Food Packaging: Antimicrobial Materials and Technologies

Food Packaging: Antimicrobial Materials and Technologies

Editor: Raven Payne

www.callistoreference.com

Callisto Reference,
118-35 Queens Blvd., Suite 400,
Forest Hills, NY 11375, USA

Visit us on the World Wide Web at:
www.callistoreference.com

ISBN: 978-1-64116-837-3 (Hardback)

Trademark Notice: Registered trademark of products or corporate names are used only for explanation and identification without intent to infringe.

Cataloging-in-Publication Data

Food packaging : antimicrobial materials and technologies / edited by Raven Payne.
   p. cm.
Includes bibliographical references and index.
ISBN 978-1-64116-837-3
1. Food--Packaging. 2. Food--Packaging--Technological innovations.
3. Food--Preservation. 4. Antibiotics in food preservation I. Payne, Raven.
TP374 .F66 2023
664.09--dc23

Table of Contents

Preface...VII

Chapter 1   **Biodegradable Antimicrobial Food Packaging: Trends and Perspectives** 1
Ludmila Motelica, Denisa Ficai, Anton Ficai, Ovidiu Cristian Oprea,
Durmuş Alpaslan Kaya and Ecaterina Andronescu

Chapter 2   **Controlling Blown Pack Spoilage using Anti-Microbial Packaging** 36
Rachael Reid, Andrey A. Tyuftin, Joe P. Kerry, Séamus Fanning,
Paul Whyte and Declan Bolton

Chapter 3   **Development and Characterization of a Biodegradable PLA Food Packaging
Hold Monoterpene–Cyclodextrin Complexes against *Alternaria alternata*** 44
Velázquez-Contreras Friné, Acevedo-Parra Hector, Nuño-Donlucas Sergio Manuel,
Núñez-Delicado Estrella and Gabaldón José Antonio

Chapter 4   **The Impact of Cross-linking Mode on the Physical and Antimicrobial
Properties of a Chitosan/Bacterial Cellulose Composite** 59
Jun Liang, Rui Wang and Ruipeng Chen

Chapter 5   **Effect of Cinnamon Extraction Oil (CEO) for Algae Biofilm
Shelf-Life Prolongation** 74
Maizatulnisa Othman, Haziq Rashid, Nur Ayuni Jamal,
Sharifah Imihezri Syed Shaharuddin, Sarina Sulaiman, H. Saffiyah Hairil,
Khalisanni Khalid and Mohd Nazarudin Zakaria

Chapter 6   **Antibacterial Nanocomposites Based on Thermosetting Polymers
Derived from Vegetable Oils and Metal Oxide Nanoparticles** 84
Ana Maria Diez-Pascual

Chapter 7   **Preparation of the Hybrids of Hydrotalcites and Chitosan by Urea
Method and their Antimicrobial Activities** 109
Bi Foua Claude Alain Gohi, Hong-Yan Zeng, Xiao-Ju Cao, Kai-Min Zou,
Wenlin Shuai and Yi Diao

Chapter 8   **Antimicrobial Activity of Lignin and Lignin-Derived Cellulose and
Chitosan Composites against Selected Pathogenic and
Spoilage Microorganisms** 132
Abla Alzagameem, Stephanie Elisabeth Klein, Michel Bergs, Xuan Tung Do,
Imke Korte, Sophia Dohlen, Carina Hüwe, Judith Kreyenschmidt,
Birgit Kamm, Michael Larkins and Margit Schulze

Chapter 9   **Use of Orange Oil Loaded Pectin Films as Antibacterial Material for
Food Packaging** 150
Tanpong Chaiwarit, Warintorn Ruksiriwanich,
Kittisak Jantanasakulwong and Pensak Jantrawut

Chapter 10   **Preparation and Characterization of Polymer Composite Materials Based on PLA/TiO$_2$ for Antibacterial Packaging** 158
Edwin A. Segura González, Dania Olmos, Miguel Ángel Lorente,
Itziar Vélaz and Javier González-Benito

Chapter 11   **Poly(3-hydroxybutyrate) Modified by Nanocellulose and Plasma Treatment for Packaging Applications** 172
Denis Mihaela Panaitescu, Eusebiu Rosini Ionita, Cristian-Andi Nicolae,
Augusta Raluca Gabor, Maria Daniela Ionita, Roxana Trusca,
Brindusa-Elena Lixandru, Irina Codita and Gheorghe Dinescu

Chapter 12   **Antimicrobial LDPE/EVOH Layered Films Containing Carvacrol Fabricated by Multiplication Extrusion** 196
Max Krepker, Cong Zhang, Nadav Nitzan, Ofer Prinz-Setter,
Naama Massad-Ivanir, Andrew Olah, Eric Baer and Ester Segal

Chapter 13   **The Influence of Accelerated UV-A and Q-SUN Irradiation on the Antibacterial Properties of Hydrophobic Coatings Containing *Eucomis comosa* Extract** 210
Małgorzata Mizielińska, Urszula Kowalska, Piotr Salachna,
Łukasz Łopusiewicz and Michał Jarosz

Chapter 14   **The Quality Evaluation of Postharvest Strawberries Stored in Nano-Ag Packages at Refrigeration Temperature** 226
Cheng Zhang, Wenhui Li, Bifen Zhu, Haiyan Chen, Hai Chi,
Lin Li, Yuyue Qin and Jing Xue

Permissions

List of Contributors

Index

Preface

The world is advancing at a fast pace like never before. Therefore, the need is to keep up with the latest developments. This book was an idea that came to fruition when the specialists in the area realized the need to coordinate together and document essential themes in the subject. That's when I was requested to be the editor. Editing this book has been an honour as it brings together diverse authors researching on different streams of the field. The book collates essential materials contributed by veterans in the area which can be utilized by students and researchers alike.

Food packaging is an important activity for restaurants, hotels and other food-related businesses. This industry is also evolving and transitioning from manufacturing of plastic based packaging to antimicrobial, biodegradable, edible, and biopolymer film packaging. Antimicrobial polymers refer to those materials that can stop or kill the growth of microorganisms in their immediate environment or on their surface. Antimicrobial polymers are appropriate for usage in the food industry for preventing bacterial contamination. This is achieved by the inactivation of microorganisms on contact or expanding their lag phase. Antimicrobial packaging significantly affects food safety and extends the shelf life of food by decreasing the growth rate of the microorganisms. They are typically made up of an antimicrobial agent and a polymer matrix, modified intentionally to provide antimicrobial properties. Apart from food packaging, these polymers also have applications in personal hygiene and medicine. This book contains some path-breaking studies related to the use of antimicrobial materials and technologies for food packaging. It will serve as a valuable source of reference for graduate and postgraduate students.

Each chapter is a sole-standing publication that reflects each author's interpretation. Thus, the book displays a multi-facetted picture of our current understanding of application, resources and aspects of the field. I would like to thank the contributors of this book and my family for their endless support.

Editor

Biodegradable Antimicrobial Food Packaging: Trends and Perspectives

Ludmila Motelica [1], Denisa Ficai [1]ⓘ, Anton Ficai [1,2]ⓘ, Ovidiu Cristian Oprea [1,*]ⓘ,
Durmuş Alpaslan Kaya [3] and Ecaterina Andronescu [1,2]ⓘ

[1] Faculty of Applied Chemistry and Materials Science, University Politehnica of Bucharest,
060042 Bucharest, Romania; motelica_ludmila@yahoo.com (L.M.); denisaficai@yahoo.ro (D.F.);
anton.ficai@upb.ro (A.F.); ecaterina.andronescu@upb.ro (E.A.)

[2] Section of Chemical Sciences, Academy of Romanian Scientists, 050045 Bucharest, Romania

[3] Department of Field Crops, Faculty of Agriculture, Hatay Mustafa Kemal University,
31030 Antakya Hatay, Turkey; dak1976@msn.com

* Correspondence: ovidiu.oprea@upb.ro

Abstract: This review presents a perspective on the research trends and solutions from recent years in the domain of antimicrobial packaging materials. The antibacterial, antifungal, and antioxidant activities can be induced by the main polymer used for packaging or by addition of various components from natural agents (bacteriocins, essential oils, natural extracts, etc.) to synthetic agents, both organic and inorganic (Ag, ZnO, TiO$_2$ nanoparticles, synthetic antibiotics etc.). The general trend for the packaging evolution is from the inert and polluting plastic waste to the antimicrobial active, biodegradable or edible, biopolymer film packaging. Like in many domains this transition is an evolution rather than a revolution, and changes are coming in small steps. Changing the public perception and industry focus on the antimicrobial packaging solutions will enhance the shelf life and provide healthier food, thus diminishing the waste of agricultural resources, but will also reduce the plastic pollution generated by humankind as most new polymers used for packaging are from renewable sources and are biodegradable. Polysaccharides (like chitosan, cellulose and derivatives, starch etc.), lipids and proteins (from vegetal or animal origin), and some other specific biopolymers (like polylactic acid or polyvinyl alcohol) have been used as single component or in blends to obtain antimicrobial packaging materials. Where the package's antimicrobial and antioxidant activities need a larger spectrum or a boost, certain active substances are embedded, encapsulated, coated, grafted into or onto the polymeric film. This review tries to cover the latest updates on the antimicrobial packaging, edible or not, using as support traditional and new polymers, with emphasis on natural compounds.

Keywords: biodegradable; polymeric nanocomposite; antimicrobial packaging; edible films; chitosan; starch; cellulose; polylactic acid; essential oils; nanoparticles toxicity

1. Introduction

The Food and Agriculture Organization of the United Nations, FAO, statistical data indicate that about one-third of the produced food is lost or wasted each year because to shelf life expiring, alteration, or spoilage due to microbial activity. About 40–50% of fruits and vegetables, 35% of fish, 30% of cereals, and 20% of dairy and meat products are lost yearly. The main culprits are the presence of bacteria and fungi, oxygen-driving processes, and presence of some enzymes. In the less developed countries 40% of the total losses occurs right after harvesting, until processing, because of poor infrastructure while in the developed countries 40% of total losses occur at retail and customer level as food waste.

At retail level large quantities are wasted because of the artificial emphasis put on the exterior look [1]. Generally speaking, the food loss occurs at production, handling, storage, and processing level, while food waste indicate the losses at distribution, market, and final consumer level [2,3].

The global food packaging market was estimated at over USD 300 billion in 2019 and has a predicted increasing rate of 5.2% yearly. The paper-based segment represents 31.9% of the total value and dominate as a biodegradable material. Unfortunately, the plastic food packaging is expected to exhibit a rapid growth globally due to superior properties and lower prices. This increase comes with environmental and pollution problems [4,5]. In developed countries the market is highly regulated by agencies such as Food and Drug Administration (FDA) or European Commission (EC). Those regulatory bodies are shifting the market toward more sustainable packaging solutions like bio-based polymers and are putting pressure on reduction of traditional plastic packaging. In the same time the economy must be transformed from a linear model to the circular economy [6]. The packaging is one of the most powerful tools in food industry as it acts also as a communication and branding medium beside its traditional role. As a marketing tool the packaging can promote a healthier lifestyle and change by itself the consumer habits [7–12].

At present, most food packaging materials are based on petrochemical polymers because of historic factors, low-cost, and good barrier performances. These polymers are nonbiodegradable and worldwide they have already raised a lot of environmental concerns regarding short- and long-term pollution [13–19]. These two factors are putting an increasing pressure on food industry to develop new types of antimicrobial packaging materials, mainly based on natural, renewable sources, or biopolymers that are environmentally safe. While the simplest way to achieve antimicrobial activity is to modify the currently used packaging materials, the environmental pressure is slowly phasing out the nonbiodegradable polymers [20–25].

The development of antimicrobial packages is a promising path for actively controlling the bacterial and fungal proliferation that leads to food spoilage [26–29]. Antimicrobial packages have a pivotal role in food safety and preservation. This type of package increases the latency phase and reduce the growth of microorganisms, enhancing the food quality and safety, ensuring a longer shelf life [30–33]. By using antimicrobial materials for food package, the shelf life is extended and the growth of microorganisms is slowed, thus ensuring a better quality and safety for meat, vegetables, fruits, and dairy products [34–37].

The methods employed in obtaining antimicrobial activity vary from adding sachets with volatile antimicrobial agents inside current packaging to incorporating the active agent directly into biopolymers and from coating or grafting antimicrobial substances on polymeric surface to the use of intrinsic antimicrobial polymers or antimicrobial pads (Figure 1) [38–42].

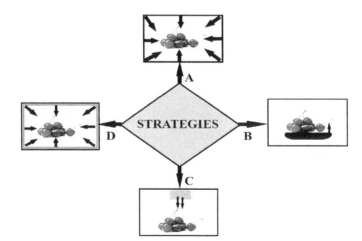

Figure 1. Main strategies employed to obtain antimicrobial packaging. The antimicrobial agents can be: incorporated into the polymeric film (**A**); in direct contact from pads (**B**); released from sachets (**C**); or coated onto polymeric film (**D**).

The antimicrobial packaging strategies can be classified in two types. First type is represented by the packaging materials with direct contact between the antimicrobial surface and the preserved food, in which the active agents can migrate into the food. Such packaging are used for food that is wrapped in foils or under vacuum. A second strategy is having the antimicrobial agent inside the package, but not in direct contact with the food, and here can be mentioned the modified atmosphere packaging (MAP) [26,43–45].

Although an increasing number of antimicrobial agents (like ethanol, carbon dioxide, silver ions or nanoparticles, chlorine dioxide, antibiotics, organic acids, essential oils etc.) [30,46–49] have been tested with the aim of inhibiting the growth of microorganisms on food, the number of antimicrobial packaging commercially available is still limited. A notable exception is in the case of silver-based antimicrobial packaging that are in use in countries like Japan or the United States [44]. The introduction of silver-based antimicrobial package solutions is expected to grow also in E.U. after the inclusion on the provisional list of permitted food additives and in the list of permitted surface biocidal products. The literature reports a wide range of antimicrobial agents like metallic nanoparticles (Ag, Cu, bimetallic etc.) [50–52], oxide nanoparticles (ZnO, TiO_2, CuO, etc.) [53–55], clay nanoparticles (cloisite, montmorillonite, bentonite) [56–58], natural extracts (essential oils or hydrophilic extracts) [59–61], natural antimicrobials (nisin, pediocin, antibiotics, etc.) [62–64], biopolymers (chitosan) [65–67], enzymes (lysozyme, peroxidase) [68–70], synthetic antimicrobial agents (including synthetic antibiotics) [71].

The natural antimicrobial compounds are obtained from vegetal sources like cloves, cinnamon, thyme, ginger, oregano, rosemary, garlic, etc. All these have a good potential for the meat industry [72,73]. In a similar way, other natural antimicrobial agents can be isolated from substances produced by bacterial or fungal activity such as pediocin, nisin, and various bacteriocins. In the meat industry, currently, some technologies based on blocking the microbial activity are in use that can enhance the organoleptic properties and packaging performance [30,74–77]. Various synthetic antimicrobial agents have been reported for bread and pastry packaging [78–80] or for fruits and vegetables [81–83].

While most of the literature present tests of antibacterial activity on strains that usually spoil the food (*Escherichia coli, Staphylococcus aureus, Salmonella enterica,* etc.) [84], there are some authors who test the packaging materials also against various fungal strains (yeasts and molds) [85]. Developing new packaging materials must pass some specific tests beside the above-mentioned antibacterial activity. Some additives (usually natural extracts or specific substances) have antioxidant activity that aids in prolonging shelf life. The new packaging material must have good mechanical resistance and barrier properties. The water vapor permeability (WVP) plays a crucial role in preserving the food. Permeability to other gases, namely oxygen, carbon dioxide, ethylene, and various aromatics is also very important when polymeric film is designed. Some other characteristics like transparency, thermo sealability, leakage of additives etc., will be described further.

The problem of food loss due to spoilage by microorganisms can be addressed by using antimicrobial packaging, be it traditional fossil-based polymers (PVC, PET, PE, PP, etc.) or biodegradable ones (cellulose, starch, chitosan, etc.). The ecological impact generated by the plastics cannot be solved unless we replace the traditional packaging materials with biopolymers from renewable sources.

The aim of the review is to present some of the latest research trends in the antimicrobial food packaging, the work horse polymers, antibacterial and antifungal agents, as well as to review the strategies presented by the literature. We searched for original articles and reviews that addressed the antimicrobial packaging subject. In order to identify the strategies and latest trends in the domain of antimicrobial food packaging we survey the Scopus, Clarivate, and Science Direct databases with emphasis on 2017–2020 interval. The search engines were used to find relevant documents by using keywords as "antimicrobial food packaging" "antimicrobial peptides" "biopolymers" "essential oils" and "antimicrobial peptides" Additional documents were identified by examining references listed in key articles. The initial number of documents obtained by searching the databases was 275,125.

By restricting the time interval, the number dropped to 88,327. Based on title and abstract information we included only the articles that were considered relevant for antimicrobial food packaging subject. Therefore, 289 articles were included in this systematic review.

2. Antimicrobial Packaging Obtained by Modification of Current Materials

Nowadays there are two main research directions for the packaging industry. First research approach is to modify the current packaging materials in order to confer antimicrobial and antioxidant activities, plus better mechanical and barrier properties. This direction is somehow closer to the manufacture sector, which will not be forced to make radical changes into the production lines, so the implementation costs are lower. While the consumer will enjoy a safer food with longer shelf life, the generated pollution is the main drawback of this approach. The nonbiodegradable plastics will still be produced and will still be an environmental hazard. The second approach, with increasing support from consumers and researchers is to replace all the nonbiodegradable polymers with biodegradable ones, obtained from renewable sources. This is more ecological but is also more costly, demanding a dramatic change of the packaging manufacture sector.

For large quantities from some foods (sugar, flour, rice, potato, beans, etc.) the storage and transport can be done in textile bags. Therefore, some researchers aim to obtain fabrics with antibacterial and antifungal properties (that can also be used outside of food industry, for clothing for example). One method presented in the literature is to load the fabric with quaternary ammonium salts that contain isocyanate moiety (CAI) and sulfopropyl betaine (SPB) by soaking-drying process. Both moieties, SPB and CAI, can be covalently bounded to the fabric surface and will give excellent bactericidal and antifungal properties. The effect is long-lasting and cannot be washed away (Figure 2). This functionalization of cotton fabric will not modify significatively the permeability to water vapor or air or its hydrophilic character. The mechanical properties of the fabric are enhanced after the treatment, the textile bag presents better tearing and breaking properties. Increasing the CAI load leads to a stronger covalent bond and the ratio SPB/CAI can be used to enhance the retention of the antibacterial agents. The fabric has not exhibited any noticeable migration of the chemical agents, the test indicating a strong retention into the cotton fiber. As a consequence there is no toxicity of stored food due to the migration of CAI or SPB [86].

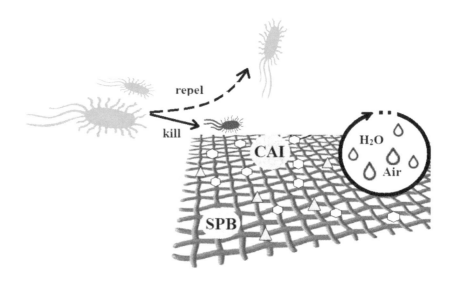

Figure 2. Antimicrobial textile loaded with nanoparticles or other bactericidal and antimycotic agents (isocyanate moiety—CAI and sulfopropyl betaine—SPB).

By same general method the antimicrobial activity can be obtained for other innovative packages. The deposition or the embedding of bactericidal and antimycotic agents into the polyvinyl chloride (PVC) [87], polyethylene terephthalate (PET) [88], polyethylene (PE) [89], or polypropylene (PP) [90] films that are used normally for food packaging, do not request a fundamental change of the production lines and the new types of packaging can easily enter production.

Mixed cellulose/PP pillow packages can be used to extend the shelf life of iceberg lettuce if emitting sachets with eugenol, carvacrol, or trans-anethole are put inside [91]. The sachets will slowly release the natural antimicrobial agent and help preserve the food. This is an easy, low-cost method to confer antibacterial activity to the packaging materials already in the market and can be classified as modified atmosphere packaging (MAP), and can be seen as an intermediate step toward obtaining antimicrobial packaging without addressing the problem of plastics pollution [92].

Small modifications for actual packaging materials are described in [93], where the authors have obtained a bilayer film from low-density polyethylene (LDPE), in which one layer has incorporated various essential oils. PVC was also modified by adding orange essential oil and the obtained films were proved to have antimicrobial activity [94]. These traditional polymeric films can have silver nanoparticles (AgNPs) or oxide nanoparticles embedded into them or deposited on the surface. The polymeric film or the nanoparticles can in turn be loaded with essential oils as antimicrobial agents. PVC film with embedded nanoparticles like AgNPs [87], TiO_2 [95], ZnO [96] are reported in literature.

The AgNPs can be loaded on silica, and these in turn can be embedded into the PP film. This kind of composite film can be obtained by hot pressing the silica powder onto the PP film. The antimicrobial film thus obtained has superior mechanical properties and the test made on fresh mackerel indicate a longer shelf life. In the samples with AgNPs/SiO_2 5% and 10% respectively concentrations of 3.06 and 3.11 log CFU/g for pathogen agents were recorded, while into the control sample the concentration reached a peak of 4.81 log CFU/g. In the same time the level of trimethyl amine increased from 1.6 to 2.5 mg/100 g in the control sample, but there was no modification in the samples packaged with composite films. The color of the fish meat stays red for samples with composite film, but changed from red to yellow for the control sample [97].

Polyethylene terephthalate (PET) can be used as antimicrobial packaging after surface modification. Because of its inertness the PET film requires an activation treatment so that functionalization/grafting can succeed. Antimicrobial peptides (AMPs) are an important class of amphiphilic molecules capable to suppress growth of microbes. They are composed of 12 to 100 amino acids and therefore are quite different structurally. Functionalization of PET with an AMP, mitochondrial-targeted peptide 1 (MTP1), is presented in by Gogliettino, M. et al. [98]. The antimicrobial packaging PET-MTP1 was used to prolong the shelf life from 4 to 10 days for meat and cheese.

Among AMPs, nisin and pediocin are used commercially, but other species such lacticin 3147 or enterocin AS-48 have been proposed for industrial applications. The prime target are the dairy products as these bacteriocins can be produced by lactic acid bacteria [99]. The main problem is to find ways to immobilize AMPs on film's surface without losing the antimicrobial activity. Direct incorporation, coating, or immobilization are the main strategies [100]. Various methods are proposed in literature, such as cold plasma treatment [101], enzyme (laccase) surface modification [102], adsorption on nanoparticles [103] etc.

Among the desirable properties that a film must possess to be used in food packaging is the heat-sealing capability. While traditional PE or PP films can be easily heat sealed, some of the biopolymers, like cellulosic films made from nanofibers, lack sealability. This problem has made the industry reluctant to market them. This shortcoming can be overcome by using the traditional PE or PP films on which cellulosic nanofibers can be deposited. This composite film will present the advantages from both materials. Peng Lu et al. [104] have demonstrated that these cellulose nanofibers can also be loaded with various antimicrobial agents, like nisin for example, to enhance the film properties (Figure 3). PE and PP films have been treated with cold plasma in order to improve hydrophilic character and to make them compatible with cellulose fibers. The cellulose layer has led to a decrease of the oxygen permeability from 67.03 to 24.02 cc/m^2·day. Adding the nisin (an AMP) has no influence on permeability, but has conferred antibacterial properties to the films. The test made against *Listeria monocytogenes* have showed a kill ratio of 94%, indicating the economic potential of this kind of packaging.

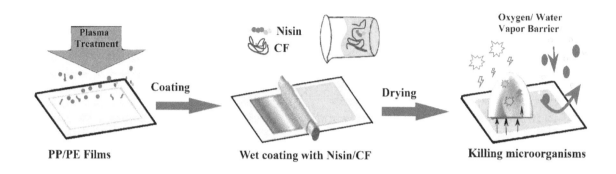

Figure 3. Coating of polypropylene/polyethylene (PP/PE) films with other polymers (cellulose fibers—CF) and loading with antibacterial agents (nisin)—adapted after information presented in [104].

Cellulose-based materials play an important role in the packaging industry, therefore, many researchers are trying to improve them. The poor mechanical performance and the lack of water resistance are some of the factors that limit the applicability domains. Zhu R. et al. [105]. have proved that by applying recycled cellulose, obtained in ionic liquids, over plain paper, the mechanical properties are improved greatly. Moreover, by using only 2% cellulose in paper composition the oxygen permeability drops by a factor of 10^6. Although the composite is hydrophilic in nature, it has good resistance to water action and keeps its form when submerged.

Traditionally, for improving the WVP values, the oil resistance, or the mechanical properties of paper packaging, at least one side is usually covered by a thin film of PE, or PET to a lesser extent. Therefore, in the quest of replacing the fossil-based polymers with biodegradable ones, researchers come with innovative composite materials that present new and superior properties. A method of enhancing the properties of paper packaging is described in [106] where the plain sheet was covered with a functionalized biopolymer coating. The film was obtained by mixing alginate with carboxymethyl cellulose and carrageenan, and was loaded with grapefruit seeds extract. The microscopy studies indicate that the biopolymer is compatible with the paper support, filling the pores and levelling the spaces between the cellulose fibers, in the end resulting a smooth surface. The whole range of properties for this type of packaging: water and oil resistance, WVP, hydrophobicity, and mechanical properties were not only superior to the simple paper support but also to paper-PE packages used at present. The obtained biopolymer film presented antibacterial properties against two pathogens *L. monocytogenes* and *E. coli*, which were destroyed in 3 and respectively 9 h of exposing. The test made on fish meat inoculated with bacteria have demonstrated the total elimination of the infection in 6–9 days, proving the economic potential of this packaging type [106].

A similar strategy for improving paper packaging is described in [107] where cellulosic paper was covered with a chitosan layer by dip-coating method. The chitosan solution also contained nisin up to 1 g/mL as antibacterial agent for the packaging material. Various compositions based also on chitosan with the addition of zein and rosemary oil coated on paper were tested by Brodnjak and Tihole [108]. As expected, the WVP values dropped sharply after coating the paper, while the new packaging exhibited antimicrobial activity.

Such experiments have demonstrated the possibility of obtaining antimicrobial packaging, usually as a nanocomposite film, based on materials that are currently in use in the food industry. The increase in production cost is one of the main drawbacks for the introduction of new technologies. The industry will adopt new technologies that can be re-engineered from the existing ones with minimal modification of the production lines. These modifications can be induced by new regulations, by predicting increasing profits or by changes in public perception and demand [109].

3. Biodegradable Polymeric Antimicrobial Packaging

Since their invention in 1959, a great number of plastic bags and other plastic packages have been produced from fossil fuels and their contribution to the pollution has become a global problem. The obvious solution is to produce bags and packages based on biodegradable materials, with low environmental impact. By adding antimicrobial properties to the biodegradable packages these new materials can offer an enhanced protection against food spoilage, extending the shelf life.

In the search of alternative materials to the plastics, the scientists have put a special emphasis on biopolymers. One of the most important conditions, the availability, is fulfilled by some abundant natural polymers like cellulose (and derivatives like carboxymethyl cellulose—CMC, methyl cellulose—MC, cellulose acetate—CA, etc.), chitosan, lignin, starch, polylactic acid (PLA), polyvinyl alcohol (PVA), alginic acid and derivatives etc. (Table 1). To be used in the food industry, these materials must be nontoxic, renewable, and present the right properties. Although many of them are included in the polysaccharides class and possess a similar monomeric structure (Figure 4), the polymeric chain configuration, hydrogen bonds, and substituents play a crucial role in shaping the specific properties of each biopolymer. The biopolymer films must be resistant, to not exfoliate and to not permit the gas or vapor exchange between food and atmosphere. Some of the most abundant biopolymers have also antimicrobial intrinsic activity (like chitosan), but usually in the native form they have low mechanical resistance and poor barrier properties. Unfortunately, cellulose, the most abundant biopolymer, lacks any antibacterial activity and therefore it must be loaded with antimicrobial agents. These agents can also become reinforcing components improving the mechanical properties and enhancing the WVP values.

Table 1. Principal film forming biodegradable classes of substances used in packaging.

Film Forming Class	Substance	Reference
Polysaccharides	Cellulose	[110,111]
	Chitosan	[112,113]
	Starch	[114,115]
	Hemicellulose	[116,117]
	Alginate	[118,119]
	Agarose	[120,121]
	Glucomannan	[122,123]
	Pullulan	[124,125]
Proteins	Soy-protein	[126,127]
	Whey-protein	[128,129]
	Zein	[126,130]
	Collagen	[131,132]
	Gelatine	[133,134]
	Casein	[135,136]
Lipids and waxes	Plant oil and animal fat	[137,138]
	Waxes	[139,140]
Others	PLA	[141,142]
	PVA	[143,144]

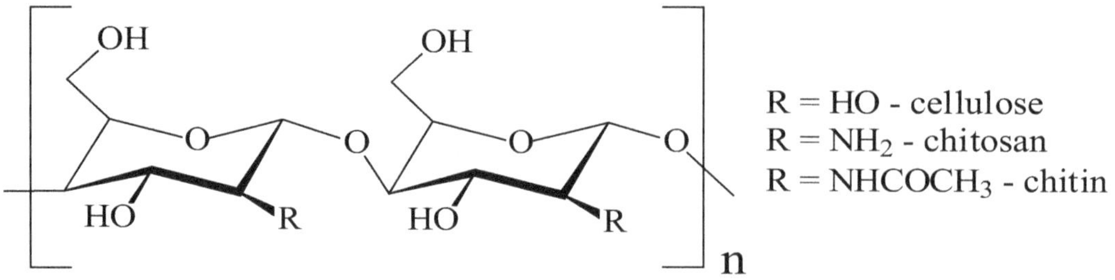

Figure 4. Comparison among structures of various polysaccharides.

PVA has high hydrophilic properties, is water soluble, with a good crystallinity, and has many hydroxyl moieties at side chains which can form hydrogen or covalent bonds with other polymers or can help loading various antimicrobial agents [145]. PLA can present good crystallinity, but is soluble only in organic solvents, and therefore reinforcing it with nanoparticles requires surface compatibilization [146]. Alginic acid and its sodium salt are easily water soluble to be used raw. Therefore they are mixed with lipidic compounds (like essential oils which also give antibacterial activity to the final material) and reinforced with various nanoparticles [147]. Chitosan is water soluble (acidic pH) and can also provide -OH and -NH$_2$ moieties for hydrogen bonds. Cellulose and its derivatives represent a class of hydrophilic biopolymers, which again present free hydroxyl moieties. The strategy for improving the water resistance of biopolymers with -OH moieties is to mix them in composite materials, where same moieties will help them cross-linking. This blending of individual polymers gives many hydrogen bonds which improve the mechanical properties of the composite. Sometimes the simple blending of two polymers does not yield automatically a better material. PVA/CMC composite films have low water resistance and present no antimicrobial activity for example. Adding inorganic nanoparticles (like Ag, ZnO, TiO$_2$, Fe$_3$O$_4$, etc.) and essential oils in such polymeric films will improve both WVP and antimicrobial activity [111,148–151].

In the food industry, the water vapor and oxygen barrier properties of the packaging are crucial, but the permeability to CO$_2$ and aromatic compounds plays an important role too. The engineering of the composite films generally aims to enhance the properties of the base biopolymer (usually a polysaccharide), in order to obtain the desired properties. Also, in the active or intelligent packaging the nanomaterials can act as antimicrobial agents, oxygen scavengers, or as signals for improper storage conditions. Compatibilization between nanomaterials and the polymeric matrix is the main challenge in obtaining antimicrobial packaging [152].

The bio-nanocomposite materials represent an important alternative to the traditional fossil-based plastics used as packaging. These are biocompatible, biodegradable, and can exhibit superior performances, both mechanical and chemical. Biopolymers like chitosan, carboxymethyl cellulose, starch, alginate, casein, carrageenan, or cellophane can solve the ecological problem of packages due to their biodegradability and non-toxic nature. But polysaccharides also have some drawbacks beside their clear advantages. Some mechanical properties are inferior, and they have a low water resistance or high permeability. Consequently, by embedding various nanomaterials into the natural polymeric film better thermal and mechanical properties can be obtained, but also the water vapor and oxygen barrier properties can be enhanced, without losing the biodegradability or the non-toxic character of the biopolymer. The most used are clay nanoparticles (like montmorillonites, kaolinite, or laponite), oxide nanoparticles (ZnO or TiO$_2$), or silver nanoparticles (AgNPs). In Table 2 are presented some types of antimicrobial packaging materials, their main polymer, antimicrobial agent, and the tested kind of food.

Table 2. Antimicrobial packages and their applicability.

Product Preserved	Packaging Material	Antimicrobial Agent	Reference
Iceberg lettuce	Cellulose	Clove and oregano oils	[91]
Cucumber	Chitosan	Limonene	[153]
Tomato	Chitosan	TiO_2 nanoparticles	[154]
Strawberries	Chitosan/CMC	Chitosan/citric acid	[155]
Strawberries	PLA	AgNPs	[150]
Strawberries	Chitosan/CMC	*Mentha spicata* oil	[156]
Strawberries	Gelatin	Butylated hydroxyanisole	[157]
Fish	PLA	Thymol	[158]
Crap fillets	Alginate/CMC	*Ziziphora clinopodioides* oil/ZnO	[159]
Rainbow trout fillet	Chitosan	Grape seed extract	[160]
Salmon	PLA	Glycerol monolaurate	[161]
Shrimps	Chitosan	Carvacrol	[162]
Shrimps	Gelatin	ZnO /clove oil	[163]
Chicken	Chitosan	Acerola residue extract	[164]
Poultry	Chitosan	Ginger oil	[165,166]
Chicken	Gelatin	Thyme oil	[167]
Chicken	Pullulan	Nisin	[125]
Chicken	Chitosan/PET	Plantaricin	[168]
Ostrich meat	Kefiran/polyurethane	*Zataria multiflora* oil	[169]
Lamb meat	Chitosan	*Satureja* plant oil	[170]
Ground beef	PLA/NC	*Mentha piperita, Bunium percicum*	[171]
Ham	Chitosan/starch	Gallic acid	[172]
Salami	Whey protein	*Cinnamomum cassia, Rosmarinus officinalis* oils	[173]
Cheese	Chitosan/PVA	TiO_2	[174]
Cheese	Cellulose /Chitosan	Monolaurin	[175]
Cheese	Starch	Clove leaf oil	[176]
Cheese	Agar	Enterocin	[177]
Cheese	Zein	Pomegranate peel extract	[178]
Peanuts (roasted)	Banana flour (starch)	Garlic essential oil	[179]

3.1. Edible Films

The edible films represent a protection packaging, with antimicrobial activity for the food safety. The literature reports a large variety of materials which can form films for food packaging, ultimately their properties indicating the exact food on which the film or coating can be used [180]. Edible packaging materials or coatings must be non-toxic, adherent to food surface, be tasteless or present an agreeable taste, have good barrier properties and prevent water depletion of the food, must have a good stability in time and prevent the mold formation, good appearance in order to be accepted by consumers, and must be economically viable.

Edible films are composed of polysaccharides, proteins, or lipids and they must present better barrier properties and antimicrobial activity in order to extend the shelf life. Among polysaccharides we can enumerate starch, cellulose derivatives, alginate, chitosan, pectin, carrageenan etc. Proteins can be extracted from animal sources like casein, whey, collagen, gelatin, egg white, etc. or can be from vegetal sources like corn zein, gluten, soy proteins, rice bran, peanut, keratin etc. The third category, lipids, is composed of saturated or unsaturated fatty acids.

The use of edible films for food packaging is not a new technology. In the past collagen-based membranes were used for various meat products like sausages, hotdogs, salami etc. At present, the food industry is searching for alternatives that can also prolongate the shelf life of the products [181].

The edible films for fruits, cheese, and meat products are usually based on natural, abundant, cheap, edible polymers, which can effectively enhance the quality of the food and at the same time can reduce the quantity of foodstuff that is altered. This kind of films can replace successfully the waxy coatings that are used on various fruits (apples, oranges, limes etc.) and that are toxic and are forbidden by an increasing number of countries. Edible biopolymers inhibit the food alteration and complies to the directives requested by EU market: organic food, healthier, and with longer shelf life.

3.2. Polysaccharides-Based Films

Chitosan, one of the most abundant natural polymers, is the work horse for such innovative edible antimicrobial packaging. It is a polysaccharide, with intrinsic antibacterial activity, usually extracted from shells of crabs and shrimps. It also has some antioxidant properties and is biocompatible, biodegradable, and most of all is edible.

A chitosan packaging film can protect the food by multiple mechanisms. The film can block the microorganism's access to the food as any physical barrier. It can hinder the respiratory activity by blocking oxygen transfer and can cause a physical obstruction so that microorganisms cannot reach the nutrients (mechanisms I and II—Figure 5). Various nutrients can be chelated by chitosan chains (mechanism III) and the outer cellular membrane of the microorganisms can be disrupted by electrostatic interactions (mechanism IV), all these adding a supplementary stress which microorganisms must endure. The antibacterial activity of chitosan can also be the result of damaged cell membrane, which will lead to leakage of intracellular electrolytes (mechanism V). Once the chitosan chains have diffused inside through cellular wall, there are multiple action pathways that leads to microorganism's death. It can chelate internal nutrients or essential metal ions from cellular plasma (mechanism VI), it can influence the gene expression (mechanism VII) or it can penetrate the nucleus and bind the DNA, thus inhibiting the replication process (mechanism VIII) [182].

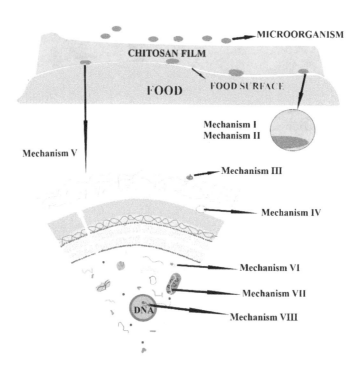

Figure 5. The mechanisms by which the chitosan films manifest antibacterial activity—adapted after information presented in [182].

Among its drawbacks are the higher cost, the acidic hydrolysis and weak mechanical properties. Therefore, in many cases it will not be used alone but in a nanocomposite film, in which the other components (polymers or nanoparticles) will compensate for the poor performances of chitosan in some areas. Usually improving the mechanical properties of polymers requires some plasticizers. Chitosan has a good compatibility with glycerol, xylitol, sorbitol etc.

The use of chitosan in edible films can raise the problem of its origin. Traditionally the chitosan is obtained from the marine crustacean's exoskeleton, which would make it undesirable for strict vegetarians or for halal and kosher food. Nevertheless, it can be obtained also from alternative sources like mushroom inferior stem (which itself is a residual product that is usually discarded) [183].

Priyadarshi, R. et al. [184] have obtained chitosan-based antimicrobial packaging by forming the polymeric film directly on the surface of the food. As binding agent, they used citric acid, that also enhanced the film stability and antioxidant properties. As plasticizer they used glycerol, which gave flexibility to the film. In fact, the elongation capacity was increased 12 folds, but this decreased the tear resistance. The as-obtained packaging material had a superior water resistance, reducing WVP by 29%. Also, the films were transparent, which is desirable from consumers' point of view, as they will always want to be able to see the food inside the packaging. Finally, the test made on green chili indicated an extended shelf life [184].

Enhancement of chitosan film properties was done by adding layers of other polymers and embedding nanoparticles. Bilayer films obtained from chitosan and polycaprolactone (PCL) have been obtained by pressing or coating. Both layers have been loaded with nanocellulose (2–5%) and grape seeds extract (15% w/w). The presence of nanocellulose (NC) have reduced significantly the WVP and the film opacity, while the grape seeds extract had the opposite action. The films obtained by pressing exhibited higher values for elastic modulus and stretch resistance than the films obtained by coating from solution. Both film types had antimicrobial activity and the grape seeds extract preserved the antioxidant activity while being loaded into the chitosan matrix [185]. Similar mechanical behavior was reported in [186] where authors obtained a composite chitosan-based film by adding cellulose nanocrystals and grape pomace extracts. Adding NC increased tensile strength and decreased elongation and WVP while the grape pomace extract had opposite effect.

Uranga J. et al. have created antibacterial films from chitosan and citric acid, but also added fish gelatin into the composition. The films have antibacterial activity on E. coli, act as a UV barrier, have good mechanical properties, and the citric acid acts like an inhibitor against swelling in the presence of water [187]. Such films can be used for products like seafood. In the case of fish or shellfish the food alteration happens because of the modification induced by enzymes activity that are naturally presented in meat, reactions like lipids oxidation or because of the microorganism's metabolism. Chitosan films can reduce the lipids oxidation reactions, although the native antioxidant properties are not so good. To overcome this problem, Uranga, J. et al. developed a new strategy by adding into the chitosan—fish gelatin mix, anthocyanins obtained from processing food wastes. Anthocyanins preserved antioxidant activity during the mold pressing process, the obtained packaging exhibiting antimicrobial and antioxidant activity, along good mechanical and WVP properties [188]. Other groups loaded the chitosan-gelatin film with procyanidin or various other antioxidants [189,190]. Enhancement of antioxidant activity of chitosan films can be done also by adding some radicals like 2,2,6,6-tetra-methylpyperidine-1-oxyl. The desired mechanical properties can be achieved by adding up to 15% cellulose nanofibers and a plasticizer like sorbitol (25%). Beside the antibacterial effect on E. coli, the films also reduced the proliferation for Salmonella enterica or L. monocytogenes [191].

Another type of edible chitosan-based film was obtained by Jiwei Yan et al. [155] and used for fresh strawberries. The fruits were covered with a bicomponent film (layer-by-layer LBL) by successive immersions in solutions of 1% chitosan and 1.5% carboxymethyl cellulose (Figure 6).

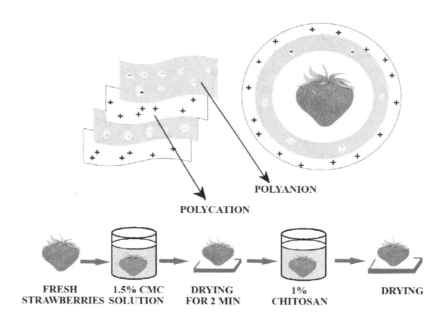

Figure 6. Obtaining the LBL films from chitosan and carboxymethyl cellulose solutions—adapted after information presented in [155].

The results have indicated a significant enhancement of the LBL film properties when compared with the simple chitosan one. The bicomponent film has conserved the fruits for longer time, preserving the consistency, the flavor and allowed only minimal changes in acidity and total soluble substance because the LBL film has decreased the metabolic degradation of carbohydrates, fatty acids, and amino acids from the fruits, preserving them for eight days [155].

The thymol, a component from thyme essential oil, has also been used as antibacterial agent in the edible chitosan films. A nano-emulsion of thymol was loaded onto a film made from chitosan and quinoa proteins. Strawberries covered with such films have presented a superior resistance against fungi. If initially the edible film has somehow altered the fruits flavor, starting from the 5th day the sensorial qualities have been enhanced when compared with control batch. In addition, the shelf life was extended by four days, mass loss was diminished, and parameters like pH, titratable acidity or soluble substances percent were not altered [192].

Another composite film that has attracted many researchers is the chitosan/ polyvinyl alcohol (PVA). This composite presents good biocompatibility and low toxicity, has broad spectrum antibacterial activity on both Gram-positive and Gram-negative strains, exhibits good thermal stability, and can be loaded with desired antimicrobial agents. Yun, Y.H. et al. [193] have prepared mixed polymeric films based on chitosan and PVA for fruit preservation. The films with 1:1 ratio between polymers, used sulfosuccinic acid as the binding agent, therefore the reticulation could be made by UV light exposure (approx. 20 min). Glycerol, xylitol, and sorbitol were used as plasticizers. The composite with sorbitol presented the best mechanical and thermal properties. All the samples had lower WVP than simple polymeric films. The reported results indicate that after 70 days storage, the fruits presented negligible decay. The packaging material is biodegradable, test showing a 40–65% disintegration after 220 days. Yang, W. et al. [194] have obtained a hydrogel by mixing chitosan and PVA in which they added 1–3% lignin nanoparticles. Lignin is the second most abundant natural polymer after cellulose, but usually is used in low value products, mostly for burning. In this hydrogel, lignin exhibited antioxidant activity and UV filter properties, while being also a reinforcing and antibacterial agent (Figure 7).

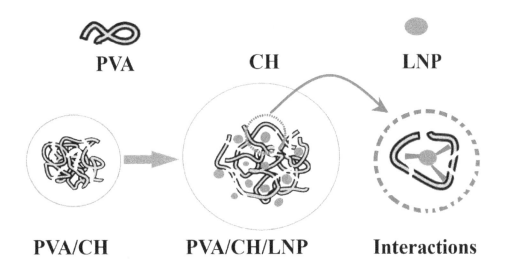

Figure 7. Schematic depiction of polyvinyl alcohol (PVA) and chitosan chains interactions with lignin nanoparticles—adapted after information presented in [194].

Best properties have been obtained for 1% lignin as higher quantity gave no further benefits because of agglomeration of nanoparticles. The hydrogel has a porous structure like honeycomb, with micron size pores which can be further loaded with antimicrobial agents. The tests made by the authors indicate a synergic activity between chitosan and lignin against *E. coli* and *S. aureus*, and also enhancing of antioxidant properties [194]. Youssef, A.M. et al. have improved same polymeric mix chitosan/PVA by using glycerol as plasticizer and adding TiO_2 nanoparticles. The coating was successfully used to preserve Karish cheese for 25 days, best results being obtained for 3% TiO_2 [174].

The chitosan-based films can also be used on fruits and vegetables that are ripening under ethylene action. Because such films can absorb the ethylene, they can postpone the unwanted ripening. The literature mentions the composition 6 g chitosan: 7 g $KMnO_4$: 2 mL sorbitol, to preserve tomatoes up to five times longer at room temperature [195]. Similar findings have been reported for other nanocomposite chitosan-based films. In order to enhance the antibacterial properties of the chitosan coating, TiO_2 nanoparticles can be embedded into the film. The coating can be applied straight from the solvent on the packaging. Such films have been used to protect the cherry tomatoes and to postpone the ripening after harvest. The authors have monitored the consistency, color changes, lycopene quantity, mass loss, total soluble substance, ascorbic acid, and concentration for CO_2 and ethylene inside the package. The conclusion was that the addition of the TiO_2 nanoparticles have enhanced the packaging quality, the cherry tomatoes presenting less changes when compared with simple chitosan packaging or with the control lot. Authors have attributed the behavior to the photocatalytic activity of TiO_2 versus ethylene gas (Figure 8) [154].

Chitosan is compatible also with ZnO nanoparticles, the composite having a more potent antibacterial activity due to synergism [54]. Good results are reported for example by [196] who manage to reduce the proliferation of *E. coli* and *S. aureus* on poultry by 4.3 and 5.32 units on log CFU/mL when compared with control. Generally speaking, the meat products have a high nutritional value and represent a good medium for microorganisms growth, thus being preserved only for a few days in the fridge. Therefore, it is no surprise that many researchers have obtained chitosan-based packaging for meat products [197,198].

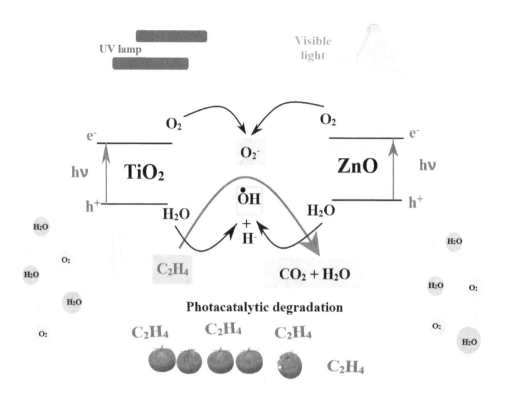

Figure 8. Scheme of the ethylene degradation due to photocatalytic activity of oxidic nanoparticles embedded into chitosan film—adapted after information presented in [154].

A chitosan/cassava starch film was obtained by Zhao, Y.J. et al. [172] by using subcritical water technology. The authors added glycerol as the plasticizer and gallic acid in order to achieve antimicrobial and antioxidant activity for the film. The formation of new hydrogen bonds, electrostatic interactions, and ester bonds between starch, chitosan, and gallic acid had the effect of decreasing permeability to water vapors. Antimicrobial tests on foodborne diseases microorganisms have indicated a prolonged shelf life for pre-inoculated ham from 7 to 25 days when compared with the control. Various edible films based on chitosan and starch (extracted from oak) have been investigated by Zheng K. et al., who have indicated that increasing the starch ration from 0 to 50% has improved the mechanical properties by 20%, has decreased the WVP by 3%, and diminished the oxygen permeability by 2.5 times. Aiming to improve the antimicrobial activity, the authors have added essential oil from *Litsea cubeba*, which further increased the barrier capabilities of the packaging [199].

As mentioned before, by incorporating essential oils into chitosan films the antimicrobial and antioxidant activity is usually greatly enhanced. The literature presents many reports of such films that are efficient against food altering pathogens, bacteria, and fungus, that spoil the harvest. The use of essential oils and plant extracts with edible natural biopolymers is one of the newest research trends in food packaging industry.

Some essential oils are produced from crops that are especially made for this purpose (lavender, thyme, cinnamon, basil, rosemary, etc.). Other essential oils or extracts are obtained as by-products processing agricultural wastes (grape seeds, apricot kernels, orange peel, etc.). One of the challenges in this type of food packaging is that the producer must match the type of essential oil with the type of the protected food. While thyme, ginger, or rosemary essential oil will be well received by consumers on the meat-based products, not many consumers will like them with fruits. Same can be said about cinnamon and vanilla which are compatible with pastry, but do not go so well with meat. In the same time the essential oils from orange or lime can be used in fruits or seafood packaging (Table 3).

Table 3. Essential oils and extracts and the usual compatibilities.

Food Type	Essential Oil/Extract	References
Meat	Acerola extract, oregano oil, ginger oil, betel oil, *mentha spicata* oil, rosemarin oil, *ziziphora clinopodioides* oil, *zataria multiflora* oil, *satureja* plant oil, thyme oil, grape seed extract, *eucalyptus* oil	[151,159,160,162,164–167,169–171, 173,200–202]
Vegetables and fruits	*Litsea cubeba* oil, clove and oregano oils, *mentha spicata*, cinnamon oil, limonene, thyme oil	[91,114,153,156,178,199,203,204]
Pastry and bread	Garlic essential oil, pomegranate peel extract, cinnamon oil, clove oil	[178,179,205]
Dairy and cheese	Clove leaf oil, ginger oil	[107,126,176]

Apricot kernels are a major agricultural waste. One possible utilization for them is to extract the essential oil from the bitter core. The chemical composition for this essential oil indicates the oleic acid as the main fatty acid, but also reveals the presence of N-methyl-2-pyrolidone which is a strong antioxidant and antimicrobial agent. By mixing chitosan with apricot essential oil in 1:1 ratio, the WVP dropped by 41% and the mechanical resistance increased by 94%. The antimicrobial activity was tested on bread slices, on which the chitosan-essential oil package managed to block the fungus growth [206]. The possible use of agricultural waste was further exemplified by Zegarra, M. et al. who have used chitosan films loaded with acerola extract from residues to extend the shelf life of meat [164].

Essential oil from *Perilla frutescens (L.)* Britt. was introduced by Zhang Z.J. et al. in a chitosan-based film, ranging from 0.2–1% *v/v*, in order to enhance the antibacterial properties against *E. coli, S. aureus,* and *Bacillus subtilis*. The essential oil also enhanced the mechanical properties, the UV blocking capacity and decreased the WVP and swelling capacity [207].

Another essential oil with good antioxidant and antimicrobial activity is extracted from cinnamon. It improves the antibacterial activity of the chitosan-based film, but also decreases the oxidation speed for lipids and reduces the WVP. Tests by Ojagh S.M. et al. have shown that if after 16 days the control sample had a bacterial load of 8.43 log CFU/g, the chitosan presented a bacterial load of 6.79 and the sample packed in chitosan-essential oil had a bacterial load of only 6.68 log CFU/g (for both chitosan-based films the bacterial load being under the limit of 7 log CFU/g) [208]. Same combination of chitosan–cinnamon essential oil has been tested for increasing the preservation time of jujube (date) fruits. Normally, these fruits have a shelf life of only 15 days and the orchards cannot be efficiently exploited as the post-harvest losses are too high. The use of fungicides is not acceptable because of health concerns and any way the ripe fruits have the highest decay rate. Xing Y. et al. have shown that chitosan–cinnamon essential oil films have lower WVP and therefore prevent the mass loss. The test indicates that the shelf life can be increased to 60 days as the decay rate is drastically reduced. The antimicrobial activity was tested on *E. coli, S. aureus, Rhizopus nigricans, Penicillium citrinum, Aspergillus flavus,* and *Penicillium expansum,* the best results being observed on *R. nigricans* and *S. aureus* strains [209].

Chitosan films have a poor performance regarding WVP, and this drawback must be removed by structural modifications. The solution is to add some lipidic compounds which are hydrophobic. Best candidates are fatty acids and essential oils or just compounds from essential oils. The use of selected compounds is a strategy to remove some of the unpleasant or too strong flavor of essential oil, or to avoid including potential toxic components into the packaging. Wang, Q. et al. have loaded the chitosan film with carvacrol and caprylic acid. Carvacrol is one of the major components of oregano and thyme essential oils with good antimicrobial and antioxidant activities and is listed as GRAS (generally recognized as safe) by FDA. The caprylic acid is also listed as GRAS, has a broad antibacterial spectrum and is naturally found in palm seed oil, coconut oil, or ruminant milk [162].

The edible films were used as packaging solution for white shrimps, *Litopenaeus vannamei*, for 10 days. During this time the growth of microorganisms was hindered, nitrogen volatile compounds were not evolved, and the pH variation was small when compared with control sample. The surface melanosis was delayed, and the organoleptic properties of the shrimps were enhanced. Research indicates that combining the chitosan with carvacrol generated a synergic effect, and adding the caprylic acid further enhanced the antimicrobial activity of the film [162]. Pabast, M. et al. have obtained better WVP properties for chitosan-based films by adding the *Satureja khuzestanica* essential oil encapsulated into nanoliposomes. The nanoliposomes of ~95 nm were obtained from a mix of soy-lecithin and had an encapsulation efficiency of 46–69%. By encapsulating the essential oil into nanoliposomes the release rate has decreased, which led to a prolonged antibacterial and antioxidant activity. The meat packed with this film was preserved for 20 days at 4 °C without spoilage [170].

Cellulose is the other most common polysaccharide used for edible films and for biodegradable packaging materials. It can be obtained from a great variety of sources (vegetal or bacterial). The poor mechanical performance and the lack of water resistance are the factors that limit the applicability domains. By loading the cellulose with an antimicrobial agent, the added value to packaging materials increases. The literature presents many cellulose-based edible films, for example obtained from CMC and glycerol mixed with essential oil from lemon, orange, or other citrus as antibacterial agent. The tests have indicated substantial increases of the antibacterial activity against *E. coli* or *S. aureus* by adding up to 2% essential oil (MIC 100, 250, and 225 mg/mL for lime, lemon, and orange) [210].

Ortiz, C.M. et al. have used microfibrillated cellulose mixed with soy proteins and glycerin as plasticizer to obtain biodegradable packaging. Further, the film was loaded with clove essential oil to enhance the antimicrobial and antioxidant activity against spoilage microorganisms. The addition of the essential oil had also the effect of a plasticizer, improved WVP, but increased oxygen permeability. The cellulose fibers had 50–60 nm diameter and ~35% crystallinity [211]. A similar composition based on soy protein, cellulose nanocrystals, and pine needle extract was reported by Yu, Z. et al. [212]. The use of cellulose nanocrystals and pine extract lowered WVP for the film by impeding on hydrogen bond formation and increased the mechanical strength.

In order to improve the mechanical and barrier properties cellulose can be reinforced with other polymers or with nanoparticles. The normal choice rests with natural, abundant polymers, which can be used as they are or after chemical modification [213]. Cellulose can be combined not only with other polysaccharides [214] but also with proteins [215,216] or lipids [217,218]. Zhang X. et al. have obtained by coagulation into a LiCl/ dimethyl acetamide/AgNO$_3$/PVP solution, in one step synthesis, a cellulosic film with embedded AgCl crystals. An important role for the embedding process is played by the presence of PVP. By light exposure some of the AgCl was decomposed and Ag@AgCl nanoparticles were generated. The shape and size of these nanoparticles can be tuned by PVP concentration. The test made on *E. coli* and *Staphylococcus aureus* have demonstrated a strong antibacterial activity, after three hours of exposure no viable bacteria being identified [219].

Zahedi, Y. et al. reports the synthesis of a composite film based on CMC with 5% montmorillonite and 1–4% ZnO [220]. The addition of these nanoparticles decreased WVP values by 53% and enhanced the mechanical resistance. The nanoparticles also were an efficient UV blocking agent. Tyagi, P. et al. obtained an improved paper-based package material by coating it with a composite made of nanocellulose, montmorillonite, soy protein, and alkyl ketene dimer [221]. When researchers obtain similar composites based on cellulose or chitosan (with montmorillonite and essential oils for example) the intrinsic antibacterial activity of chitosan will always make the difference between them [222]. Elsewhere [223], AgNPs immobilized on laponite were obtained with quaternized chitosan and used for litchi packaging. Turning chitosan into a quaternary ammonium salt or subjecting it to other chemical modifications is an efficient way to enhance its properties. CMC can be mixed in any proportion with such modified chitosan as the two substances are miscible because of the formation of hydrogen bonds. The composite film has better mechanical and thermal performances. The WVP is decreased, but the oxygen permeability is increased in such composites. The CMC-chitosan composite has antibacterial activity against *S. aureus* and *E. coli*, and the test made on fresh bananas indicated an increased shelf life for the coated fruits [224].

Bacterial cellulose (BC) from *Gluconacetobacter xylinus* can be transformed into cellulose nanocrystals by acidic hydrolysis. These BC nanocrystals with size of 20–30 nm, together with silver nanoparticles (AgNPs) with size of 35–50 nm can be embedded into chitosan films. The embedding process has a significative influence on the color and transparency of the films, but it greatly enhances the mechanical and barrier properties. The forming of new hydrogen bonds between chitosan and BC indicate that the embedding is not only physical, but also implying a chemical process. The antimicrobial properties of these films vs. food specific pathogenic agents, suggest a synergic action of chitosan and BC-incorporated nanoparticles [225]. Incorporation of AgNPs into the methylcellulose films can be achieved with *Lippia alba* extract. The films have a smaller mechanical resistance and elastic modulus than the control lot, but they are easier to stretch and have a higher hydrophobicity. In the same time the antioxidant and antimicrobial activity are higher when compared with same control films [226].

The literature reports also nanocomposite films based on cellulose that contain not only AgNPs but also oxide nanoparticles from transition metals like CuO or ZnO. During production of regenerated cellulose fibers from cotton or microcrystalline cellulose from BC, oxide nanoparticles can be attached to the fiber surface by multiple hydrogen bonds (generated by hydroxyl moieties), without any additional functionalization. The number of oxide nanoparticles attached to the cellulose is nevertheless smaller than the AgNPs that can be loaded. The thermal stability of the composite films with AgNPs and ZnONPs is higher if compared with bare cellulose film. Antibacterial tests done on *E. coli* and *L. monocytogenes* indicate a strong inhibition of this strains [227].

Starch is also a common polysaccharide used for edible films, and unlike cellulose, it has also a nutritional value for humans. Simple starch films mixed with various natural compounds can form efficient antimicrobial packaging materials [228]. Starch films loaded with clove leaf oil can preserve *Listeria* inoculated cheese for 24 days. The clove leaf oil fulfills multiple roles, as it increases tensile strength and elongation break, but also inhibited *L. monocytogenes* proliferation, acted as UV barrier and radical scavenger [176]. An innovative design based on cheese whey and starch from agricultural waste is presented by Dinika et al. [229]. In this case the AMP from cheese whey are responsible for the activity of the final edible antimicrobial packaging film. Trongchuen, K. et al., have also obtained a starch-based composite that can be used as antimicrobial packaging material. Tests indicated a good activity against *E. coli* and *S. aureus* for composition containing starch foam, 10% extracted spent coffee ground (SCG), and 8% oregano essential oil [230]. A similar composition but with enhanced antimicrobial and antioxidant properties was developed by Ounkaew A. et al. who mixed starch with PVA and in the resulting nanocomposite-incorporated SCG and citric acid [231].

Other authors used bentonite clay to reinforce the starch-based film (cassava starch) in order to improve the mechanical properties. As plasticizer 2% glycerol was used. To ensure antimicrobial activity, 2.5% cinnamon essential oil was added. Antibacterial tests on *E. coli*, *S. aureus*, and *Salmonella typhimurium* confirmed the capacity of this packaging material to protect the food against such strains. Test made on meat balls indicates that the film was able to extend the shelf life up to 96 h (keeping the bacterial population under the FDA recommended value of 10^6 CFU/g) [232].

3.3. Other Biopolymers-Based Films

Another biopolymer that has attracted much attention is the PLA, which can be easily produced by bacterial fermentation of some renewable resources like corn starch or sugar beet. At the end of economic life, the PLA products are easily decomposed as they are biodegradable. The main problem for PLA films to be accepted as a good candidate for food packages is the poor barrier properties, as oxygen can easily permeate thru PLA films.

Composite based on PLA and nanocellulose loaded with plant essential oils (from *Mentha piperita* or *Bunium percicum*) were used to extend shelf life for grounded beef up to 12 days in refrigerated condition [171]. While the authors report no significant differences of WVP when adding nanocellulose to PLA, the values decrease to 50% when essential oils are added. Therefore some authors studied the possibility of using PLA films loaded only with some major components of essential oils as antimicrobials [233]. Better properties are reported when PLA is mixed with proteins before being loaded with essential oils [234].

Another way to improve the WVP is presented by Zhu J.Y. et al., who have obtained bi and tri-layer films from PLA and gliadin, without surface modification [158]. These films have a good mechanical resistance, do not exfoliate, and most important, the gliadin improves the barrier properties of the PLA toward oxygen, while the PLA improves the barrier properties of gliadin toward water vapor. By loading the films with antimicrobial agents such thymol, the authors have obtained a biodegradable polymeric film with antibacterial activity, which is suitable for food packaging. Moreover, the film has exhibit different antimicrobial activity on each side, the gliadin side being more potent than the PLA, indicating a preferential release (Figure 9).

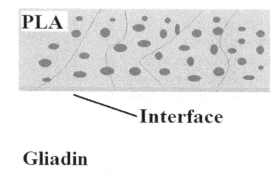

Figure 9. The interface between polylactic acid (PLA) and gliadin layers—adapted after information presented in [158].

Yin, X. et al. have also synthetized PLA films in which they added salicylic acid (SA). SA is heavily utilized after harvesting for fruits and vegetables preservation. The authors have demonstrated that if PLA film is loaded with up to 6% SA the antibacterial activity against *E. coli* can reach 97.63% while the WVP drops to 11.4×10^{-14} g cm/(cm^2 s Pa) [235].

Three-components composite films, based on PLA mixed with chitosan and loaded with tea polyphenols, in various molar rapports have been obtained by Ye, J. et al. [236]. The three-component film has superior mechanical properties, lower WVP, and better heat-sealability. While the pure PLA has a strength of heat-seal of 2.75 ± 0.41 N/15 mm, the composite film with PLA-tea polyphenols-chitosan presents a strength up to three times higher. The large number of -OH and -NH$_2$ moieties from the chitosan polymeric chains and its viscosity ensures an easy and durable welding of the membrane. The tests made on cherries have indicated that the optimum ratio between tea polyphenols and chitosan is 3:7 for prolonging the shelf life from 2 to 8 days at room temperature. In this case was observed the lowest degradation speed and was obtained the best preservation of the vitamins and total solid substance. The vitamin C content was higher for all samples packed with films with tea polyphenols as these substances have antioxidant activity and thus protect vitamin C from interacting with free radicals, decreasing the decay speed.

Like other polymers used for food packaging, the PLA films can also be reinforced with nanoparticles, which can exhibit antimicrobial activity [237]. Zhang, C. et al. have added up to 10% *w/w* AgNPs into PLA obtaining a film that has the capacity to prolong freshness of fruits (strawberries). The optimum results were reported for the film containing 5% AgNPs [150]. Li, W. et al. [238] have reported that both TiO$_2$ and Ag nanoparticles can be embedded into PLA. The obtained films have lower WVP when compared with the bare polymer, lower transparency, and lower elastic modulus. The FTIR analysis has failed to evidence any new bonds, indicating the physical embedding of nanoparticles into the PLA. Same results were obtained by DSC, as the vitreous transition and melting temperatures remained unchanged. The only observable modification is the crystallinity degree due to the presence of nanoparticles. Although there is only a physical embedding, the migration tests revealed a good nanocomposite stability, with no detectable levels of nanoparticles migrating into the food. The antimicrobial activity has been tested against *E. coli* and *L. monocytogenes* as typical microorganisms infecting food. As the TiO_2 percent increased from 1 to 5% the antimicrobial activity improved, the bacterial concentration decreased from 9.12 to 3.45 log CFU/mL. By adding the AgNPs along with TiO_2 the bacterial concentration further decreased to 3.06 log CFU/mL. Antibacterial activity against *E. coli* for PLA-based films reinforced with aluminum-doped ZnO nanoparticles was reported by Valerini, D. et al. [239] and PLA films loaded with ZnO nanoparticles were investigated for preserving fresh cut apple in [240].

Gelatin films have also been tested in conjunction with various antimicrobial agents as active packaging materials [157,167,241]. In order to slow the release kinetics of the antimicrobials, the active substance can be encapsulated in a copolymer, in beta-cyclodextrin, in clays nanoparticles (like montmorillonite, halloysite etc.) or can be included as a nanoemulsion [241,242]. In order to improve the mechanical and barrier properties, gelatin can be reinforced with other polymers like polycaprolactone [133], sodium alginate [243], chitosan [244], starch [245], cellulose, and derivatives [246]. The mechanical properties are improved also by adding nanoparticles like ZnO [163], TiO_2 [247], Ag [248].

The bionanocomposites based on poly-(butylene adipate-co-terephthalate)/ Ag_2O (PBAT/Ag_2O) obtained directly from solvent ($CHCl_3$) have presented promising properties for antimicrobial packaging. The best properties were obtained for 1–10% Ag_2O load on the PBAT film. Over 10% Ag_2O the nanoparticles present a tendency to form agglomerates. The mechanical properties of composites were superior to the matrix polymer film up to 7% Ag_2O. Most importantly, the permeability of oxygen and water vapor was lower for the nanocomposite films than for the simple PBAT film. The antimicrobial activity against *Klebsiella pneumonia* and *S. aureus* strains indicate that these nanocomposite films can be used in food industry [249]. Nisin can also be incorporated into PBAT polymeric matrix to produce an antimicrobial packaging [250], with good results against *L. monocytogenes*.

4. Toxicity Studies

The materials used for the innovative antimicrobial food packaging films are usually biopolymers that are biocompatible and safe for human use. Most of them are classified as GRAS and are edible. For example, chitosan or cellulose are considered GRAS in micron scale. While nanocellulose indicate minimal cytotoxicity it still may have an impact over gut microbial population and alter intestinal function, by reducing nutrient absorption [251–253]. The use of nanomaterials in food packaging brings a series of advantages as direct antimicrobial activity, support for other antibacterial agents or sensors to detect food contaminants (mostly metabolic products of bacterial activity), but at the same time there is a need to evaluate the risks implied by the use of metallic and oxide nanoparticles. The nanoparticles interact with bacterial cells and can even inhibit biofilms formation, but at the same time can migrate into food and interact with human cells [254–257]. Therefore, before marketing any novel antimicrobial food packaging, a careful study of nanoparticles migration into food and their associated toxicology must be undertaken. From the available literature studies, it is clear that nanoparticles present toxicological effects in living organisms and may have a profound impact on environmental ecosystems [258–260].

AgNPs are among the most studied and used antibacterial agents and therefore many research are devoted to the AgNPs effects and implications for human health [261–263]. Studies on zebrafish indicate that AgNPs and AuNPs enhance the secretion of alanine aminotransferase and aspartate aminotransferase. The increased levels of these enzymes lead to increased level generation of ROS, that can cause oxidative stress and immunotoxicity. AgNPs also induced formation of micronuclei and nuclear abnormalities [264]. Unfortunately, with the increased application of nanomaterials, metal nanoparticles have been widely detected in the environment. There are alarming reports about nanoparticles accumulation in seafood [265–267], but also about their cytotoxic and genotoxic effect in plants [268–270].

Toxicological concerns exist also for copper and copper oxide nanoparticles as studies indicate a strong relation between decreasing nanoparticles size and increasing phytotoxicity [271–273]. Some recent studies also indicate effects on plastids, mitochondria, protoplasm, and membranes for *Hordeum sativum distichum* [274] or *Abelmoschus esculentus* [275].

ZnO and TiO$_2$ nanoparticles are widely used not only in antimicrobial packaging but also in cosmetics and paints. ZnO is classified by FDA as GRAS and its toxicological effects are still under investigation, but some reports on ZnO nanoparticles are already alarming [276–278]. ZnO nanoparticles can pass the stomach and enter in the digestion process in intestine [279] and can induce hyperproliferation of malignant cells [280]. Studies made on mice indicated that TiO$_2$ nanoparticles alter gastrointestinal homeostasis, induce cytotoxicity and genotoxicity [281–285].

For essential oils the studies report no toxicity at the concentration used in food packaging [286–288]. There are some concerns that at inhibitory concentration for pathogenic bacteria the essential oil can negatively impact also the beneficial bacteria [289].

5. Conclusions

As we presented here, there are many strategies that can be adopted to combat the food loss. The use of antimicrobial packaging is gaining ground as both regulatory bodies authorities and general public are taking steps toward acceptance. The general trend is to implement new biodegradable polymers so both problems (antimicrobial activity and plastic pollution) can be addressed. Emphasis is on natural, abundant, and cheap polymers, but they require modification and carefully chosen additives in order to confer the properties required by the packaging industry.

Beside the main polymer, the composite usually contains at least one additive as a plasticizer. This additive plays an important role in achieving the desired mechanical properties. Researchers have a wide range of compounds to choose from: xylitol, mannitol, glycerol, sucrose, polyethylene glycol (PEG), sorbitol, fatty acids, urea, corn syrup, soy proteins, etc. Plasticizers can represent a weight percent between 10 and 65%, depending on the polymer rigidity. Beside the plasticizers, other additives might be included into the polymeric mix, such as antioxidants, antimicrobials, emulsifiers, nutraceuticals, probiotics, etc. Sometimes these additives influence the mechanical properties. While nanoparticles are also reinforcing agents that usually increase the mechanical resistance, essential oils increase the elasticity of the film. Some additives modify the barrier properties, like essential oils or other lipids that decrease WVP.

The replacement of traditional non-compostable oil-derived plastics with antimicrobial biodegradable packaging materials brought new challenges due to incorporation of the antimicrobials into the polymeric matrix, compatibilities between various components and easier degradation by heat and light. Some of the antimicrobials suffer a high loss rate because of inherent volatilization (essential oils for example) and future studies are required to improve the durability and efficiency of novel antimicrobial packaging materials.

The increased demand for cleaner and less processed food, with lower additives content also led to the development of antimicrobial packaging. Beyond the desired antimicrobial activity, the additives used in these innovative packaging can be responsible for the contamination of the food. Migration of undesirable substances must be under the limits established by regulations to ensure the consumer's safety. For nanoparticles specific toxicological test must be carried out to indicate whether they are safe or not for humans in long term. Also, one must take into account the possible long-term environmental impact of these antimicrobial substances once they end up to landfill. Therefore, reducing the toxicity and environmental impact of synthetic antimicrobial agents is an important subject for further research.

The presence of various additives into the polymeric film must also not affect in a negative way the organoleptic properties of the food, otherwise consumers will reject the new packaging. The packaging must be as transparent as possible because the customers want to see what they are buying. Wrong tints that will give the impression of dusty, dirty, or oxidized will not be received well. Once this kind of antimicrobial packaging is available at a larger scale, the extension of shelf life will become a reality, and since the producers usually mark only the expiry date, customers will not be able to judge the exact freshness of the food. Nevertheless, both regulatory bodies and consumer pressure will shape the future of antimicrobial packaging toward novel, cost efficient, bio-degradable materials that can ensure food safety, quality, and longer shelf life with fewer additives.

Author Contributions: Conceptualization, O.C.O. and D.A.K.; resources, D.A.K.; writing—original draft preparation, L.M. and D.F.; writing—review and editing, A.F. and O.C.O.; visualization, L.M.; supervision, E.A.; project administration, E.A.; funding acquisition, O.C.O. All authors have read and agreed to the published version of the manuscript.

References

1. Technical Platform on the Measurement and Reduction of Food Loss and Waste. Available online: http://www.fao.org/platform-food-loss-waste/en/ (accessed on 25 August 2020).
2. Ishangulyyev, S.; Kim, S.; Lee, S.H. Understanding Food Loss and Waste-Why Are We Losing and Wasting Food? *Foods* **2019**, *8*, 297. [CrossRef] [PubMed]
3. Jiang, B.; Na, J.X.; Wang, L.L.; Li, D.M.; Liu, C.H.; Feng, Z.B. Reutilization of Food Waste: One-Step Extration, Purification and Characterization of Ovalbumin from Salted Egg White by Aqueous Two-Phase Flotation. *Foods* **2019**, *8*, 286. [CrossRef] [PubMed]
4. Food Packaging Market Size, Share&Trends Analysis Report By Type (Rigid, Flexible), By Material (Paper, Plastic), By Application (Bakery and Confectionery, Dairy Products), By Region, And Segment Forecasts, 2020–2027. Available online: https://www.grandviewresearch.com/industry-analysis/food-packaging-market (accessed on 26 September 2020).
5. Wohner, B.; Gabriel, V.H.; Krenn, B.; Krauter, V.; Tacker, M. Environmental and economic assessment of food-packaging systems with a focus on food waste. Case study on tomato ketchup. *Sci. Total Environ.* **2020**, *738*. [CrossRef]
6. EU Circular Economy Action Plan. A New Circular Economy Action Plan for a Cleaner and More Competitive Europe. Available online: https://ec.europa.eu/environment/circular-economy/index_en.htm (accessed on 26 September 2020).
7. Kuster-Boluda, I.; Vila, N. Can Health Perceptions, Credibility, and Physical Appearance of Low-Fat Foods Stimulate Buying Intentions? *Foods* **2020**, *9*, 866. [CrossRef] [PubMed]
8. Kuster, I.; Vila, N.; Sarabia, F. Food packaging cues as vehicles of healthy information: Visions of millennials (early adults and adolescents). *Food Res. Int.* **2019**, *119*, 170–176. [CrossRef]
9. Hernandez, A.; Vila, N.; Kuster, I.; Rodriguez, C. Clustering Spanish alcoholic beverage shoppers to focus marketing strategies. *Int. J. Wine Bus. Res.* **2019**, *31*, 362–384. [CrossRef]
10. Vila-Lopez, N.; Kuster-Boluda, I. Consumers' physiological and verbal responses towards product packages: Could these responses anticipate product choices? *Physiol. Behav.* **2019**, *200*, 166–173. [CrossRef]
11. Vila-Lopez, N.; Küster-Boluda, I. A bibliometric analysis on packaging research: Towards sustainable and healthy packages. *Br. Food J.* **2020**. ahead of print. [CrossRef]
12. Bou-Mitri, C.; Abdessater, M.; Zgheib, H.; Akiki, Z. Food packaging design and consumer perception of the product quality, safety, healthiness and preference. *Nutr. Food Sci.* **2020**. ahead-of-print. [CrossRef]
13. Makaremi, M.; Yousefi, H.; Cavallaro, G.; Lazzara, G.; Goh, C.B.S.; Lee, S.M.; Solouk, A.; Pasbakhsh, P. Safely Dissolvable and Healable Active Packaging Films Based on Alginate and Pectin. *Polymers* **2019**, *11*, 1594. [CrossRef]
14. Kaladharan, P.; Singh, V.V.; Asha, P.S.; Edward, L.; Sukhadane, K.S. Marine plastic litter in certain trawl grounds along the peninsular coasts of India. *Mar. Pollut. Bull.* **2020**, *157*, 111299. [CrossRef] [PubMed]
15. Chen, F.Y.; Chen, H.; Yang, J.H.; Long, R.Y.; Li, W.B. Impact of regulatory focus on express packaging waste recycling behavior: Moderating role of psychological empowerment perception. *Environ. Sci. Pollut. Res.* **2019**, *26*, 8862–8874. [CrossRef] [PubMed]

16.   White, A.; Lockyer, S. Removing plastic packaging from fresh produce—What's the impact? *Nutr. Bull.* **2020**, *45*, 35–50. [CrossRef]

17.   Guo, J.J.; Huang, X.P.; Xiang, L.; Wang, Y.Z.; Li, Y.W.; Li, H.; Cai, Q.Y.; Mo, C.H.; Wong, M.H. Source, migration and toxicology of microplastics in soil. *Environ. Int.* **2020**, *137*, 105263. [CrossRef] [PubMed]

18.   Boz, Z.; Korhonen, V.; Sand, C.K. Consumer Considerations for the Implementation of Sustainable Packaging: A Review. *Sustainability* **2020**, *12*, 2192. [CrossRef]

19.   Clark, N.; Trimingham, R.; Wilson, G.T. Incorporating Consumer Insights into the UK Food Packaging Supply Chain in the Transition to a Circular Economy. *Sustainability* **2020**, *12*, 6106. [CrossRef]

20.   Zemljic, L.F.; Plohl, O.; Vesel, A.; Luxbacher, T.; Potrc, S. Physicochemical Characterization of Packaging Foils Coated by Chitosan and Polyphenols Colloidal Formulations. *Int. J. Mol. Sci.* **2020**, *21*, 495. [CrossRef] [PubMed]

21.   Esmailzadeh, H.; Sangpour, P.; Shahraz, F.; Eskandari, A.; Hejazi, J.; Khaksar, R. CuO/LDPE nanocomposite for active food packaging application: A comparative study of its antibacterial activities with ZnO/LDPE nanocomposite. *Polym. Bull.* **2020**. published online. [CrossRef]

22.   Zia, J.; Paul, U.C.; Heredia-Guerrero, J.A.; Athanassiou, A.; Fragouli, D. Low-density polyethylene/curcumin melt extruded composites with enhanced water vapor barrier and antioxidant properties for active food packaging. *Polymer* **2019**, *175*, 137–145. [CrossRef]

23.   Habib, S.; Lehocky, M.; Vesela, D.; Humpolicek, P.; Krupa, I.; Popelka, A. Preparation of Progressive Antibacterial LDPE Surface via Active Biomolecule Deposition Approach. *Polymers* **2019**, *11*, 1704. [CrossRef]

24.   Giannakas, A.; Salmas, C.; Leontiou, A.; Tsimogiannis, D.; Oreopoulou, A.; Braouhli, J. Novel LDPE/Chitosan Rosemary and Melissa Extract Nanostructured Active Packaging Films. *Nanomaterials* **2019**, *9*, 1105. [CrossRef] [PubMed]

25.   Kuster-Boluda, A.; Vila, N.V.; Kuster, I. Managing international distributors' complaints: An exploratory study. *J. Bus. Ind. Mark.* **2020**. [CrossRef]

26.   Kerry, J.P.; O'Grady, M.N.; Hogan, S.A. Past, current and potential utilisation of active and intelligent packaging systems for meat and muscle-based products: A review. *Meat Sci.* **2006**, *74*, 113–130. [CrossRef] [PubMed]

27.   O'Callaghan, K.A.M.; Kerry, J.P. Assessment of the antimicrobial activity of potentially active substances (nanoparticled and non-nanoparticled) against cheese-derived micro-organisms. *Int. J. Dairy Technol.* **2014**, *67*, 483–489. [CrossRef]

28.   Rai, M.; Ingle, A.P.; Gupta, I.; Pandit, R.; Paralikar, P.; Gade, A.; Chaud, M.V.; dos Santos, C.A. Smart nanopackaging for theenhancement of foodshelf life. *Environ. Chem. Lett.* **2019**, *17*, 277–290. [CrossRef]

29.   Vilas, C.; Mauricio-Iglesias, M.; Garcia, M.R. Model-based design of smart active packaging systems with antimicrobial activity. *Food Packag. Shelf* **2020**, *24*, 100446. [CrossRef]

30.   Jayasena, D.D.; Jo, C. Essential oils as potential antimicrobial agents in meat and meat products: A review. *Trends Food Sci. Technol.* **2013**, *34*, 96–108. [CrossRef]

31.   Szabo, K.; Teleky, B.E.; Mitrea, L.; Calinoiu, L.F.; Martau, G.A.; Simon, E.; Varvara, R.A.; Vodnar, D.C. Active Packaging-Poly(Vinyl Alcohol) Films Enriched with Tomato By-Products Extract. *Coatings* **2020**, *10*, 141. [CrossRef]

32.   Motelica, L.; Ficai, D.; Oprea, O.C.; Ficai, A.; Andronescu, E. Smart Food Packaging Designed by Nanotechnological and Drug Delivery Approaches. *Coatings* **2020**, *10*, 806. [CrossRef]

33.   Shruthy, R.; Jancy, S.; Preetha, R. Cellulose nanoparticles synthesised from potato peel for the development of active packaging film for enhancement of shelf life of raw prawns (*Penaeus monodon*) during frozen storage. *Int. J. Food Sci. Technol.* **2020**. [CrossRef]

34.   Han, J.H. Antimicrobial food packaging. *Food Technol. Chic.* **2000**, *54*, 56–65.

35.   Surendhiran, D.; Li, C.Z.; Cui, H.Y.; Lin, L. Fabrication of high stability active nanofibers encapsulated with pomegranate peel extract using chitosan/PEO for meat preservation. *Food Packag. Shelf Life* **2020**, *23*, 100439. [CrossRef]

36.   Settier-Ramirez, L.; Lopez-Carballo, G.; Gavara, R.; Hernandez-Munoz, P. PVOH/protein blend films embedded with lactic acid bacteria and their antilisterial activity in pasteurized milk. *Int. J. Food Microbiol.* **2020**, *322*, 108545. [CrossRef]

37. Ramos, M.; Beltran, A.; Fortunati, E.; Peltzer, M.; Cristofaro, F.; Visai, L.; Valente, A.J.M.; Jimenez, A.; Kenny, J.M.; Garrigos, M.C. Controlled Release of Thymol from Poly(Lactic Acid)-Based Silver Nanocomposite Films with Antibacterial and Antioxidant Activity. *Antioxidants* **2020**, *9*, 395. [CrossRef] [PubMed]

38. Chiabrando, V.; Garavaglia, L.; Giacalone, G. The Postharvest Quality of Fresh Sweet Cherries and Strawberries with an Active Packaging System. *Foods* **2019**, *8*, 335. [CrossRef] [PubMed]

39. Ojogbo, E.; Ward, V.; Mekonnen, T.H. Functionalized starch microparticles for contact-active antimicrobial polymer surfaces. *Carbohyd. Polym.* **2020**, *229*, 115442. [CrossRef] [PubMed]

40. Pan, Y.F.; Xia, Q.Y.; Xiao, H.N. Cationic Polymers with Tailored Structures for Rendering Polysaccharide-Based Materials Antimicrobial: An Overview. *Polymers* **2019**, *11*, 1283. [CrossRef]

41. Brezoiu, A.M.; Prundeanu, M.; Berger, D.; Deaconu, M.; Matei, C.; Oprea, O.; Vasile, E.; Negreanu-Pirjol, T.; Muntean, D.; Danciu, C. Properties of Salvia officinalis L. and Thymus serpyllum L. Extracts Free and Embedded into Mesopores of Silica and Titania Nanomaterials. *Nanomaterials* **2020**, *10*, 820. [CrossRef] [PubMed]

42. Zhang, Z.J.; Wang, X.J.; Gao, M.; Zhao, Y.L.; Chen, Y.Z. Sustained release of an essential oil by a hybrid cellulose nanofiber foam system. *Cellulose* **2020**, *27*, 2709–2721. [CrossRef]

43. Fang, Z.X.; Zhao, Y.Y.; Warner, R.D.; Johnson, S.K. Active and intelligent packaging in meat industry. *Trends Food Sci. Technol.* **2017**, *61*, 60–71. [CrossRef]

44. Realini, C.E.; Marcos, B. Active and intelligent packaging systems for a modern society. *Meat Sci.* **2014**, *98*, 404–419. [CrossRef] [PubMed]

45. Yildirim, S.; Rocker, B.; Pettersen, M.K.; Nilsen-Nygaard, J.; Ayhan, Z.; Rutkaite, R.; Radusin, T.; Suminska, P.; Marcos, B.; Coma, V. Active Packaging Applications for Food. *Compr. Rev. Food Sci. Food Saf.* **2018**, *17*, 165–199. [CrossRef]

46. Malhotra, B.; Keshwani, A.; Kharkwal, H. Antimicrobial food packaging: Potential and pitfalls. *Front. Microbiol.* **2015**, *6*. [CrossRef] [PubMed]

47. Suppakul, P.; Miltz, J.; Sonneveld, K.; Bigger, S.W. Antimicrobial properties of basil and its possible application in food packaging. *J. Agric. Food Chem.* **2003**, *51*, 3197–3207. [CrossRef] [PubMed]

48. Saadat, S.; Pandey, G.; Tharmavaram, M.; Braganza, V.; Rawtani, D. Nano-interfacial decoration of Halloysite Nanotubes for the development of antimicrobial nanocomposites. *Adv. Colloid Interface Sci.* **2020**, *275*, 102063. [CrossRef]

49. Arsenie, L.V.; Lacatusu, I.; Oprea, O.; Bordei, N.; Bacalum, M.; Badea, N. Azelaic acid-willow bark extract-panthenol—Loaded lipid nanocarriers improve the hydration effect and antioxidant action of cosmetic formulations. *Ind. Crops Prod.* **2020**, *154*, 112658. [CrossRef]

50. Arfat, Y.A.; Ejaz, M.; Jacob, H.; Ahmed, J. Deciphering the potential of guar gum/Ag-Cu nanocomposite films as an active food packaging material. *Carbohyd. Polym.* **2017**, *157*, 65–71. [CrossRef]

51. Nedelcu, I.A.; Ficai, A.; Sonmez, M.; Ficai, D.; Oprea, O.; Andronescu, E. Silver Based Materials for Biomedical Applications. *Curr. Org. Chem.* **2014**, *18*, 173–184. [CrossRef]

52. Pica, A.; Guran, C.; Andronescu, E.; Oprea, O.; Ficai, D.; Ficai, A. Antimicrobial performances of some film forming materials based on silver nanoparticles. *J. Optoelectron. Adv. Mater.* **2012**, *14*, 863–868.

53. Oprea, O.; Andronescu, E.; Ficai, D.; Ficai, A.; Oktar, F.N.; Yetmez, M. ZnO Applications and Challenges. *Curr. Org. Chem.* **2014**, *18*, 192–203. [CrossRef]

54. Vasile, B.S.; Oprea, O.; Voicu, G.; Ficai, A.; Andronescu, E.; Teodorescu, A.; Holban, A. Synthesis and characterization of a novel controlled release zinc oxide/gentamicin-chitosan composite with potential applications in wounds care. *Int. J. Pharm.* **2014**, *463*, 161–169. [CrossRef] [PubMed]

55. Siripatrawan, U.; Kaewklin, P. Fabrication and characterization of chitosan-titanium dioxide nanocomposite film as ethylene scavenging and antimicrobial active food packaging. *Food Hydrocoll.* **2018**, *84*, 125–134. [CrossRef]

56. de Araujo, M.J.G.; Barbosa, R.C.; Fook, M.V.L.; Canedo, E.L.; Silva, S.M.L.; Medeiros, E.S.; Leite, I.F. HDPE/Chitosan Blends Modified with Organobentonite Synthesized with Quaternary Ammonium Salt Impregnated Chitosan. *Materials* **2018**, *11*, 291. [CrossRef] [PubMed]

57. Nouri, A.; Yaraki, M.T.; Ghorbanpour, M.; Agarwal, S.; Gupta, V.K. Enhanced Antibacterial effect of chitosan film using Montmorillonite/CuO nanocomposite. *Int. J. Biol. Macromol.* **2018**, *109*, 1219–1231. [CrossRef]

58. Salmas, C.; Giannakas, A.; Katapodis, P.; Leontiou, A.; Moschovas, D.; Karydis-Messinis, A. Development of ZnO/Na-Montmorillonite Hybrid Nanostructures Used for PVOH/ZnO/Na-Montmorillonite Active Packaging Films Preparation via a Melt-Extrusion Process. *Nanomaterials* **2020**, *10*, 1079. [CrossRef]

59. Lacatusu, I.; Badea, N.; Badea, G.; Brasoveanu, L.; Stan, R.; Ott, C.; Oprea, O.; Meghea, A. Ivy leaves extract based—Lipid nanocarriers and their bioefficacy on antioxidant and antitumor activities. *RSC Adv.* **2016**, *6*, 77243–77255. [CrossRef]

60. Lacatusu, I.; Badea, N.; Badea, G.; Oprea, O.; Mihaila, M.A.; Kaya, D.A.; Stan, R.; Meghea, A. Lipid nanocarriers based on natural oils with high activity against oxygen free radicals and tumor cell proliferation. *Mater. Sci. Eng. C* **2015**, *56*, 88–94. [CrossRef]

61. Yahaya, W.A.W.; Abu Yazid, N.; Azman, N.A.M.; Almajano, M.P. Antioxidant Activities and Total Phenolic Content of Malaysian Herbs as Components of Active Packaging Film in Beef Patties. *Antioxidants* **2019**, *8*, 204. [CrossRef]

62. Yu, H.H.; Kim, Y.J.; Park, Y.J.; Shin, D.M.; Choi, Y.S.; Lee, N.K.; Paik, H.D. Application of mixed natural preservatives to improve the quality of vacuum skin packaged beef during refrigerated storage. *Meat Sci.* **2020**, *169*, 108219. [CrossRef]

63. Huang, T.Q.; Qian, Y.S.; Wei, J.; Zhou, C.C. Polymeric Antimicrobial Food Packaging and Its Applications. *Polymers* **2019**, *11*, 560. [CrossRef]

64. Liang, S.M.; Wang, L.J. A Natural Antibacterial-Antioxidant Film from Soy Protein Isolate Incorporated with Cortex Phellodendron Extract. *Polymers* **2018**, *10*, 71. [CrossRef] [PubMed]

65. Gingasu, D.; Mindru, I.; Patron, L.; Ianculescu, A.; Vasile, E.; Marinescu, G.; Preda, S.; Diamandescu, L.; Oprea, O.; Popa, M.; et al. Synthesis and Characterization of Chitosan-Coated Cobalt Ferrite Nanoparticles and Their Antimicrobial Activity. *J. Inorg. Organomet. Polym.* **2018**, *28*, 1932–1941. [CrossRef]

66. Radulescu, M.; Ficai, D.; Oprea, O.; Ficai, A.; Andronescu, E.; Holban, A.M. Antimicrobial Chitosan based Formulations with Impact on Different Biomedical Applications. *Curr. Pharm. Biotechnol.* **2015**, *16*, 128–136. [CrossRef] [PubMed]

67. Nguyen, T.T.; Dao, U.T.T.; Bui, Q.P.T.; Bach, G.L.; Thuc, C.N.H.; Thuc, H.H. Enhanced antimicrobial activities and physiochemical properties of edible film based on chitosan incorporated with Sonneratia caseolaris (L.) Engl. leaf extract. *Prog. Org. Coat.* **2020**, *140*, 105487. [CrossRef]

68. Sofi, S.A.; Singh, J.; Rafiq, S.; Ashraf, U.; Dar, B.N.; Nayik, G.A. A Comprehensive Review on Antimicrobial Packaging and its Use in Food Packaging. *Curr. Nutr. Food Sci.* **2018**, *14*, 305–312. [CrossRef]

69. Mirabelli, V.; Salehi, S.M.; Angiolillo, L.; Belviso, B.D.; Conte, A.; Del Nobile, M.A.; Di Profio, G.; Caliandro, R. Enzyme Crystals and Hydrogel Composite Membranes as New Active Food Packaging Material. *Glob. Chall.* **2018**, *2*. [CrossRef]

70. Galante, Y.M.; Merlini, L.; Silvetti, T.; Campia, P.; Rossi, B.; Viani, F.; Brasca, M. Enzyme oxidation of plant galactomannans yielding biomaterials with novel properties and applications, including as delivery systems. *Appl. Microbiol. Biotechnol.* **2018**, *102*, 4687–4702. [CrossRef]

71. Avramescu, S.M.; Butean, C.; Popa, C.V.; Ortan, A.; Moraru, I.; Temocico, G. Edible and Functionalized Films/Coatings—Performances and Perspectives. *Coatings* **2020**, *10*, 687. [CrossRef]

72. Feng, K.; Wen, P.; Yang, H.; Li, N.; Lou, W.Y.; Zong, M.H.; Wu, H. Enhancement of the antimicrobial activity of cinnamon essential oil-loaded electrospun nanofilm by the incorporation of lysozyme. *RSC Adv.* **2017**, *7*, 1572–1580. [CrossRef]

73. He, S.K.; Yang, Q.M.; Ren, X.Y.; Zi, J.H.; Lu, S.S.; Wang, S.Q.; Zhang, Y.B.; Wang, Y.F. Antimicrobial Efficiency of Chitosan Solutions and Coatings Incorporated with Clove Oil and/or Ethylenediaminetetraacetate. *J. Food Saf.* **2014**, *34*, 345–352. [CrossRef]

74. Arvanitoyannis, I.S.; Stratakos, A.C. Application of Modified Atmosphere Packaging and Active/Smart Technologies to Red Meat and Poultry: A Review. *Food Bioprocess Technol.* **2012**, *5*, 1423–1446. [CrossRef]

75. Fratianni, F.; De Martino, L.; Melone, A.; De Feo, V.; Coppola, R.; Nazzaro, F. Preservation of Chicken Breast Meat Treated with Thyme and Balm Essential Oils. *J. Food Sci.* **2010**, *75*, M528–M535. [CrossRef] [PubMed]

76. Mulla, M.; Ahmed, J.; Al-Attar, H.; Castro-Aguirre, E.; Arfat, Y.A.; Auras, R. Antimicrobial efficacy of clove essential oil infused into chemically modified LLDPE film for chicken meat packaging. *Food Control* **2017**, *73*, 663–671. [CrossRef]

77. Radulescu, M.; Popescu, S.; Ficai, D.; Sonmez, M.; Oprea, O.; Spoiala, A.; Ficai, A.; Andronescu, E. Advances in Drug Delivery Systems, from 0 to 3D Superstructures. *Curr. Drug Targets* **2018**, *19*, 393–405. [CrossRef] [PubMed]

78. Lopes, F.A.; Soares, N.D.F.; Lopes, C.D.P.; da Silva, W.A.; Baffa, J.C.; Medeiros, E.A.A. Conservation of Bakery Products Through Cinnamaldehyde Antimicrobial Films. *Packag. Technol. Sci.* **2014**, *27*, 293–302. [CrossRef]

79. Rivero, S.; Giannuzzi, L.; Garcia, M.A.; Pinotti, A. Controlled delivery of propionic acid from chitosan films for pastry dough conservation. *J. Food Eng.* **2013**, *116*, 524–531. [CrossRef]

80. Mihaly-Cozmuta, A.; Peter, A.; Craciun, G.; Falup, A.; Mihaly-Cozmuta, L.; Nicula, C.; Vulpoi, A.; Baia, M. Preparation and characterization of active cellulose-based papers modified with TiO_2, Ag and zeolite nanocomposites for bread packaging application. *Cellulose* **2017**, *24*, 3911–3928. [CrossRef]

81. Al-Naamani, L.; Dutta, J.; Dobretsov, S. Nanocomposite Zinc Oxide-Chitosan Coatings on Polyethylene Films for Extending Storage Life of Okra (*Abelmoschus esculentus*). *Nanomaterials* **2018**, *8*, 479. [CrossRef]

82. Xing, Y.G.; Li, X.L.; Guo, X.L.; Li, W.X.; Chen, J.W.; Liu, Q.; Xu, Q.L.; Wang, Q.; Yang, H.; Shui, Y.R.; et al. Effects of Different TiO_2 Nanoparticles Concentrations on the Physical and Antibacterial Activities of Chitosan-Based Coating Film. *Nanomaterials* **2020**, *10*, 1365. [CrossRef]

83. Xing, Y.G.; Li, W.X.; Wang, Q.; Li, X.L.; Xu, Q.L.; Guo, X.L.; Bi, X.F.; Liu, X.C.; Shui, Y.R.; Lin, H.B.; et al. Antimicrobial Nanoparticles Incorporated in Edible Coatings and Films for the Preservation of Fruits and Vegetables. *Molecules* **2019**, *24*, 1695. [CrossRef]

84. Sun, L.J.; Yang, S.S.; Qian, X.R.; An, X.H. High-efficacy and long term antibacterial cellulose material: Anchored guanidine polymer via double "click chemistry". *Cellulose* **2020**, *27*, 8799–8812. [CrossRef]

85. Jouneghani, R.S.; Castro, A.H.F.; Panda, S.K.; Swennen, R.; Luyten, W. Antimicrobial Activity of Selected Banana Cultivars Against Important Human Pathogens, Including Candida Biofilm. *Foods* **2020**, *9*, 435. [CrossRef] [PubMed]

86. Zhang, S.B.; Yang, X.H.; Tang, B.; Yuan, L.J.; Wang, K.; Liu, X.Y.; Zhu, X.L.; Li, J.N.; Ge, Z.C.; Chen, S.G. New insights into synergistic antimicrobial and antifouling cotton fabrics via dually finished with quaternary ammonium salt and zwitterionic sulfobetaine. *Chem. Eng. J.* **2018**, *336*, 123–132. [CrossRef]

87. Braga, L.R.; Rangel, E.T.; Suarez, P.A.Z.; Machado, F. Simple synthesis of active films based on PVC incorporated with silver nanoparticles: Evaluation of the thermal, structural and antimicrobial properties. *Food Packag. Shelf Life* **2018**, *15*, 122–129. [CrossRef]

88. Bastante, C.C.; Cardoso, L.C.; Ponce, M.T.F.; Serrano, C.M.; de la Ossa-Fernandez, E.J.M. Characterization of olive leaf extract polyphenols loaded by supercritical solvent impregnation into PET/PP food packaging films. *J. Supercrit. Fluid* **2018**, *140*, 196–206. [CrossRef]

89. Ahmed, J.; Mulla, M.; Arfat, Y.A.; Bher, A.; Jacob, H.; Auras, R. Compression molded LLDPE films loaded with bimetallic (Ag-Cu) nanoparticles and cinnamon essential oil for chicken meat packaging applications. *LWT Food Sci. Technol.* **2018**, *93*, 329–338. [CrossRef]

90. Egodage, D.P.; Jayalath, H.T.S.; Samarasekara, A.M.P.B.; Amarasinghe, D.A.S.; Madushani, S.P.A.; Senerath, S.M.N.S. Novel Antimicrobial Nano Coated Polypropylene Based Materials for Food Packaging Systems. In Proceedings of the 2017 3rd International Moratuwa Engineering Research Conference (Mercon), Moratuwa, Sri Lanka, 29–31 May 2017; pp. 88–92.

91. Wieczynska, J.; Cavoski, I. Antimicrobial, antioxidant and sensory features of eugenol, carvacrol and trans-anethole in active packaging for organic ready-to-eat iceberg lettuce. *Food Chem.* **2018**, *259*, 251–260. [CrossRef] [PubMed]

92. Llana-Ruiz-Cabello, M.; Pichardo, S.; Bermudez, J.M.; Banos, A.; Ariza, J.J.; Guillamon, E.; Aucejo, S.; Camean, A.M. Characterisation and antimicrobial activity of active polypropylene films containing oregano essential oil and Allium extract to be used in packaging for meat products. *Food Addit. Contam. A* **2018**, *35*, 782–791. [CrossRef] [PubMed]

93. Dong, Z.; Xu, F.J.; Ahmed, I.; Li, Z.X.; Lin, H. Characterization and preservation performance of active polyethylene films containing rosemary and cinnamon essential oils for Pacific white shrimp packaging. *Food Control* **2018**, *92*, 37–46. [CrossRef]

94. da Silva, C.F.; de Oliveira, F.S.M.; Gaetano, V.F.; Vinhas, G.M.; Cardoso, S.A. Orange essential oil as antimicrobial additives in poly(vinyl chloride) films. *Polimeros* **2018**, *28*, 332–338. [CrossRef]

95. Krehula, L.K.; Papic, A.; Krehula, S.; Gilja, V.; Foglar, L.; Hrnjak-Murgic, Z. Properties of UV protective films of poly(vinyl-chloride)/TiO₂ nanocomposites for food packaging. *Polym. Bull.* **2017**, *74*, 1387–1404. [CrossRef]

96. Li, X.H.; Xing, Y.; Jiang, Y.H.; Ding, Y.L.; Li, W.L. Antimicrobial activities of ZnO powder-coated PVC film to inactivate food pathogens. *Int. J. Food Sci. Technol.* **2009**, *44*, 2161–2168. [CrossRef]

97. Singh, S.; Lee, M.; Gaikwad, K.K.; Lee, Y.S. Antibacterial and amine scavenging properties of silver-silica composite for post-harvest storage of fresh fish. *Food Bioprod. Process.* **2018**, *107*, 61–69. [CrossRef]

98. Gogliettino, M.; Balestrieri, M.; Ambrosio, R.L.; Anastasio, A.; Smaldone, G.; Proroga, Y.T.R.; Moretta, R.; Rea, I.; De Stefano, L.; Agrillo, B.; et al. Extending the Shelf-Life of Meat and Dairy Products via PET-Modified Packaging Activated With the Antimicrobial Peptide MTP1. *Front. Microbiol.* **2020**, *10*. [CrossRef] [PubMed]

99. Silva, C.C.G.; Silva, S.P.M.; Ribeiro, S.C. Application of Bacteriocins and Protective Cultures in Dairy Food Preservation. *Front. Microbiol.* **2018**, *9*. [CrossRef]

100. Santos, J.C.P.; Sousa, R.C.S.; Otoni, C.G.; Moraes, A.R.F.; Souza, V.G.L.; Medeiros, E.A.A.; Espitia, P.J.P.; Pires, A.C.S.; Coimbra, J.S.R.; Soares, N.F.F. Nisin and other antimicrobial peptides: Production, mechanisms of action, and application in active food packaging. *Innov. Food Sci. Emerg.* **2018**, *48*, 179–194. [CrossRef]

101. Agrillo, B.; Balestrieri, M.; Gogliettino, M.; Palmieri, G.; Moretta, R.; Proroga, Y.T.R.; Rea, I.; Cornacchia, A.; Capuano, F.; Smaldone, G.; et al. Functionalized Polymeric Materials with Bio-Derived Antimicrobial Peptides for "Active" Packaging. *Int. J. Mol. Sci.* **2019**, *20*, 601. [CrossRef]

102. Corrales-Urena, Y.R.; Souza-Schiaber, Z.; Lisboa, P.N.; Marquenet, F.; Michael Noeske, P.L.; Gatjen, L.; Rischka, K. Functionalization of hydrophobic surfaces with antimicrobial peptides immobilized on a bio-interfactant layer. *RSC Adv.* **2020**, *10*, 376–386. [CrossRef]

103. Torres, L.M.F.C.; Braga, N.A.; Gomes, I.P.; Almeida, M.T.; Santos, T.L.; de Mesquita, J.P.; da Silva, L.M.; Martins, H.R.; Kato, K.C.; dos Santos, W.T.P.; et al. Nanobiostructure of fibrous-like alumina functionalized with an analog of the BP100 peptide: Synthesis, characterization and biological applications. *Colloid Surf. B* **2018**, *163*, 275–283. [CrossRef]

104. Lu, P.; Guo, M.Y.; Xu, Z.J.; Wu, M. Application of Nanofibrillated Cellulose on BOPP/LDPE Film as Oxygen Barrier and Antimicrobial Coating Based on Cold Plasma Treatment. *Coatings* **2018**, *8*, 207. [CrossRef]

105. Zhu, R.N.; Liu, X.; Song, P.P.; Wang, M.; Xu, F.; Jiang, Y.J.; Zhang, X.M. An approach for reinforcement of paper with high strength and barrier properties via coating regenerated cellulose. *Carbohyd. Polym.* **2018**, *200*, 100–105. [CrossRef] [PubMed]

106. Shankar, S.; Rhim, J.W. Antimicrobial wrapping paper coated with a ternary blend of carbohydrates (alginate, carboxymethyl cellulose, carrageenan) and grapefruit seed extract. *Carbohyd. Polym.* **2018**, *196*, 92–101. [CrossRef] [PubMed]

107. Divsalar, E.; Tajik, H.; Moradi, M.; Forough, M.; Lotfi, M.; Kuswandi, B. Characterization of cellulosic paper coated with chitosan-zinc oxide nanocomposite containing nisin and its application in packaging of UF cheese. *Int. J. Biol. Macromol.* **2018**, *109*, 1311–1318. [CrossRef] [PubMed]

108. Brodnjak, U.V.; Tihole, K. Chitosan Solution Containing Zein and Essential Oil as Bio Based Coating on Packaging Paper. *Coatings* **2020**, *10*, 497. [CrossRef]

109. Mandolini, M.; Campi, F.; Favi, C.; Germanil, M. Manufacturing processes re-engineering for cost reduction: The investment casting case study. In *Proceedings of the Asme International Design Engineering Technical Conferences and Computers and Information in Engineering Conference, 2019, Vol 4*; American Society of Mechanical Engineers: New York, NY, USA, 2020; Volume 4.

110. Khezrian, A.; Shahbazi, Y. Application of nanocompostie chitosan and carboxymethyl cellulose films containing natural preservative compounds in minced camel's meat. *Int. J. Biol. Macromol.* **2018**, *106*, 1146–1158. [CrossRef] [PubMed]

111. Zhao, S.W.; Guo, C.R.; Hu, Y.Z.; Guo, Y.R.; Pan, Q.J. The preparation and antibacterial activity of cellulose/ZnO composite: A review. *Open Chem.* **2018**, *16*, 9–20. [CrossRef]

112. Kaya, M.; Khadem, S.; Cakmak, Y.S.; Mujtaba, M.; Ilk, S.; Akyuz, L.; Salaberria, A.M.; Labidi, J.; Abdulqadir, A.H.; Deligoz, E. Antioxidative and antimicrobial edible chitosan films blended with stem, leaf and seed extracts of Pistacia terebinthus for active food packaging. *RSC Adv.* **2018**, *8*, 3941–3950. [CrossRef]

113. Sun, L.J.; Sun, J.J.; Liu, D.J.; Fu, M.H.; Yang, X.; Guo, Y.R. The preservative effects of chitosan film incorporated with thinned young apple polyphenols on the quality of grass carp (*Ctenopharyngodon idellus*) fillets during cold storage: Correlation between the preservative effects and the active properties of the film. *Food Packag. Shelf Life* **2018**, *17*, 1–10. [CrossRef]

114. Beltran, C.A.U.; Florez, I.F. Preparation of bio-packages from native starches and essential oils prolonging shelf life in strawberries. *Rev. Colomb. Investig.* **2018**, *5*, 71–86. [CrossRef]

115. Issa, A.T.; Schimmel, K.A.; Worku, M.; Shahbazi, A.; Ibrahim, S.A.; Tahergorabi, R. Sweet Potato Starch-Based Nanocomposites: Development, Characterization, and Biodegradability. *Starch-Starke* **2018**, *70*. [CrossRef]

116. Tedeschi, G.; Guzman-Puyol, S.; Ceseracciu, L.; Paul, U.C.; Picone, P.; Di Carlo, M.; Athanassiou, A.; Heredia-Guerrero, J.A. Multifunctional Bioplastics Inspired by Wood Composition: Effect of Hydrolyzed Lignin Addition to Xylan-Cellulose Matrices. *Biomacromolecules* **2020**, *21*, 910–920. [CrossRef] [PubMed]

117. Nechita, P.; Roman, M. Review on Polysaccharides Used in Coatings for Food Packaging Papers. *Coatings* **2020**, *10*, 566. [CrossRef]

118. Fabra, M.J.; Falco, I.; Randazzo, W.; Sanchez, G.; Lopez-Rubio, A. Antiviral and antioxidant properties of active alginate edible films containing phenolic extracts. *Food Hydrocoll.* **2018**, *81*, 96–103. [CrossRef]

119. Aziz, M.S.A.; Salama, H.E.; Sabaa, M.W. Biobased alginate/castor oil edible films for active food packaging. *LWT Food Sci. Technol.* **2018**, *96*, 455–460. [CrossRef]

120. Hu, Z.; Hong, P.Z.; Liao, M.N.; Kong, S.Z.; Huang, N.; Ou, C.Y.; Li, S.D. Preparation and Characterization of Chitosan-Agarose Composite Films. *Materials* **2016**, *9*, 816. [CrossRef]

121. Kusznierewicz, B.; Staroszczyk, H.; Malinowska-Panczyk, E.; Parchem, K.; Bartoszek, A. Novel ABTS-dot-blot method for the assessment of antioxidant properties of food packaging. *Food Packag. Shelf Life* **2020**, *24*, 100478. [CrossRef]

122. Lei, J.; Zhou, L.; Tang, Y.J.; Luo, Y.; Duan, T.; Zhu, W.K. High-Strength Konjac Glucomannan/Silver Nanowires Composite Films with Antibacterial Properties. *Materials* **2017**, *10*, 524. [CrossRef]

123. Wu, C.H.; Li, Y.L.; Sun, J.S.; Lu, Y.Z.; Tong, C.L.; Wang, L.; Yan, Z.M.; Pang, J. Novel konjac glucomannan films with oxidized chitin nanocrystals immobilized red cabbage anthocyanins for intelligent food packaging. *Food Hydrocoll.* **2020**, *98*, 105245. [CrossRef]

124. Li, S.B.; Yi, J.J.; Yu, X.M.; Wang, Z.Y.; Wang, L. Preparation and characterization of pullulan derivative/chitosan composite film for potential antimicrobial applications. *Int. J. Biol. Macromol.* **2020**, *148*, 258–264. [CrossRef]

125. Hassan, A.H.A.; Cutter, C.N. Development and evaluation of pullulan-based composite antimicrobial films (CAF) incorporated with nisin, thymol and lauric arginate to reduce foodborne pathogens associated with muscle foods. *Int. J. Food Microbiol.* **2020**, *320*, 108519. [CrossRef]

126. da Silva, F.T.; da Cunha, K.F.; Fonseca, L.M.; Antunes, M.D.; El Halal, S.L.M.; Fiorentini, A.M.; Zavareze, E.D.; Dias, A.R.G. Action of ginger essential oil (*Zingiber officinale*) encapsulated in proteins ultrafine fibers on the antimicrobial control in situ. *Int. J. Biol. Macromol.* **2018**, *118*, 107–115. [CrossRef] [PubMed]

127. Echeverria, I.; Lopez-Caballero, M.E.; Gomez-Guillen, M.C.; Mauri, A.N.; Montero, M.P. Active nanocomposite films based on soy proteins-montmorillonite- clove essential oil for the preservation of refrigerated bluefin tuna (*Thunnus thynnus*) fillets. *Int. J. Food Microbiol.* **2018**, *266*, 142–149. [CrossRef] [PubMed]

128. Kouravand, F.; Jooyandeh, H.; Barzegar, H.; Hojjati, M. Characterization of cross-linked whey protein isolate-based films containing Satureja Khuzistanica Jamzad essential oil. *J. Food Process. Perservation* **2018**, *42*. [CrossRef]

129. Aziz, S.G.G.; Almasi, H. Physical Characteristics, Release Properties, and Antioxidant and Antimicrobial Activities of Whey Protein Isolate Films Incorporated with Thyme (*Thymus vulgaris* L.) Extract-Loaded Nanoliposomes. *Food Bioprocess. Technol.* **2018**, *11*, 1552–1565. [CrossRef]

130. Altan, A.; Aytac, Z.; Uyar, T. Carvacrol loaded electrospun fibrous films from zein and poly(lactic acid) for active food packaging. *Food Hydrocoll.* **2018**, *81*, 48–59. [CrossRef]

131. Ahmad, M.; Nirmal, N.P.; Danish, M.; Chuprom, J.; Jafarzedeh, S. Characterisation of composite films fabricated from collagen/chitosan and collagen/soy protein isolate for food packaging applications. *RSC Adv.* **2016**, *6*, 82191–82204. [CrossRef]

132. Ocak, B. Film-forming ability of collagen hydrolysate extracted from leather solid wastes with chitosan. *Environ. Sci. Pollut. Res.* **2018**, *25*, 4643–4655. [CrossRef]

133. Figueroa-Lopez, K.J.; Castro-Mayorga, J.L.; Andrade-Mahecha, M.M.; Cabedo, L.; Lagaron, J.M. Antibacterial and Barrier Properties of Gelatin Coated by Electrospun Polycaprolactone Ultrathin Fibers Containing Black Pepper Oleoresin of Interest in Active Food Biopackaging Applications. *Nanomaterials* **2018**, *8*, 199. [CrossRef]

134. Kim, H.; Beak, S.E.; Song, K.B. Development of a hagfish skin gelatin film containing cinnamon bark essential oil. *LWT Food Sci. Technol.* **2018**, *96*, 583–588. [CrossRef]

135. Abdollahzadeh, E.; Ojagh, S.M.; Fooladi, A.A.I.; Shabanpour, B.; Gharahei, M. Effects of Probiotic Cells on the Mechanical and Antibacterial Properties of Sodium-Caseinate Films. *Appl. Food Biotechnol.* **2018**, *5*, 155–162. [CrossRef]

136. Chevalier, E.; Assezat, G.; Prochazka, F.; Oulahal, N. Development and characterization of a novel edible extruded sheet based on different casein sources and influence of the glycerol concentration. *Food Hydrocoll.* **2018**, *75*, 182–191. [CrossRef]

137. Akyuz, L.; Kaya, M.; Ilk, S.; Cakmak, Y.S.; Salaberria, A.M.; Labidi, J.; Yilmaz, B.A.; Sargin, I. Effect of different animal fat and plant oil additives on physicochemical, mechanical, antimicrobial and antioxidant properties of chitosan films. *Int. J. Biol. Macromol.* **2018**, *111*, 475–484. [CrossRef] [PubMed]

138. Tonyali, B.; McDaniel, A.; Amamcharla, J.; Trinetta, V.; Yucel, U. Release kinetics of cinnamaldehyde, eugenol, and thymol from sustainable and biodegradable active packaging films. *Food Packag. Shelf Life* **2020**, *24*, 100484. [CrossRef]

139. Saurabh, C.K.; Gupta, S.; Variyar, P.S. Development of guar gum based active packaging films using grape pomace. *J. Food Sci. Technol.* **2018**, *55*, 1982–1992. [CrossRef]

140. Zhang, D.; Yu, G.H.; Long, Z.; Xiao, H.N.; Qian, L.Y. Bio-Wax Latex-Modified Paper as Antimicrobial and Water-Vapor-Resistant Packaging Material. *J. Wood Chem. Technol.* **2016**, *36*, 182–191. [CrossRef]

141. Swaroop, C.; Shukla, M. Nano-magnesium oxide reinforced polylactic acid biofilms for food packaging applications. *Int. J. Biol. Macromol.* **2018**, *113*, 729–736. [CrossRef] [PubMed]

142. Lopusiewicz, L.; Jedra, F.; Mizielinska, M. New Poly(lactic acid) Active Packaging Composite Films Incorporated with Fungal Melanin. *Polymers* **2018**, *10*, 386. [CrossRef]

143. Shao, P.; Yan, Z.P.; Chen, H.J.; Xiao, J. Electrospun poly(vinyl alcohol)/permutite fibrous film loaded with cinnamaldehyde for active food packaging. *J. Appl. Polym. Sci.* **2018**, *135*. [CrossRef]

144. Arrieta, M.P.; de Dicastillo, C.L.; Garrido, L.; Roa, K.; Galotto, M.J. Electrospun PVA fibers loaded with antioxidant fillers extracted from Durvillaea antarctica algae and their effect on plasticized PLA bionanocomposites. *Eur. Polym. J.* **2018**, *103*, 145–157. [CrossRef]

145. Chen, C.W.; Xu, Z.W.; Ma, Y.R.; Liu, J.L.; Zhang, Q.J.; Tang, Z.P.; Fu, K.J.; Yang, F.X.; Xie, J. Properties, vapour-phase antimicrobial and antioxidant activities of active poly(vinyl alcohol) packaging films incorporated with clove oil. *Food Control* **2018**, *88*, 105–112. [CrossRef]

146. Baek, N.; Kim, Y.T.; Marcy, J.E.; Duncan, S.E.; O'Keefe, S.F. Physical properties of nanocomposite polylactic acid films prepared with oleic acid modified titanium dioxide. *Food Packag. Shelf Life* **2018**, *17*, 30–38. [CrossRef]

147. Alboofetileh, M.; Rezaei, M.; Hosseini, H.; Abdollahi, M. Morphological, physico-mechanical, and antimicrobial properties of sodium alginate-montmorillonite nanocomposite films incorporated with marjoram essential oil. *J. Food Process. Perservation* **2018**, *42*, e13596. [CrossRef]

148. Youssef, A.M.; Abdel-Aziz, M.E.; El-Sayed, E.S.A.; Abdel-Aziz, M.S.; El-Hakim, A.A.A.; Kamel, S.; Turky, G. Morphological, electrical & antibacterial properties of trilayered Cs/PAA/PPy bionanocomposites hydrogel based on Fe3O4-NPs. *Carbohyd. Polym.* **2018**, *196*, 483–493. [CrossRef]

149. Yousuf, B.; Qadri, O.S.; Srivastava, A.K. Recent developments in shelf-life extension of fresh-cut fruits and vegetables by application of different edible coatings: A review. *LWT Food Sci. Technol.* **2018**, *89*, 198–209. [CrossRef]

150. Zhang, C.; Li, W.H.; Zhu, B.F.; Chen, H.Y.; Chi, H.; Li, L.; Qin, Y.Y.; Xue, J. The Quality Evaluation of Postharvest Strawberries Stored in Nano-Ag Packages at Refrigeration Temperature. *Polymers* **2018**, *10*, 894. [CrossRef] [PubMed]

151. Singh, S.; Gaikwad, K.K.; Lee, Y.S. Antimicrobial and antioxidant properties of polyvinyl alcohol bio composite films containing seaweed extracted cellulose nano-crystal and basil leaves extract. *Int. J. Biol. Macromol.* **2018**, *107*, 1879–1887. [CrossRef]

152. Youssef, A.M.; El-Sayed, S.M. Bionanocomposites materials for food packaging applications: Concepts and future outlook. *Carbohyd. Polym.* **2018**, *193*, 19–27. [CrossRef]

153. Maleki, G.; Sedaghat, N.; Woltering, E.J.; Farhoodi, M.; Mohebbi, M. Chitosan-limonene coating in combination with modified atmosphere packaging preserve postharvest quality of cucumber during storage. *J. Food Meas. Charact.* **2018**, *12*, 1610–1621. [CrossRef]

154. Kaewklin, P.; Siripatrawan, U.; Suwanagul, A.; Lee, Y.S. Active packaging from chitosan-titanium dioxide nanocomposite film for prolonging storage life of tomato fruit. *Int. J. Biol. Macromol.* **2018**, *112*, 523–529. [CrossRef]

155. Yan, J.; Luo, Z.; Ban, Z.; Lu, H.; Li, D.; Yang, D.; Aghdam, M.S.; Li, L. The effect of the layer-by-layer (LBL) edible coating on strawberry quality and metabolites during storage. *Postharvest Biol. Technol.* **2019**, *147*, 29–38. [CrossRef]

156. Shahbazi, Y. Application of carboxymethyl cellulose and chitosan coatings containing Mentha spicata essential oil in fresh strawberries. *Int. J. Biol. Macromol.* **2018**, *112*, 264–272. [CrossRef] [PubMed]

157. Li, L.L.; Wang, H.L.; Chen, M.M.; Jiang, S.W.; Jiang, S.T.; Li, X.J.; Wang, Q.Y. Butylated hydroxyanisole encapsulated in gelatin fiber mats: Volatile release kinetics, functional effectiveness and application to strawberry preservation. *Food Chem.* **2018**, *269*, 142–149. [CrossRef] [PubMed]

158. Zhu, J.Y.; Tang, C.H.; Yin, S.W.; Yang, X.Q. Development and characterisation of polylactic acid-gliadin bilayer/trilayer films as carriers of thymol. *Int. J. Food Sci. Technol.* **2018**, *53*, 608–618. [CrossRef]

159. Rezaei, F.; Shahbazi, Y. Shelf-life extension and quality attributes of sauced silver carp fillet: A comparison among direct addition, edible coating and biodegradable film. *LWT Food Sci. Technol.* **2018**, *87*, 122–133. [CrossRef]

160. Hassanzadeh, P.; Moradi, M.; Vaezi, N.; Moosavy, M.H.; Mahmoudi, R. Effects of chitosan edible coating containing grape seed extract on the shelf-life of refrigerated rainbow trout fillet. *Vet. Res. Forum.* **2018**, *9*, 73–79.

161. Ma, Y.C.; Li, L.; Wang, Y.F. Development of PLA-PHB-based biodegradable active packaging and its application to salmon. *Packag. Technol. Sci.* **2018**, *31*, 739–746. [CrossRef]

162. Wang, Q.Y.; Lei, J.; Ma, J.J.; Yuan, G.F.; Sun, H.Y. Effect of chitosan-carvacrol coating on the quality of Pacific white shrimp during iced storage as affected by caprylic acid. *Int. J. Biol. Macromol.* **2018**, *106*, 123–129. [CrossRef]

163. Ejaz, M.; Arfat, Y.A.; Mulla, M.; Ahmed, J. Zinc oxide nanorods/clove essential oil incorporated Type B gelatin composite films and its applicability for shrimp packaging. *Food Packag. Shelf Life* **2018**, *15*, 113–121. [CrossRef]

164. Zegarra, M.D.C.P.; Santos, A.M.P.; Silva, A.M.A.D.; Melo, E.D. Chitosan films incorporated with antioxidant extract of acerola agroindustrial residue applied in chicken thigh. *J. Food Process. Perservation* **2018**, *42*. [CrossRef]

165. Souza, V.G.L.; Pires, J.R.A.; Vieira, E.T.; Coelhoso, I.M.; Duarte, M.P.; Fernando, A.L. Shelf Life Assessment of Fresh Poultry Meat Packaged in Novel Bionanocomposite of Chitosan/Montmorillonite Incorporated with Ginger Essential Oil. *Coatings* **2018**, *8*, 177. [CrossRef]

166. Pires, J.R.A.; de Souza, V.G.L.; Fernando, A.L. Chitosan/montmorillonite bionanocomposites incorporated with rosemary and ginger essential oil as packaging for fresh poultry meat. *Food Packag. Shelf Life* **2018**, *17*, 142–149. [CrossRef]

167. Lin, L.; Zhu, Y.L.; Cui, H.Y. Electrospun thyme essential oil/gelatin nanofibers for active packaging against Campylobacter jejuni in chicken. *LWT Food Sci. Technol.* **2018**, *97*, 711–718. [CrossRef]

168. Yang, W.E.; Xie, Y.H.; Jin, J.H.; Liu, H.; Zhang, H.X. Development and Application of an Active Plastic Multilayer Film by Coating a Plantaricin BM-1 for Chilled Meat Preservation. *J. Food Sci.* **2019**, *84*, 1864–1870. [CrossRef] [PubMed]

169. Rad, F.H.; Sharifan, A.; Asadi, G. Physicochemical and antimicrobial properties of kefiran /waterborne polyurethane film incorporated with essential oils on refrigerated ostrich meat. *LWT Food Sci. Technol.* **2018**, *97*, 794–801. [CrossRef]

170. Pabast, M.; Shariatifar, N.; Beikzadeh, S.; Jahed, G. Effects of chitosan coatings incorporating with free or nano-encapsulated Satureja plant essential oil on quality characteristics of lamb meat. *Food Control* **2018**, *91*, 185–192. [CrossRef]

171. Talebi, F.; Misaghi, A.; Khanjari, A.; Kamkar, A.; Gandomi, H.; Rezaeigolestani, M. Incorporation of spice essential oils into poly-lactic acid film matrix with the aim of extending microbiological and sensorial shelf life of ground beef. *LWT Food Sci. Technol.* **2018**, *96*, 482–490. [CrossRef]

172. Zhao, Y.J.; Teixeira, J.S.; Ganzle, M.M.; Saldana, M.D.A. Development of antimicrobial films based on cassava starch, chitosan and gallic acid using subcritical water technology. *J. Supercrit. Fluid* **2018**, *137*, 101–110. [CrossRef]

173. Ribeiro-Santos, R.; de Melo, N.R.; Andrade, M.; Azevedo, G.; Machado, A.V.; Carvalho-Costa, D.; Sanches-Silva, A. Whey protein active films incorporated with a blend of essential oils: Characterization and effectiveness. *Packag. Technol. Sci.* **2018**, *31*, 27–40. [CrossRef]

174. Youssef, A.M.; El-Sayed, S.M.; El-Sayed, H.S.; Salama, H.H.; Assem, F.M.; Abd El-Salam, M.H. Novel bionanocomposite materials used for packaging skimmed milk acid coagulated cheese (Karish). *Int. J. Biol. Macromol.* **2018**, *115*, 1002–1011. [CrossRef]

175. Lotfi, M.; Tajik, H.; Moradi, M.; Forough, M.; Divsalar, E.; Kuswandi, B. Nanostructured chitosan/monolaurin film: Preparation, characterization and antimicrobial activity against Listeria monocytogenes on ultrafiltered white cheese. *LWT Food Sci. Technol.* **2018**, *92*, 576–583. [CrossRef]

176. Yang, S.Y.; Cao, L.; Kim, H.; Beak, S.E.; Bin Song, K. Utilization of Foxtail Millet Starch Film Incorporated with Clove Leaf Oil for the Packaging of Queso Blanco Cheese as a Model Food. *Starch-Starke* **2018**, *70*. [CrossRef]

177. Guitian, M.V.; Ibarguren, C.; Soria, M.C.; Hovanyecz, P.; Banchio, C.; Audisio, M.C. Anti-Listeria monocytogenes effect of bacteriocin-incorporated agar edible coatings applied on cheese. *Int. Dairy J.* **2019**, *97*, 92–98. [CrossRef]

178. Mushtaq, M.; Gani, A.; Gani, A.; Punoo, H.A.; Masoodi, F.A. Use of pomegranate peel extract incorporated zein film with improved properties for prolonged shelf life of fresh Himalayan cheese (Kalari/kradi). *Innov. Food Sci. Emerg.* **2018**, *48*, 25–32. [CrossRef]

179. Orsuwan, A.; Sothornvit, R. Active Banana Flour Nanocomposite Films Incorporated with Garlic Essential Oil as Multifunctional Packaging Material for Food Application. *Food Bioprocess. Technol.* **2018**, *11*, 1199–1210. [CrossRef]

180. Pavli, F.; Tassou, C.; Nychas, G.J.E.; Chorianopoulos, N. Probiotic Incorporation in Edible Films and Coatings: Bioactive Solution for Functional Foods. *Int. J. Mol. Sci.* **2018**, *19*, 150. [CrossRef]

181. Yousefi, M.; Azizi, M.; Ehsani, A. Antimicrobial coatings and films on meats: A perspective on the application of antimicrobial edible films or coatings on meats from the past to future. *Bali Med. J.* **2018**, *7*, 87–96. [CrossRef]

182. Wang, H.X.; Qan, J.; Ding, F.Y. Emerging Chitosan-Based Films for Food Packaging Applications. *J. Agric. Food Chem.* **2018**, *66*, 395–413. [CrossRef]

183. Dutta, A.S.P.K. Extraction of Chitin-Glucan Complex from Agaricus bisporus: Characterization and Antibacterial Activity. *J. Polym. Mater.* **2017**, *34*, 1–9.

184. Priyadarshi, R.; Sauraj; Kumar, B.; Negi, Y.S. Chitosan film incorporated with citric acid and glycerol as an active packaging material for extension of green chilli shelf life. *Carbohyd. Polym.* **2018**, *195*, 329–338. [CrossRef]

185. Sogut, E.; Seydim, A.C. Development of Chitosan and Polycaprolactone based active bilayer films enhanced with nanocellulose and grape seed extract. *Carbohyd. Polym.* **2018**, *195*, 180–188. [CrossRef]

186. Xu, Y.X.; Willis, S.; Jordan, K.; Sismour, E. Chitosan nanocomposite films incorporating cellulose nanocrystals and grape pomace extracts. *Packag. Technol. Sci.* **2018**, *31*, 631–638. [CrossRef]

187. Uranga, J.; Puertas, A.I.; Etxabide, A.; Duenas, M.T.; Guerrero, P.; de la Caba, K. Citric acid-incorporated fish gelatin/chitosan composite films. *Food Hydrocoll.* **2019**, *86*, 95–103. [CrossRef]

188. Uranga, J.; Etxabide, A.; Guerrero, P.; de la Caba, K. Development of active fish gelatin films with anthocyanins by compression molding. *Food Hydrocoll.* **2018**, *84*, 313–320. [CrossRef]

189. Ramziia, S.; Ma, H.; Yao, Y.Z.; Wei, K.R.; Huang, Y.Q. Enhanced antioxidant activity of fish gelatin-chitosan edible films incorporated with procyanidin. *J. Appl. Polym. Sci.* **2018**, *135*. [CrossRef]

190. Perez-Cordoba, L.J.; Norton, I.T.; Batchelor, H.K.; Gkatzionis, K.; Spyropoulos, F.; Sobral, P.J.A. Physico-chemical, antimicrobial and antioxidant properties of gelatin-chitosan based films loaded with nanoemulsions encapsulating active compounds. *Food Hydrocoll.* **2018**, *79*, 544–559. [CrossRef]

191. Soni, B.; Mahmoud, B.; Chang, S.; El-Giar, E.M.; Hassan, E.B. Physicochemical, antimicrobial and antioxidant properties of chitosan/TEMPO biocomposite packaging films. *Food Packag. Shelf Life* **2018**, *17*, 73–79. [CrossRef]

192. Robledo, N.; Lopez, L.; Bunger, A.; Tapia, C.; Abugoch, L. Effects of Antimicrobial Edible Coating of Thymol Nanoemulsion/Quinoa Protein/Chitosan on the Safety, Sensorial Properties, and Quality of Refrigerated Strawberries (*Fragaria × ananassa*) Under Commercial Storage Environment. *Food Bioprocess. Technol.* **2018**, *11*, 1566–1574. [CrossRef]

193. Yun, Y.H.; Lee, C.M.; Kim, Y.S.; Yoon, S.D. Preparation of chitosan/polyvinyl alcohol blended films containing sulfosuccinic acid as the crosslinking agent using UV curing process. *Food Res. Int.* **2017**, *100*, 377–386. [CrossRef]

194. Yang, W.; Fortunati, E.; Bertoglio, F.; Owczarek, J.S.; Bruni, G.; Kozanecki, M.; Kenny, J.M.; Torre, L.; Visai, L.; Puglia, D. Polyvinyl alcohol/chitosan hydrogels with enhanced antioxidant and antibacterial properties induced by lignin nanoparticles. *Carbohyd. Polym.* **2018**, *181*, 275–284. [CrossRef]

195. Warsiki, E. Application of chitosan as biomaterial for active packaging of ethylene absorber. In Proceedings of the 2nd International Conference on Biomass: Toward Sustainable Biomass Utilization for Industrial and Energy Applications, Bogor, Indonesia, 24–25 July 2017; Volume 141. [CrossRef]

196. Souza, V.G.L.; Rodrigues, C.; Valente, S.; Pimenta, C.; Pires, J.R.A.; Alves, M.M.; Santos, C.F.; Coelhoso, I.M.; Fernando, A.L. Eco-Friendly ZnO/Chitosan Bionanocomposites Films for Packaging of Fresh Poultry Meat. *Coatings* **2020**, *10*, 110. [CrossRef]

197. Suo, B.; Li, H.R.; Wang, Y.X.; Li, Z.; Pan, Z.L.; Ai, Z.L. Effects of ZnO nanoparticle-coated packaging film on pork meat quality during cold storage. *J. Sci. Food Agric.* **2017**, *97*, 2023–2029. [CrossRef] [PubMed]

198. Kumar, S.; Mudai, A.; Roy, B.; Basumatary, I.B.; Mukherjee, A.; Dutta, J. Biodegradable Hybrid Nanocomposite of Chitosan/Gelatin and Green Synthesized Zinc Oxide Nanoparticles for Food Packaging. *Foods* **2020**, *9*, 1143. [CrossRef] [PubMed]

199. Zheng, K.W.; Li, W.; Fu, B.Q.; Fu, M.F.; Ren, Q.L.; Yang, F.; Qin, C.Q. Physical, antibacterial and antioxidant properties of chitosan films containing hardleaf oatchestnut starch and Litsea cubeba oil. *Int. J. Biol. Macromol.* **2018**, *118*, 707–715. [CrossRef] [PubMed]

200. Sharma, H.; Mendiratta, S.K.; Agrawal, R.K.; Talukder, S.; Kumar, S. Studies on the potential application of various blends of essential oils as antioxidant and antimicrobial preservatives in emulsion based chicken sausages. *Br. Food J.* **2018**, *120*, 1398–1411. [CrossRef]

201. Shahbazi, Y.; Karami, N.; Shavisi, N. Effect of Mentha spicata essential oil on chemical, microbial, and sensory properties of minced camel meat during refrigerated storage. *J. Food Saf.* **2018**, *38*. [CrossRef]

202. Azadbakht, E.; Maghsoudlou, Y.; Khomiri, M.; Kashiri, M. Development and structural characterization of chitosan films containing Eucalyptus globulus essential oil: Potential as an antimicrobial carrier for packaging of sliced sausage. *Food Packag. Shelf Life* **2018**, *17*, 65–72. [CrossRef]

203. Niu, B.; Yan, Z.P.; Shao, P.; Kang, J.; Chen, H.J. Encapsulation of Cinnamon Essential Oil for Active Food Packaging Film with Synergistic Antimicrobial Activity. *Nanomaterials* **2018**, *8*, 598. [CrossRef]

204. Dolea, D.; Rizo, A.; Fuentes, A.; Barat, J.M.; Fernandez-Segovia, I. Effect of thyme and oregano essential oils on the shelf life of salmon and seaweed burgers. *Food Sci. Technol. Int.* **2018**, *24*, 394–403. [CrossRef]

205. Ju, J.; Xu, X.M.; Xie, Y.F.; Guo, Y.H.; Cheng, Y.L.; Qian, H.; Yao, W.R. Inhibitory effects of cinnamon and clove essential oils on mold growth on baked foods. *Food Chem.* **2018**, *240*, 850–855. [CrossRef]

206. Priyadarshi, R.; Sauraj; Kumar, B.; Deeba, F.; Kulshreshtha, A.; Negi, Y.S. Chitosan films incorporated with Apricot (*Prunus armeniaca*) kernel essential oil as active food packaging material. *Food Hydrocoll.* **2018**, *85*, 158–166. [CrossRef]

207. Zhang, Z.J.; Li, N.; Li, H.Z.; Li, X.J.; Cao, J.M.; Zhang, G.P.; He, D.L. Preparation and characterization of biocomposite chitosan film containing *Perilla frutescens* (L.) Britt. essential oil. *Ind. Crops Prod.* **2018**, *112*, 660–667. [CrossRef]

208. Ojagh, S.M.; Rezaei, M.; Razavi, S.H.; Hosseini, S.M.H. Effect of chitosan coatings enriched with cinnamon oil on the quality of refrigerated rainbow trout. *Food Chem.* **2010**, *120*, 193–198. [CrossRef]

209. Xing, Y.G.; Lin, H.B.; Cao, D.; Xu, Q.L.; Han, W.F.; Wang, R.R.; Che, Z.M.; Li, X.H. Effect of Chitosan Coating with Cinnamon Oil on the Quality and Physiological Attributes of China Jujube Fruits. *Biomed. Res. Int.* **2015**, 835151. [CrossRef] [PubMed]

210. Srikandace, Y.; Indrarti, L.; Indriyati; Sancoyorini, M.K. Antibacterial activity of bacterial cellulose-based edible film incorporated with *Citrus spp* essential oil. *IOP Conf. Ser. Earth Environ.* **2018**, *160*. [CrossRef]

211. Ortiz, C.M.; Salgado, P.R.; Dufresne, A.; Mauri, A.N. Microfibrillated cellulose addition improved the physicochemical and bioactive properties of biodegradable films based on soy protein and clove essential oil. *Food Hydrocoll.* **2018**, *79*, 416–427. [CrossRef]

212. Yu, Z.L.; Sun, L.; Wang, W.; Zeng, W.C.; Mustapha, A.; Lin, M.S. Soy protein-based films incorporated with cellulose nanocrystals and pine needle extract for active packaging. *Ind. Crops Prod.* **2018**, *112*, 412–419. [CrossRef]

213. Fardioui, M.; Kadmiri, I.M.; Qaiss, A.E.; Bouhfid, R. Bio-active nanocomposite films based on nanocrystalline cellulose reinforced styrylquinoxalin-grafted-chitosan: Antibacterial and mechanical properties. *Int. J. Biol. Macromol.* **2018**, *114*, 733–740. [CrossRef]

214. Freire, C.S.R.; Fernandes, S.C.M.; Silvestre, A.J.D.; Neto, C.P. Novel cellulose-based composites based on nanofibrillated plant and bacterial cellulose: Recent advances at the University of Aveiro—A review. *Holzforschung* **2013**, *67*, 603–612. [CrossRef]

215. Shabanpour, B.; Kazemi, M.; Ojagh, S.M.; Pourashouri, P. Bacterial cellulose nanofibers as reinforce in edible fish myofibrillar protein nanocomposite films. *Int. J. Biol. Macromol.* **2018**, *117*, 742–751. [CrossRef]

216. Gilbert, J.; Cheng, C.J.; Jones, O.G. Vapor Barrier Properties and Mechanical Behaviors of Composite Hydroxypropyl Methylcelluose/Zein Nanoparticle Films. *Food Biophys.* **2018**, *13*, 25–36. [CrossRef]

217. Shen, Z.; Kamdem, D.P. Antimicrobial activity of sugar beet lignocellulose films containing tung oil and cedarwood essential oil. *Cellulose* **2015**, *22*, 2703–2715. [CrossRef]

218. Coma, V.; Sebti, I.; Pardon, P.; Deschamps, A.; Pichavant, F.H. Antimicrobial edible packaging based on cellulosic ethers, fatty acids, and nisin incorporation to inhibit Listeria innocua and Staphylococcus aureus. *J. Food Protect.* **2001**, *64*, 470–475. [CrossRef] [PubMed]

219. Zhang, X.M.; Shu, Y.; Su, S.P.; Zhu, J. One-step coagulation to construct durable anti-fouling and antibacterial cellulose film exploiting Ag@AgCl nanoparticle- triggered photo-catalytic degradation. *Carbohyd. Polym.* **2018**, *181*, 499–505. [CrossRef] [PubMed]

220. Zahedi, Y.; Fathi-Achachlouei, B.; Yousefi, A.R. Physical and mechanical properties of hybrid montmorillonite/zinc oxide reinforced carboxymethyl cellulose nanocomposites. *Int. J. Biol. Macromol.* **2018**, *108*, 863–873. [CrossRef] [PubMed]

221. Tyagi, P.; Hubbe, M.A.; Lucia, L.; Pal, L. High performance nanocellulose-based composite coatings for oil and grease resistance. *Cellulose* **2018**, *25*, 3377–3391. [CrossRef]

222. Shahbazi, Y. Characterization of nanocomposite films based on chitosan and carboxymethylcellulose containing Ziziphora clinopodioides essential oil and methanolic Ficus carica extract. *J. Food Process. Perservation* **2018**, *42*. [CrossRef]

223. Wu, Z.G.; Huang, X.J.; Li, Y.C.; Xiao, H.Z.; Wang, X.Y. Novel chitosan films with laponite immobilized Ag nanoparticles for active food packaging. *Carbohyd. Polym.* **2018**, *199*, 210–218. [CrossRef]

224. Hu, D.Y.; Wang, H.X.; Wang, L.J. Physical properties and antibacterial activity of quaternized chitosan/carboxymethyl cellulose blend films. *LWT Food Sci. Technol.* **2016**, *65*, 398–405. [CrossRef]

225. Salari, M.; Khiabani, M.S.; Mokarram, R.R.; Ghanbarzadeh, B.; Kafil, H.S. Development and evaluation of chitosan based active nanocomposite films containing bacterial cellulose nanocrystals and silver nanoparticles. *Food Hydrocoll.* **2018**, *84*, 414–423. [CrossRef]

226. Nunes, M.R.; Castilho, M.D.M.; Veeck, A.P.D.; da Rosa, C.G.; Noronha, C.M.; Maciel, M.V.O.B.; Barreto, P.M. Antioxidant and antimicrobial methylcellulose films containing Lippia alba extract and silver nanoparticles. *Carbohyd. Polym.* **2018**, *192*, 37–43. [CrossRef]

227. Shankar, S.; Oun, A.A.; Rhim, J.W. Preparation of antimicrobial hybrid nano-materials using regenerated cellulose and metallic nanoparticles. *Int. J. Biol. Macromol.* **2018**, *107*, 17–27. [CrossRef] [PubMed]

228. Caetano, K.D.; Lopes, N.A.; Costa, T.M.H.; Brandelli, A.; Rodrigues, E.; Flores, S.H.; Cladera-Olivera, F. Characterization of active biodegradable films based on cassava starch and natural compounds. *Food Packag. Shelf Life* **2018**, *16*, 138–147. [CrossRef]

229. Dinika, I.; Utama, G.L. Cheese whey as potential resource for antimicrobial edible film and active packaging production. *Food Raw Mater.* **2019**, *7*, 229–239. [CrossRef]

230. Trongchuen, K.; Ounkaew, A.; Kasemsiri, P.; Hiziroglu, S.; Mongkolthanaruk, W.; Wannasutta, R.; Pongsa, U.; Chindaprasirt, P. Bioactive Starch Foam Composite Enriched With Natural Antioxidants from Spent Coffee Ground and Essential Oil. *Starch-Starke* **2018**, *70*. [CrossRef]

231. Ounkaew, A.; Kasemsiri, P.; Kamwilaisak, K.; Saengprachatanarug, K.; Mongkolthanaruk, W.; Souvanh, M.; Pongsa, U.; Chindaprasirt, P. Polyvinyl Alcohol (PVA)/Starch Bioactive Packaging Film Enriched with Antioxidants from Spent Coffee Ground and Citric Acid. *J. Polym. Environ.* **2018**, *26*, 3762–3772. [CrossRef]

232. Iamareerat, B.; Singh, M.; Sadiq, M.B.; Anal, A.K. Reinforced cassava starch based edible film incorporated with essential oil and sodium bentonite nanoclay as food packaging material. *J. Food Sci. Technol.* **2018**, *55*, 1953–1959. [CrossRef]

233. Jash, A.; Paliyath, G.; Lim, L.T. Activated release of bioactive aldehydes from their precursors embedded in electrospun poly(lactic acid) nonwovens. *RSC Adv.* **2018**, *8*, 19930–19938. [CrossRef]

234. Beak, S.; Kim, H.; Song, K.B. Sea Squirt Shell Protein and Polylactic Acid Laminated Films Containing Cinnamon Bark Essential Oil. *J. Food Sci.* **2018**, *83*, 1896–1903. [CrossRef]

235. Yin, X.; Sun, C.; Tian, M.Y.; Wang, Y. Preparation and Characterization of Salicylic Acid/Polylactic Acid Composite Packaging Materials. *Lect. Notes Electr. Eng.* **2018**, *477*, 811–818. [CrossRef]

236. Ye, J.S.; Wang, S.Y.; Lan, W.J.; Qin, W.; Liu, Y.W. Preparation and properties of polylactic acid-tea polyphenol-chitosan composite membranes. *Int. J. Biol. Macromol.* **2018**, *117*, 632–639. [CrossRef]

237. Ahmed, J.; Arfat, Y.A.; Bher, A.; Mulla, M.; Jacob, H.; Auras, R. Active Chicken Meat Packaging Based on Polylactide Films and Bimetallic Ag-Cu Nanoparticles and Essential Oil. *J. Food Sci.* **2018**, *83*, 1299–1310. [CrossRef] [PubMed]

238. Li, W.H.; Zhang, C.; Chi, H.; Li, L.; Lan, T.Q.; Han, P.; Chen, H.Y.; Qin, Y.Y. Development of Antimicrobial Packaging Film Made from Poly(Lactic Acid) Incorporating Titanium Dioxide and Silver Nanoparticles. *Molecules* **2017**, *22*, 1170. [CrossRef] [PubMed]

239. Valerini, D.; Tammaro, L.; Di Benedetto, F.; Vigliotta, G.; Capodieci, L.; Terzi, R.; Rizzo, A. Aluminum-doped zinc oxide coatings on polylactic acid films for antimicrobial food packaging. *Thin Solid Films* **2018**, *645*, 187–192. [CrossRef]

240. Li, W.H.; Li, L.; Cao, Y.; Lan, T.Q.; Chen, H.Y.; Qin, Y.Y. Effects of PLA Film Incorporated with ZnO Nanoparticle on the Quality Attributes of Fresh-Cut Apple. *Nanomaterials* **2017**, *7*, 207. [CrossRef] [PubMed]

241. Li, M.X.; Zhang, F.; Liu, Z.D.; Guo, X.F.; Wu, Q.; Qiao, L.R. Controlled Release System by Active Gelatin Film Incorporated with beta-Cyclodextrin-Thymol Inclusion Complexes. *Food Bioprocess. Technol.* **2018**, *11*, 1695–1702. [CrossRef]

242. Raquel, B.; Cristina, N.; Filomena, S. Encapsulation Systems for Antimicrobial Food Packaging Components: An Update. *Molecules* **2020**, *25*, 1134.

243. Dou, L.X.; Li, B.F.; Zhang, K.; Chu, X.; Hou, H. Physical properties and antioxidant activity of gelatin-sodium alginate edible films with tea polyphenols. *Int. J. Biol. Macromol.* **2018**, *118*, 1377–1383. [CrossRef]

244. Bonilla, J.; Poloni, T.; Lourenco, R.V.; Sobral, P.J.A. Antioxidant potential of eugenol and ginger essential oils with gelatin/chitosan films. *Food Biosci.* **2018**, *23*, 107–114. [CrossRef]

245. Moreno, O.; Atares, L.; Chiralt, A.; Cruz-Romero, M.C.; Kerry, J. Starch-gelatin antimicrobial packaging materials to extend the shelf life of chicken breast fillets. *LWT Food Sci. Technol.* **2018**, *97*, 483–490. [CrossRef]

246. Liu, Y.Y.; Li, Y.; Deng, L.L.; Zou, L.; Feng, F.Q.; Zhang, H. Hydrophobic Ethylcellulose/Gelatin Nanofibers Containing Zinc Oxide Nanoparticles for Antimicrobial Packaging. *J. Agric. Food Chem.* **2018**, *66*, 9498–9506. [CrossRef]

247. He, Q.Y.; Zhang, Y.C.; Cai, X.X.; Wang, S.Y. Fabrication of gelatin-TiO$_2$ nanocomposite film and its structural, antibacterial and physical properties. *Int. J. Biol. Macromol.* **2016**, *84*, 153–160. [CrossRef] [PubMed]

248. Kumar, S.; Shukla, A.; Baul, P.P.; Mitra, A.; Halder, D. Biodegradable hybrid nanocomposites of chitosan/gelatin and silver nanoparticles for active food packaging applications. *Food Packag. Shelf Life* **2018**, *16*, 178–184. [CrossRef]

249. Venkatesan, R.; Rajeswari, N.; Tamilselvi, A. Antimicrobial, mechanical, barrier, and thermal properties of bio-based poly (butylene adipate-co-terephthalate) (PBAT)/Ag2O nanocomposite films for packaging application. *Polym. Adv. Technol* **2018**, *29*, 61–68. [CrossRef]

250. Zehetmeyer, G.; Meira, S.M.M.; Scheibel, J.M.; da Silva, C.D.; Rodembusch, F.S.; Brandelli, A.; Soares, R.M.D. Biodegradable and antimicrobial films based on poly(butylene adipate-co-terephthalate) electrospun fibers. *Polym. Bull.* **2017**, *74*, 3243–3268. [CrossRef]

251. Khare, S.; DeLoid, G.M.; Molina, R.M.; Gokulan, K.; Couvillion, S.P.; Bloodsworth, K.J.; Eder, E.K.; Wong, A.R.; Hoyt, D.W.; Bramer, L.M.; et al. Effects of ingested nanocellulose on intestinal microbiota and homeostasis in Wistar Han rats. *Nanoimpact* **2020**, *18*. [CrossRef] [PubMed]

252. DeLoid, G.M.; Cao, X.Q.; Molina, R.M.; Silva, D.I.; Bhattacharya, K.; Ng, K.W.; Loo, S.C.J.; Brain, J.D.; Demokritou, P. Toxicological effects of ingested nanocellulose in in vitro intestinal epithelium and in vivo rat models. *Environ. Sci. Nano* **2019**, *6*, 2105–2115. [CrossRef]

253. Chen, Y.J.; Lin, Y.J.; Nagy, T.; Kong, F.B.; Guo, T.L. Subchronic exposure to cellulose nanofibrils induces nutritional risk by non-specifically reducing the intestinal absorption. *Carbohyd. Polym.* **2020**, *229*. [CrossRef]

254. Dimitrijevic, M.; Karabasil, N.; Boskovic, M.; Teodorovic, V.; Vasilev, D.; Djordjevic, V.; Kilibarda, N.; Cobanovic, N. Safety aspects of nanotechnology applications in food packaging. In Proceedings of the 58th International Meat Industry Conference (Meatcon2015), Mt. Zlatibor, Serbia, 4–7 October 2015; Volume 5, pp. 57–60. [CrossRef]

255. dos Santos, C.A.; Ingle, A.P.; Rai, M. The emerging role of metallic nanoparticles in food. *Appl. Microbiol. Biotechnol.* **2020**, *104*, 2373–2383. [CrossRef]

256. Doskocz, N.; Zaleska-Radziwill, M.; Affek, K.; Lebkowska, M. Ecotoxicity of selected nanoparticles in relation to micro-organisms in the water ecosystem. *Desalin Water Treat.* **2020**, *186*, 50–55. [CrossRef]

257. Noori, A.; Ngo, A.; Gutierrez, P.; Theberge, S.; White, J.C. Silver nanoparticle detection and accumulation in tomato (Lycopersicon esculentum). *J. Nanopart Res.* **2020**, *22*. [CrossRef]

258. Dash, S.R.; Kundu, C.N. Promising opportunities and potential risk of nanoparticle on the society. *IET Nanobiotechnol.* **2020**, *14*, 253–260. [CrossRef]

259. Garcia, C.V.; Shin, G.H.; Kim, J.T. Metal oxide-based nanocomposites in food packaging: Applications, migration, and regulations. *Trends Food Sci. Technol.* **2018**, *82*, 21–31. [CrossRef]

260. Souza, V.G.L.; Fernando, A.L. Nanoparticles in food packaging: Biodegradability and potential migration to food-A review. *Food Packag. Shelf Life* **2016**, *8*, 63–70. [CrossRef]

261. Istiqola, A.; Syafiuddin, A. A review of silver nanoparticles in food packaging technologies: Regulation, methods, properties, migration, and future challenges. *J. Chin. Chem. Soc.* **2020**. [CrossRef]

262. Sahoo, R.K. Packaging: Polymer-Metal-Based Micro- and Nanocomposites. *Encycl. Polym. Appl.* **2019**, *I–III*, 2021–2040. [CrossRef]

263. Morais, L.D.; Macedo, E.V.; Granjeiro, J.M.; Delgado, I.F. Critical evaluation of migration studies of silver nanoparticles present in food packaging: A systematic review. *Crit. Rev. Food Sci.* **2019**. [CrossRef] [PubMed]

264. Ramachandran, R.; Krishnaraj, C.; Kumar, V.K.A.; Harper, S.L.; Kalaichelvan, T.P.; Yun, S.I. In vivo toxicity evaluation of biologically synthesized silver nanoparticles and gold nanoparticles on adult zebrafish: A comparative study. *3 Biotech* **2018**, *8*, 441. [CrossRef]

265. Zhou, Q.F.; Liu, L.H.; Liu, N.A.; He, B.; Hu, L.G.; Wang, L.N. Determination and characterization of metal nanoparticles in clams and oysters. *Ecotoxicol. Environ. Saf.* **2020**, *198*. [CrossRef]

266. Gong, Y.; Chai, M.W.; Ding, H.; Shi, C.; Wang, Y.; Li, R.L. Bioaccumulation and human health risk of shellfish contamination to heavy metals and As in most rapid urbanized Shenzhen, China. *Environ. Sci. Pollut. Res.* **2020**, *27*, 2096–2106. [CrossRef]

267. Shah, N.; Khan, A.; Ali, R.; Marimuthu, K.; Uddin, M.N.; Rizwan, M.; Rahman, K.U.; Alam, M.; Adnan, M.; Jawad, S.M.; et al. Monitoring Bioaccumulation (in Gills and Muscle Tissues), Hematology, and Genotoxic Alteration in Ctenopharyngodon idella Exposed to Selected Heavy Metals. *Biomed. Res. Int.* **2020**, *2020*. [CrossRef]

268. Wu, J.; Wang, G.Y.; Vijver, M.G.; Bosker, T.; Peijnenburg, W.J.G.M. Foliar versus root exposure of AgNPs to lettuce: Phytotoxicity, antioxidant responses and internal translocation. *Environ. Pollut.* **2020**, *261*. [CrossRef] [PubMed]

269. Becaro, A.A.; Siqueira, M.C.; Puti, F.C.; de Moura, M.R.; Correa, D.S.; Marconcini, J.M.; Mattoso, L.H.C.; Ferreira, M.D. Cytotoxic and genotoxic effects of silver nanoparticle/carboxymethyl cellulose on Allium cepa. *Environ. Monit. Assess.* **2017**, *189*. [CrossRef] [PubMed]

270. Ma, C.X.; Liu, H.; Chen, G.C.; Zhao, Q.; Guo, H.Y.; Minocha, R.; Long, S.; Tang, Y.Z.; Saad, E.M.; DeLaTorreRoche, R.; et al. Dual roles of glutathione in silver nanoparticle detoxification and enhancement of nitrogen assimilation in soybean (*Glycine max* (L.) Merrill). *Environ. Sci. Nano* **2020**, *7*, 1954–1966. [CrossRef]

271. Yang, Z.Z.; Xiao, Y.F.; Jiao, T.T.; Zhang, Y.; Chen, J.; Gao, Y. Effects of Copper Oxide Nanoparticles on the Growth of Rice (*Oryza sativa* L.) Seedlings and the Relevant Physiological Responses. *Int. J. Environ. Res. Public Health* **2020**, *17*, 1260. [CrossRef]

272. Yusefi-Tanha, E.; Fallah, S.; Rostamnejadi, A.; Pokhrel, L.R. Particle size and concentration dependent toxicity of copper oxide nanoparticles (CuONPs) on seed yield and antioxidant defense system in soil grown soybean (Glycine max cv. Kowsar). *Sci. Total Environ.* **2020**, *715*. [CrossRef]

273. Rajput, V.; Minkina, T.; Sushkova, S.; Behal, A.; Maksimov, A.; Blicharska, E.; Ghazaryan, K.; Movsesyan, H.; Barsova, N. ZnO and CuO nanoparticles: A threat to soil organisms, plants, and human health. *Environ. Geochem. Health* **2020**, *42*, 147–158. [CrossRef]

274. Rajput, V.; Minkina, T.; Fedorenko, A.; Sushkova, S.; Mandzhieva, S.; Lysenko, V.; Duplii, N.; Fedorenko, G.; Dvadnenko, K.; Ghazaryan, K. Toxicity of copper oxide nanoparticles on spring barley (Hordeum sativum distichum). *Sci Total Environ.* **2018**, *645*, 1103–1113. [CrossRef]

275. Baskar, V.; Safia, N.; Preethy, K.S.; Dhivya, S.; Thiruvengadam, M.; Sathishkumar, R. A comparative study of phytotoxic effects of metal oxide (CuO, ZnO and NiO) nanoparticles on in-vitro grown Abelmoschus esculentus. *Plant. Biosyst.* **2020**. [CrossRef]

276. Jeon, Y.R.; Yu, J.; Choi, S.J. Fate Determination of ZnO in Commercial Foods and Human Intestinal Cells. *Int. J. Mol. Sci.* **2020**, *21*, 433. [CrossRef]

277. Pereira, F.F.; Paris, E.C.; Bresolin, J.D.; Mitsuyuki, M.C.; Ferreira, M.D.; Correa, D.S. The Effect of ZnO Nanoparticles Morphology on the Toxicity Towards Microalgae Pseudokirchneriella subcapitata. *J. Nanosci. Nanotechnol.* **2020**, *20*, 48–63. [CrossRef]

278. Vasile, O.R.; Serdaru, I.; Andronescu, E.; Trusca, R.; Surdu, V.A.; Oprea, O.; Ilie, A.; Vasile, B.S. Influence of the size and the morphology of ZnO nanoparticles on cell viability. *CR Chim.* **2015**, *18*, 1335–1343. [CrossRef]

279. Voss, L.; Saloga, P.E.J.; Stock, V.; Bohmert, L.; Braeuning, A.; Thunemann, A.F.; Lampen, A.; Sieg, H. Environmental Impact of ZnO Nanoparticles Evaluated by in Vitro Simulated Digestion. *ACS Appl. Nano Mater.* **2020**, *3*, 724–733. [CrossRef]

280. Meng, J.; Zhou, X.L.; Yang, J.; Qu, X.J.; Cui, S.X. Exposure to low dose ZnO nanoparticles induces hyperproliferation and malignant transformation through activating the CXCR2/NF-kappa B/STAT3/ERK and AKT pathways in colonic mucosal cells. *Environ. Pollut.* **2020**, *263*. [CrossRef] [PubMed]

281. Musial, J.; Krakowiak, R.; Mlynarczyk, D.T.; Goslinski, T.; Stanisz, B.J. Titanium Dioxide Nanoparticles in Food and Personal Care Products-What Do We Know about Their Safety? *Nanomaterials* **2020**, *10*, 1110. [CrossRef] [PubMed]

282. Kurtz, C.C.; Mitchell, S.; Nielsen, K.; Crawford, K.D.; Mueller-Spitz, S.R. Acute high-dose titanium dioxide nanoparticle exposure alters gastrointestinal homeostasis in mice. *J. Appl. Toxicol.* **2020**, *40*, 1384–1395. [CrossRef] [PubMed]

283. Hashem, M.M.; Abo-El-Sooud, K.; Abd-Elhakim, Y.M.; Badr, Y.A.H.; El-Metwally, A.E.; Bahy-El-Diena, A. The long-term oral exposure to titanium dioxide impaired immune functions and triggered cytotoxic and genotoxic impacts in rats. *J. Trace Elem. Med. Biol.* **2020**, *60*. [CrossRef] [PubMed]

284. Cao, X.Q.; Han, Y.H.; Gu, M.; Du, H.J.; Song, M.Y.; Zhu, X.A.; Ma, G.X.; Pan, C.; Wang, W.C.; Zhao, E.M.; et al. Foodborne Titanium Dioxide Nanoparticles Induce Stronger Adverse Effects in Obese Mice than Non-Obese Mice: Gut Microbiota Dysbiosis, Colonic Inflammation, and Proteome Alterations. *Small* **2020**. [CrossRef]

285. Bettencourt, A.; Goncalves, L.M.; Gramacho, A.C.; Vieira, A.; Rolo, D.; Martins, C.; Assuncao, R.; Alvito, P.; Silva, M.J.; Louro, H. Analysis of the Characteristics and Cytotoxicity of Titanium Dioxide Nanomaterials Following Simulated In Vitro Digestion. *Nanomaterials* **2020**, *10*, 1516. [CrossRef]

286. Najnin, H.; Alam, N.; Mujeeb, M.; Ahsan, H.; Siddiqui, W.A. Biochemical and toxicological analysis of Cinnamomum tamala essential oil in Wistar rats. *J. Food Process. Perservation* **2020**, *44*. [CrossRef]

287. Costa, W.K.; de Oliveira, J.R.S.; de Oliveira, A.M.; Santos, I.B.D.; da Cunha, R.X.; de Freitas, A.F.S.; da Silva, J.W.L.M.; Silva, V.B.G.; de Aguiar, J.C.R.D.F.; da Silva, A.G.; et al. Essential oil from Eugenia stipitata McVaugh leaves has antinociceptive, anti-inflammatory and antipyretic activities without showing toxicity in mice. *Ind. Crops Prod.* **2020**, *144*. [CrossRef]

288. Bonin, E.; Carvalho, V.M.; Avila, V.D.; dos Santos, N.C.A.; Benassi-Zanqueta, E.; Lancheros, C.A.C.; Previdelli, I.T.S.; Ueda-Nakamura, T.; de Abreu, B.A.; do Prado, I.N. Baccharis dracunculifolia: Chemical constituents, cytotoxicity and antimicrobial activity. *LWT Food Sci. Technol.* **2020**, *120*. [CrossRef]

289. Abou Baker, D.H.; Al-Moghazy, M.; ElSayed, A.A.A. The in vitro cytotoxicity, antioxidant and antibacterial potential of Satureja hortensis L. essential oil cultivated in Egypt. *Bioorg. Chem.* **2020**, *95*. [CrossRef] [PubMed]

Controlling Blown Pack Spoilage using Anti-Microbial Packaging

Rachael Reid [1], Andrey A. Tyuftin [2], Joe P. Kerry [3], Séamus Fanning [3] ⓘ, Paul Whyte [3] and Declan Bolton [1],*

[1] Food Safety Department, Teagasc Food Research Centre, Ashtown, Dublin 15, Ireland; rachel.reid@teagasc.ie
[2] Food Packaging Group, School of Food & Nutritional Sciences, University College Cork, College Road, Cork, Ireland; a.tiuftin@ucc.ie
[3] School of Public Health, University College Dublin, Belfield, Dublin 4, Ireland; joe.kerry@ucc.ie (J.P.K.); sfanning@ucd.ie (S.F.); paul.whyte@ucd.ie (P.W.)
* Correspondence: declan.bolton@teagasc.ie

Abstract: Active (anti-microbial) packaging was prepared using three different formulations; Auranta FV; Inbac-MDA and sodium octanoate at two concentrations (2.5 and 3.5 times their minimum inhibitory concentration (MIC, the lowest concentration that will inhibit the visible growth of the organisms) against *Clostridium estertheticum*, DSMZ 8809). Inoculated beef samples were packaged using the active packaging and monitored for 100 days storage at 2 °C for blown pack spoilage. The time to the onset of blown pack spoilage was significantly ($p < 0.01$) increased using Auranta FV and sodium octanoate (caprylic acid sodium salt) at both concentrations. Moreover, sodium octanoate packs had significantly ($p < 0.01$) delayed blown pack spoilage as compared to Auranta FV. It was therefore concluded that Auranta FV or sodium octanoate, incorporated into the packaging materials used for vacuum packaged beef, would inhibit blown pack spoilage and in the case of the latter, well beyond the 42 days storage period currently required for beef primals.

Keywords: blown pack spoilage; *C. estertheticum*; antimicrobials; gelatine films; edible coatings; active food packaging

1. Introduction

Blown pack spoilage (BPS), characterised by a putrid smell (H_2S) with a metallic sheen on the meat with or without gas production, occurs in correctly chilled batches (0 to 2 °C) after four to six weeks and is caused by psychrophilic *Clostridium* spp. [1]. Although *Clostridium algidicarnis*, *Clostridium frigoris*, *Clostridium bowmanii*, *Clostridium frigidicarmis* and *Clostridium ruminantium* have been associated with meat spoilage, they do not produce gas [2–4]. Blown pack spoilage is usually caused by other *Clostridium* spp., including *C. estertheticum* and *C. gasigenes*, which produce large volumes of gas, primarily carbon dioxide [3,5–7].

A low percentage of beef primals are consistently contaminated with *C. estertheticum* or *C. gasigenes* spores [1] and previous research by Moschonas et al. [6] showed that low contamination levels (as low as 1 spore per cm^2) are sufficient to cause spoilage. When beef is vacuum packaged, the shrinkage step (e.g., 90 °C for 3 s) activates the spores [8] which germinate and grow in the anaerobic and low temperature environment in which beef primals and sub-primals are typically stored. Spoilage may occur as soon as two weeks but typically after four to six weeks.

As there are no specific interventions available to prevent BPS, control is currently reliant on reducing contamination using sporicidal agents, such as peroxyacetic acid, to disinfect the plant and equipment. Moreover, apart from lactic acid treatment of carcasses, processors are not legally permitted to apply chemical treatments to beef products. Active packaging provides a vehicle by which anti-microbials can be applied to inhibit bacterial growth on beef. Previous research has demonstrated a reduction in *Lactobacillus helveticus* and *Brochothrix thermosphacta* counts on vacuum-packed beef using a polyethylene-based plastic film incorporating nisin [9]. Oregano and garlic have also been incorporated into whey-protein based films to control *Salmonella enteritidis*, *Listeria monocytogenes*, *Escherichia coli*, *Staphylococcus aureus* and *Lactobacillus plantarum* [10].

Antimicrobial compounds in active packaging films may be incorporated into the film or coated in a carrier matrix onto the inner surface of the film [11]. Carrier matrices include edible polymers such as gelatine. Commercially available antimicrobials Auranta FV (AFV) (composed of bioflavonoids, citric, malic, lactic, and caprylic acids), Inbac-MDA (IMDA) (composed of sodium diacetate, malic acid, mono and diglycerides of fatty acids, salt and excipients) and sodium octanoate (SO) are considered to be safe, with potential application in active packaging systems [12]. Moreover, they are odourless and do not adversely affect other sensory attributes of food, such as taste or texture. Their application in a gelatine carrier to inhibit anaerobic bacteria has been previously demonstrated [13]. The objective of this study was to test three different formulations; AFV, IMDA and SO at two concentrations, 2.5% and 3.5% times their MIC against *C. estertheticum*, as agents in active packaging to prevent the growth of this bacteria.

2. Materials and Methods

2.1. Materials

The antimicrobials used in this study included AFVand SO which were obtained fromSigma-Aldrich, Gillingham, Dorset, UK and IMDA which was purchased from Envirotech Innovative Products Ltd, Ardee, County Louth, Ireland. Glycerol (KB Scientific Ltd, Cork, Ireland) was used as a plasticizer and beef gelatine 100 bloom (Healan ingredients, York, UK) was used as the matrix material for all film forming solutions. Beef sub-primal striploins were purchased from a local beef supplier. Conventional vacuum heat shrinking pouches (265 × 290 mm, 50 μm; water vapour transmission rate of 50 g/m² day) were supplied by Cryovac, Trade Name BB3055X (Sealed Air W.R. Grace Europe Inc., Lausanne, Switzerland) and used as industry standard materials for coatings and meat packaging trials.

2.2. Plasma Treatment

In order to increase the hydrophilicity of the Low-density polyethylene (LDPE) inner part of the vacuum pouches; cold plasma treatment was carried out using a Dielectric-Barrier Discharge plasma system prior to the application of the antimicrobial coatings. Briefly, pouches were cut to a size of 190 × 500 mm and the surface of the LDPE side of the laminate pouches were plasma treated at atmospheric pressure using atmospheric air. The plasma source consisted of two circular aluminium plate electrodes (outer diameter = 158 mm). The top dielectric barrier was a perspex dielectric barrier (10 mm thickness) and the bottom dielectric barrier was a polypropylene sheet (5 mm thickness). When the potential across the gap reached the breakdown voltage, the dielectric barrier prevented the arc transition and homogenised the micro-discharges. The voltage applied was 75 ± 0.2 kV which was obtained from a step-up transformer (Phoenix Technologies, Inc., Campbell, CA, USA) using a variac. The input of 230 V, 50 Hz was given to the primary winding of high voltage step-up transformer from the mains supply. The samples were plasma treated for 60 s in three different places to cover the entire film area, leaving only approximately 5 cm from the edge of the film (high voltage electrode was placed 1 cm above the film).

Plasma treatment was carried out on the film samples. Following treatment, plasma treated samples were placed in Ziploc® plastic bags to protect the films from antistatic and dust particles. The water droplet test was used to determine the activation of the surface. Plasma treated film samples were then coated with water-based gelatine coatings containing the test antimicrobials.

2.3. Coatings Preparations and Packaging of Beef

2.3.1. Preparation of Film Forming Solutions and Coatings

Exactly 25 g of dry beef gelatine was dissolved in 475 mL of distilled water (5% w/w) in a 500 mL flask by heating at 90 °C in a shaking water bath (SW23, Julabo USA INC., Allentown, PA, USA) for 30 min during which 8.25 g (33% w/w) of glycerol was added under constant stirring. This solution was cooled to 40 °C in a waterbath, before the addition of the antimicrobials. The antimicrobial solutions were prepared as follows; 25.41 mL and 35.55 mL of AFV (liquid), 19.05 g and 26.65 g of IMDA (solids) and 19.0 g and 26.65 g of SO (solids) dissolved in 50 mL of distilled water before addition to the gelatine solutions to give final concentrations for each treatment of 2.5 and 3.5 times the MIC against *C. estertheticum*. These concentrations were selected to ensure the antimicrobials were present at concentrations at which they were effective against *C. estertheticum*, allowing for a dilution effect, etc. when working in food systems. Each solution was then cast on conventional polyamide/Low-density polyethylene (PA/LDPE) films using a Micron II film applicator (Gardco, Pompano Beach, FL, USA), sealed and dried at 20 °C for 48 h. The thickness of each resultant gelatine coated film was measured using a digital micrometer—Käfer Digital Thickness gauge (Käfer Messuhrenfabrik GmbH & Co., Villingen-Schwenningen, Germany) and ranged from 5 to 25 µm.

2.3.2. Vacuum Packaging of Beef

Conventional PA/LDPE films (BB3055X, Cryovac, Sealed Air Ltd, St Neots, UK) coated with the active gelatine-based antimicrobials were detached from the flat surface on which they were coated, the edges of each laminate sample were cleaned with water and/or ethanol and dried. Each film was then heat-sealed to form a pouch (approx. 170 × 220 mm) using a Webomatic type D463 (Webomatic Vacuum Packaging Systems, Bochum, Germany) with the sealing time set at 2.7 s. In order to avoid adhesion between the coated films, sterile food grade aluminium foil was placed between the films prior to sealing. Exactly 15 samples (5 in triplicate) were prepared for each antimicrobial-concentration combination. Untreated PA/LDPE film was used for the control pouches.

2.4. Inoculation of Beef Samples and Monitoring for Blown Pack Spoilage

2.4.1. Preparation of Blown Pack Spoilage *C. estertheticum*

Reference strain *C.estertheticum* subsp. *estertheticum* (DSMZ 8809[T]), was purchased as a freeze dried culture from the Deutsche Sammlung von Mikroorganismen und Zellkulturen GmbH (DSMZ, Braunchweig, Germany). The strain was revived under anaerobic conditions in 10 mL pre-reduced Peptone Yeast Extract Glucose Starch (PYGS) broth [14] and incubated for 3 weeks at 4 °C.

2.4.2. Preparation of Spore Inocula

Spore concentrates were prepared by transferring 5 mL of exponentially growing culture to 100 mL of pre-reduced peptone yeast extract glucose starch (PYGS) broth [14] and incubating at 4 °C for a minimum of 3 months to promote sporulation. Prior to inoculation all media were pre-reduced in an anaerobic cabinet for 24 h (Don Whitley Scientific Ltd, Shipley, UK) under an atmosphere of 100% carbon dioxide at 20 °C. Spores were harvested using the method described by Moschonas et al. [6]. Briefly, spore suspensions were recovered by centrifugation (7500 g, 4 °C, 10 min) and washed with saline (0.85% NaCl in sterile water). This was repeated 3 times. The washed spore suspension was then sonicated (40 kHz, 15 min) in an ultrasonic waterbath (VWR International, Dublin, Ireland) and centrifuged/washed as described above (three sonication/centrifugation/wash cycles) before final suspension in 10 mL saline and storage at −20 °C. Final spore numbers were estimated by preparing serial dilutions of the spore suspensions in saline (0.85% NaCl) and plating 0.1 mL aliquots on Columbia blood agar (CBA) supplemented with 5% defibrinated horse blood and incubating anaerobically for 3 weeks at 4 °C.

2.4.3. Preparation of Meat Samples, *C. estertheticum* Inoculation and Packing

Exactly 90 (10 × 10 × 1 cm) samples were prepared from *Biceps femoris* muscles (Charolais Cross heifers), purchased from a commercial beef processing plant. In a laminar flow unit, samples were spread inoculated with the prepared inocula to a final mean concentration of 10^3 cfu·cm^{-2} and allowed to dry for 30 min at room temperature. The samples were then placed in individual bags with antimicrobial treatment or control bags containing a hydrogen sulphide strip (Sigma Aldrich, Gillingham, UK) and vacuum packed using the Vac Star S220 (Vac Star Shop, Sugiez, Switzerland). All samples were heat shrunk at 90 °C for 3 s and stored at 2 °C for 100 days in cardboard boxes in a chilling unit located in the on-site abattoir in Teagasc Food Research Centre (Dublin) The chiller temperature was monitored using an Easylog USB data logger (Lascar Electronics Ltd, Salisbury, UK) and the surface temperature of the samples was monitored using a Testo T-175 data logger.

2.5. Monitoring Vacuum Packs

Packs were visually examined every four days for the presence of gas and scored against the following criteria as described by Boerema et al. [15]; 0 (no gas bubbles in drip), 1 (gas bubbles in drip), 2 (loss of vacuum, considered to be the start of blown pack spoilage), 3 ("blown"), 4 (presence of sufficient gas inside the packs to produce pack distension) and 5 (tightly stretched, overblown packs or packs that are leaking).

2.6. Statistical Analysis

To obtain sufficient data for statistical analysis, five replicate samples were used for each antimicrobial treatment and five samples were used as treatment controls. The experiment was repeated on three separate occasions. Data on the time to the onset of blown pack spoilage, defined as the first day when each pack was assigned the score of 2, was analysed using GenStat Release 14.1 (VSN International Ltd, Hemel Hempstead, UK). Since all individual and pooled data failed the normality tests, data was analysed using the Mann—Whitney U (Wilcoxon rank-sum) test.

3. Results

The results are presented in Figures 1–3. AFV active packs took significantly longer ($p < 0.01$) to spoil than the corresponding controls (Figure 1). This was primarily due to the onset of blown pack spoilage (score = 2) being delayed from approximately 28 days (control packs) to 48 days in the treated packs. Interestingly, there was no significant difference ($p > 0.01$) between the different concentrations of AFV used (2.5 and 3.5 times the MIC). In contrast, there was no significant difference in the IMDA treated films when compared to the control (Figure 2). Moreover, the time to the onset of blown pack spoilage was similar to that observed in the AFV control packs. The inoculated samples in SO treated packs showed a similar pattern to the AFV packs, as the time to spoilage in product wrapped in the treated films was significantly longer ($p < 0.01$) than the corresponding controls (Figure 3) and there was no significant difference ($p > 0.01$) between the different concentrations of SO used (2.5 and 3.5 times the MIC). Moreover, SO packs had significantly ($p < 0.01$) delayed blown pack spoilage as compared to AFV.

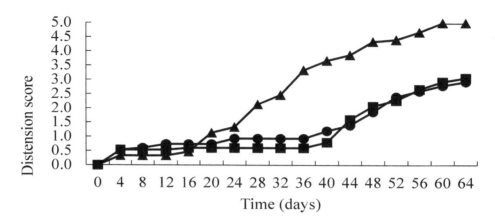

Figure 1. Distension status over time (days) of vacuum packs inoculated with spores of *C. estertheticum* and packaged in films containing 0 × MIC (▲), 2.5 × MIC (●) and 3.5 × MIC (■) AFV.

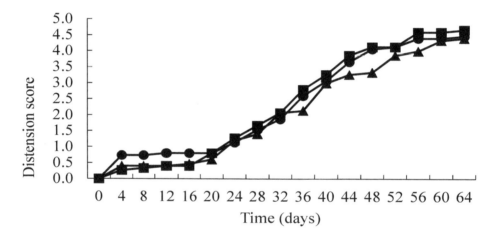

Figure 2. Distension status over time (days) of vacuum packs inoculated with spores of *C. estertheticum* and packaged in films containing 0 × MIC (▲), 2.5 × MIC (●) and 3.5 × MIC (■) IMDA.

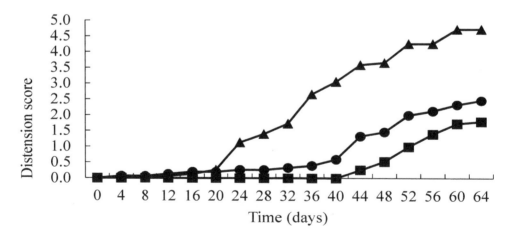

Figure 3. Distension status over time (days) of vacuum packs inoculated with spores of *C. estertheticum* and packaged in films containing 0 × MIC (▲), 2.5 × MIC (●) and 3.5 × MIC (■) SO.

4. Discussion

Blown pack spoilage (BPS) is a global issue for the beef sector [16–18], including in Ireland where 0.8% of beef primals are contaminated with *C. estertheticum* [1]. Although meat spoiled in this way has no commercial value, control is reliant on sanitation of beef plants and equipment with a sporicidal agent such as peroxyacetic acid which is highly corrosive and often ineffective. Active packaging is a potential solution if suitable antimicrobials can be found.

Antimicrobial packaging incorporates an antimicrobial agent into a polymer film that prevents the growth of target microorganisms by extending the lag period, decreasing the live counts of microorganisms and/or reduces growth rate [19]. The antimicrobials used include organic acids, enzymes, bacteriocins, fungicides, polymers, natural extracts and essential oils [20]. However, the packaging methods and/or materials used are also important. Nisin, for example, incorporated into low-density polyethylene (LDPE) will suppress the growth of *Staphylococcus aureus* and *Listeria monocytogenes* [21], *Lactobacillus plantarum* when incorporated into soy protein and corn zein based films and *Salmonella Typhimurium* when coated onto polymeric films like PVC and nylon [22].

In this study, AFV and SO incorporated into active packaging films inhibited the growth of *C. estertheticum*, significantly retarding blown pack spoilage of beef primals. AFV contains bioflavanoids, citric, malic, lactic and caprylic acid, all of which have previously demonstrated antibacterial activity against Gram-positive bacteria [12,23,24]. Moreover, bioflavanoids are known to have antimicrobial activity against Clostridium spp. [25], possibly from the inhibition of membrane bound or intracellular proteins [26]. Although the exact antibacterial mechanisms of organic acids is not fully understood, it is assumed the undissociated form penetrates the cell, dissociates into anions and protons resulting in a decrease in cytoplasmic pH which inhibits a range of cellular functions [27]. Caprylic acid may also lower the pH of the cytoplasm disrupting the normal activity of intracellular enzymes [28] and has been shown to have antimicrobial activity against a range of foodborne bacterial pathogens including *Escherichia coli* O157, *Enterobacter sakazakii* and *L. monocytogenes* [29,30]. Interestingly, sodium octanoate ($C_8H_{15}NaO_2$), is a derivative of caprylic acid ($C_8H_{16}O_2$) and has similar antimicrobial properties [29,30].

In contrast, IMDA did not demonstrate anti-*estertheticum* properties when incorporated into the packaging film. This was unexpected as IMDA is composed of sodium diacetate, malic acid, mono and diglycerides of fatty acids, salt and excipients, all of which have previously been demonstrated to have anti-bacterial, including anti-Clostridium properties [27,31]. However, the effectiveness, or otherwise, of an anti-microbial compound incorporated into an active packaging film is dependent on a range of factors including the properties of the film/matrix and the characteristics of the food (pH, moisture, temperature, etc.). Thus, the apparent ineffectiveness of IMDA may be attributed to differences in important parameters such as release rate and reaction with the matrix (gelatine) [32].

5. Conclusions

In conclusion, the results of this study suggest that Auranta FV (AFV) and sodium octanoate (SO), incorporated in a gelatine matrix at concentrations of 2.5% or 3.5 times their MIC against *C. estertheticum* could be used in an active packaging system to prevent blown pack spoilage of beef primals.

Acknowledgments: The funding for this research was provided by the Food Institutional Research Measure (FIRM) administered by the Department of Agriculture, Food and the Marine, Ireland. The authors would also like to thank Paula Reid (Teagasc) for her help with statistical analysis of the data and Paula Bourke and Patrick J. Cullen (both Dublin Institute of Technology) for their assistance in the plasma pre-treatment of the packaging films.

Author Contributions: Declan Bolton conceived and designed the experiments; Rachael Reid performed the experiments; Rachael Reid and Declan Bolton analysed the data; Declan Bolton and Rachael Reid wrote the paper; Declan Bolton, Séamus Fanning, Paul Whyte, Andrey A. Tiuftin and Joe P. Kerry supervised the study. All authors read and approved the manuscript.

References

1. Bolton, D.J.; Carroll, J.; Walsh, D. A four-year survey of blown pack spoilage *Clostridium estertheticum* and *Clostridium gasigenes* on beef primal cuts. *Lett. Appl. Microbiol.* **2015**, *61*, 153–157. [CrossRef] [PubMed]

2.   Broda, D.M.; Saul, D.J.; Lawson, P.A.; Bell, R.G.; Musgrave, D.R. *Clostridium gasigenes* sp. nov., a psychrophile causing spoilage of vacuum-packed meat. *Int. J. Syst. Evol. Microbiol.* **2000**, *1*, 107–118. [CrossRef] [PubMed]

3.   Adam, K.H.; Flint, S.H.; Brightwell, G. Psychrophilic and psychrotrophic clostridia: Sporulation and germination processes and their role in the spoilage of chilled, vacuum-packaged beef, lamb and venison. *Int. J. Food Sci. Technol.* **2010**, *45*, 1539–1544. [CrossRef]

4.   Cavill, L.; Renteria-Monterrubio, A.L.; Helps, C.R.; Corry, J.E.L. Detection of cold-tolerant clostridia other than Clostridium estertheticum in raw vacuum-packed chill-stored meat. *Food Microbiol.* **2011**, *28*, 957–963. [CrossRef] [PubMed]

5.   Dainty, R.H.; Edwards, R.A.; Hibbard, C.M. Spoilage of vacuum-packed beef by a *clostridium sp.* *J. Sci. Food Agric.* **1989**, *49*, 473–486. [CrossRef]

6.   Moschonas, G.; Bolton, D.J.; Sheridan, J.J.; McDowell, D.A. The effect of storage temperature and inoculum level on the time of onset of 'blown pack' spoilage. *J. Appl. Microbol.* **2010**, *108*, 532–539. [CrossRef] [PubMed]

7.   Yang, X.; Gill, C.O.; Balamurugan, S. Enumeration of *Clostridium estertheticum* spores in samples from meat plant conveyors and silage stacks by conventional and real time PCR procedures. *Int. J. Food Saf.* **2010**, *12*, 115–121.

8.   Moschonas, G.; Bolton, D.J.; Sheridan, J.J.; McDowell, D.A. The effect of heat shrink treatment and storage temperature on the time of onset of "blown pack" spoilage. *Meat Sci.* **2011**, *87*, 115–118. [CrossRef] [PubMed]

9.   Siragusa, G.; Cutter, C.; Willett, J. Incorporation of bacteriocin in plastic retains activity and inhibits surface growth of bacteria on meat. *Food Microbiol.* **1999**, *16*, 229–235. [CrossRef]

10. Seydim, A.; Sarikus, G. Antimicrobial activity of whey protein based edible films incorporated with oregano, rosemary and garlic essential oils. *Food Res. Int.* **2006**, *39*, 639–644. [CrossRef]

11. Cooksey, K. Antimicrobial food packaging materials. *Addit. Polym.* **2001**, *8*, 6–10. [CrossRef]

12. Cruz-Romero, M.; Murphy, T.; Morris, M.; Cummins, E.; Kerry, J. Antimicrobial activity of chitosan, organic acids and nano-sized solubilisates for potential use in smart antimicrobially-active packaging for potential food applications. *Food Control* **2013**, *34*, 393–397. [CrossRef]

13. Clarke, D.; Molinaro, S.; Tyuftin, A.; Bolton, D.; Fanning, S.; Kerry, J.P. Incorporation of commercially-derived antimicrobials into gelatin-based films and assessment of their antimicrobial activity and impact on physical film properties. *Food Control* **2016**, *64*, 202–211. [CrossRef]

14. Lund, B.M.; Graham, A.F.; George, S.M.; Brown, D. The combined effect of incubation temperature, pH and sorbic acid on the probability of growth of nonproteolytic type B *Clostridium botulinum*. *J. Appl. Bacteriol.* **1990**, *89*, 481–492. [CrossRef]

15. Boerema, J.A.; Broda, D.M.; Penney, N.; Brightwell, G. Influence of peroxyacetic acid-based carcass rinse on the onset of "blown pack" spoilage in artificially inoculated vacuum-packed chilled beef. *J. Food Prot.* **2007**, *70*, 1434–1439. [CrossRef] [PubMed]

16. Lawson, P.; Dainty, R.H.; Kristiansen, N.; Berg, J.; Collins, M.D. Characterization of a psychrotrophic Clostridium causing spoilage in vacuum-packed cooked pork: Description of *Clostridium algidicarnis* sp. nov. *Lett. Appl. Microbiol.* **1994**, *19*, 153–157. [CrossRef] [PubMed]

17. Yang, X.; Gill, C.O.; Balamurugan, S. Effects of temperature and pH on the growth of bacteria isolated from blown packs of vacuum-packaged beef. *J. Food Prot.* **2009**, *72*, 2380–2385. [CrossRef]

18. Silva, A.R.; Paulo, E.N.; Sant'ana, A.S.; Chaves, R.D.; Massaguer, P.R. Involvement of *Clostridium gasigenes* and *Clostridium algidicarnis* in 'blown pack' spoilage of Brazilian vacuum-packed beef. *Int. J. Food Microbiol.* **2011**, *148*, 156–163. [CrossRef] [PubMed]

19. Han, J.H. Antimicrobial food packaging. *Food Technol.* **2000**, *54*, 56–65.

20. Malhotra, B.; Keshwani, A.; Harsha, K. Antimicrobial food packaging: Potential and pitfalls. *Front. Microbiol.* **2015**, *6*, 611. [CrossRef] [PubMed]

21. Cooksey, K. Utilization of antimicrobial packaging films for the inhibition of selected microorganisms. In *Food Packaging: Testing Methods and Applications*; ACS Publications: Washington, DC, USA, 2000; pp. 17–25.

22. Natrajan, N.; Sheldon, B.W. Efficacy of Nisin-Coated polymer films to inactivate Salmonella Typhimurium on fresh broiler skin. *J. Food Prot.* **2000**, *63*, 1189–1196. [CrossRef] [PubMed]

23. Burt, S. Essential oils: Their antibacterial properties and potential applications in foods—A review. *Int. J. Food Microbiol.* **2004**, *94*, 223–253. [CrossRef] [PubMed]

24. Batovska, D.; Parushev, S.; Stamboliyska, B.; Tsvetkova, I.; Ninova, M.; Najdenski, H. Examination of growth inhibitory properties of synthetic chalcones for which antibacterial activity was predicted. *Eur. J. Med. Chem.* **2009**, *44*, 2211–2218. [CrossRef] [PubMed]

25. Wu, X.; Alam, Z.; Feng, L.; Tsutsumi, L.S.; Sun, D.; Hurdle, J.G. Prospects for flavonoid and related phytochemicals as nature-inspired treatments for Clostridium difficile infection. *J. Appl. Microbiol.* **2014**, *116*, 23–31. [CrossRef] [PubMed]

26. Cushnie, T.P.; Lamb, A.J. Recent advances in undersatdning the antibacterial properties of flavonoids. *Int. J. Antimicrobiol. Agents* **2011**, *38*, 99–107. [CrossRef] [PubMed]

27. Ricke, S.C. Perspectives on the use of organic acids and short chain fatty acids as antimicrobials. *Poult. Sci.* **2002**, *82*, 632–639. [CrossRef]

28. Sun, C.Q.; O' Conner, C.J.; Robertson, C.J. The antimicrobial properties of milk fat after partial hydrolysis of calf pregastric lipase. *Chem. Biol. Interact.* **2002**, *40*, 185–198. [CrossRef]

29. Annamalai, T.; Nair, M.K.M.; Marek, P.; Vasudevan, P.; Schreiber, D.; Knight, R.; Hoagland, T.; Venkitanarayanan, K. In vitro inactivation of enterohemorrhagic *Escherichia coli* O157:H7 in bovine rumen fluid by caprylic acid. *J. Food Prot.* **2004**, *67*, 884–888. [CrossRef] [PubMed]

30. Nair, M.K.M.; Vasudevan, P.; Hoagland, T.; Venkitanarayanan, K. Inactivation of *Escherichia coli* O157:H7 and *Listeria monocytogenes* in milk by caprylic acid and monocaprylin. *Food Microbiol.* **2004**, *21*, 611–616. [CrossRef]

31. Juneja, V.K.; Thippareddi, H. Inhibitory effects of organic acid salts on growth of *Clostridium perfringens* from spore inocula during chilling of marinated ground turkey breast. *Int. J. Food Microbiol.* **2004**, *93*, 155–163. [CrossRef] [PubMed]

32. Appendini, P.; Hotchkiss, J.H. Review of antimicrobial food packaging. *Innov. Food Sci. Emerg. Technol.* **2002**, *3*, 113–126.

3

Development and Characterization of a Biodegradable PLA Food Packaging Hold Monoterpene–Cyclodextrin Complexes against *Alternaria alternata*

Velázquez-Contreras Friné [1,2], Acevedo-Parra Hector [2], Nuño-Donlucas Sergio Manuel [3], Núñez-Delicado Estrella [1] and Gabaldón José Antonio [1,*]

[1] Campus de los Jerónimos, Universidad Católica San Antonio de Murcia, 135, 30107 Guadalupe, Murcia, Spain; fvelazqu@up.edu.mx (V.-C.F.); enunez@ucam.edu (N.-D.E.)

[2] Universidad Panamericana ESDAI, Álvaro del Portillo 49, Zapopan 45010, Jalisco, Mexico; hector.acevedo@up.edu.mx

[3] Departamento de Ingeniería Química, Universidad de Guadalajara, Blvd, Marcelino García Barragán 1421, Guadalajara 44430, Jalisco, Mexico; gigio@cencar.udg.mx

* Correspondence: jagabaldon@ucam.edu.

Abstract: The fungi of the genus *Alternaria* are among the main pathogens causing post-harvest diseases and significant economic losses. The consumption of *Alternaria* contaminated foods may be a major risk to human health, as many *Alternaria* species produce several toxic mycotoxins and secondary metabolites. To protect consumer health and extend the shelf life of food products, the development of new ways of packaging is of outmost importance. The aim of this work was to investigate the antifungal capacity of a biodegradable poly(lactic acid) (PLA) package filled with thymol or carvacrol complexed in β-cyclodextrins (β-CDs) by the solubility method. Once solid complexes were obtained by spray drying, varying proportions (0.0%, 1.5%, 2.5%, and 5.0 wt%) of β-CD–thymol or β-CD–carvacrol were mixed with PLA for packaging development by injection process. The formation of stable complexes between β-CDs and carvacrol or thymol molecules was assessed by Fourier-transform infrared spectroscopy (FTIR). Mechanical, structural, and thermal characterization of the developed packaging was also carried out. The polymer surface showed a decrease in the number of cuts and folds as the amount of encapsulation increased, thereby reducing the stiffness of the packaging. In addition, thermogravimetric analysis (TGA) revealed a slight decrease in the temperature of degradation of PLA package as the concentration of the complexes increased, with β-CD–carvacrol or β-CDs–thymol complexes acting as plasticisers that lowered the intermolecular forces of the polymer chains, thereby improving the breaking point. Packages containing 2.5% and 5% β-CD–carvacrol, or 5% β-CD–thymol showed *Alternaria alternata* inhibition after 10 days of incubation revealing their potential uses in agrofood industry.

Keywords: food packaging; poly(lactic acid); thymol; carvacrol; β-cyclodextrin; antifungal activity

1. Introduction

Nowadays, the increasing consumer demand for healthy, freshly prepared, and convenient fruits and vegetables has driven the rapid growth of the fresh-cut produce industry worldwide, with benefits in multi-billion dollars [1]. In addition, new lifestyle drivers such as health and aging of population has stimulated the agrofood industry to enhance the offer and delivery of value-added products, such as minimally processed fruits and vegetables packaged in sealed polymeric films or on trays, ready for immediate consumption or direct cooking.

However, this trend has disturbed the scenario of foodborne diseases worldwide caused by pathogenic microorganisms, with important economic and social impacts [2], since fresh and minimally processed foods may undergo negative qualitative changes related to high respiratory rate, moisture loss, rapid enzymatic browning, and microbial contamination which lead to the rapid deterioration of the products [3]. In addition, fungal contamination of crops through latent infections usually occurs in the fields; nevertheless, the rotting arises later, during the storage and transport before marketing. The fungi of the genus *Alternaria* are among the main pathogens causing post-harvest diseases and significant economic losses. These fungi also represent a serious toxicological risk as they produce a broad spectrum of mycotoxins and secondary metabolites, which can cause problems in humans and animals [4].

This issue has raised considerable challenges for food packaging companies and researchers that specifically use biodegradable materials or bio-based packaging for food preservation. Despite the good properties of petroleum-based plastics, their widespread use and accumulation cause serious environmental problems and dependence on fossil resources. In fact, packaging applications contribute to 63% of the current plastic waste, and it is estimated that less than 14% are recyclable [5].

To overcome the described drawbacks, different approaches have been carried out to obtain bioplastics with analogous functionalities to petrochemical polymers. Poly(lactic acid) (PLA), a biodegradable aliphatic polyester which can be obtained by fermentation of renewable resources such as corn, tapioca, and sugarcane [6], meets several requirements such as high mechanical strength, biodegradability, biocompatibility, bio-absorbability, transparency, low toxicity, and easy process ability [7] to be thoroughly employed in agricultural films, biomedical devices, and food packaging [8,9] and used as a suitable carrier of active compounds to yield antioxidant or antimicrobial effects [10,11].

Currently, consumer concerns about the potential toxicity to humans of synthetic antimicrobials such as butylated hydroxytoluene (BHT) or butylated hydroxyanisole (BHA) have resulted in the increased use of natural antimicrobials, which receive a good deal of attention for a number of microorganism control issues [12]. As a result, different antimicrobials have been added to different packaging materials. In particular, essential oils and their bioactive molecules such as carvacrol and thymol have been thoroughly tested in vitro [13,14] or in different food systems such as meat, dairy or vegetable samples [15] due to their insecticidal, antiviral, antimicrobial, and antifungal activities [16]; however, their high volatility and reactivity limits their application as food preservatives. In fact, long storage time and temperature could magnify volatilization and drastically lessen their activity, requiring as consequence high concentrations to ensure antimicrobial activity, which is a detrimental praxis for organoleptic attributes (flavor, taste, and aroma) and acceptability of foods, so this strategy is not considered in practice.

In order to increase the applicability of natural antimicrobial formulations, these drawbacks could be overcome by microencapsulation or complexation techniques using cyclodextrins (CDS), which are cyclic oligosaccharides derived from starch made up of 6, 7, or 8 units of D-glucose monomers linked by $\alpha(1,4)$ bonds, shaped as a truncated hollow cone [17] that presents an internal hydrophobic cavity to interact with non-polar active constituents of essential oils or their bioactive molecules such as carvacrol and thymol, whereas the external face is hydrophilic, improving their water solubility and gradually increasing their effectiveness using lower concentrations of these compounds.

As a preliminary stage to subsequently evaluate the antifungal capacity of a biodegradable poly(lactic acid) (PLA) package carrying as preservatives carvacrol or thymol complexed in CDs (as described here), their complexation was carried out with native and modified CDs [18] and the antimicrobial and antifungal effects of their respective complexes was verified by comparison with hydroxypropyl-β-cyclodextrins (selected due to their highest Kc values) against *Escherichia coli*, *Staphylococcus aureus*, and *Galactomyces citri-aurantii* [19,20]. However, only native CDs (α, β, and γ) are considered as GRAS (generally recognized as safe) and are the only ones authorized to come into contact with foods.

Therefore, the present study focuses on the design and optimization of a controlled release system of antifungal carvacrol or thymol volatiles encapsulated in β-CD to be incorporated into a biodegradable polymeric matrix of PLA by industrial injection. The optimization of stable complexes between β-CDs and carvacrol or thymol molecules and characterization by Fourier-transform infrared spectroscopy (FTIR) were carried out. Mechanical, structural, and thermal characterization of developed packaging was carried out and materials behavior against *Alternaria alternata* growth was also investigated.

2. Materials and Methods

2.1. Materials

Carvacrol (CAS: 499-75-2, 99.5% purity), thymol (CAS: 89-83-8, 98.7% purity), and β-cyclodextrin (β-CD >95%, food grade) were purchased from Sigma-Aldrich Corp (Saint Louis, MO, USA). The chemical structures of the two monoterpenes are shown in Figure 1. Poly(lactic acid) (PLA, Ingeo™ Biopolymer ref. code: 3251D) with a weight-average molecular weight ($\overline{M}$w) of 5.5×10^4 g/mol, polydispersity index (PI) of 1.62, and low D-isomer content (99% L-lactide/1% D-lactide), provided by PromaPlast Co (Guadalajara, Jalisco, Mexico) and manufactured by Nature Works LLC (Blair, NE, USA), was selected for injection moulding applications since it has a higher flow capability (relative viscosity 2.5, glass transition temperature Tg = 55–60 °C, melting temperature Tm = 155–170 °C, and processing temperature 188–210 °C) than other resins currently available in the marketplace. *Alternaria alternata* strain ATCC 42761 (isolated from blackberries in Georgia, USA) was purchased from SENNA laboratories, Mexico City. Potato dextrose agar (PDA) was provided by Bioxon, Mexico. The rest of the chemical products were of analytical grade.

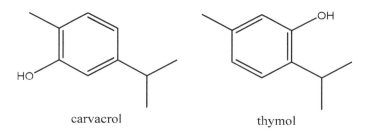

Figure 1. Chemical structures of carvacrol and thymol monoterpenes.

2.2. Preparation of β-CD Inclusion Complexes

Both β-CD–carvacrol and β-CD–thymol inclusion complexes were prepared using the solubility method [18]. For that, aqueous solutions of increasing concentrations of β-cyclodextrin (0–15 mM) were prepared in sodium phosphate buffer (100 mM, pH of 7.0) in a total volume of 100 mL. A saturating amount of carvacrol or thymol was added to each one of the solutions and kept in an ultrasound bath (Ultrasons HP, Selecta, Spain) for 60 min in the dark at 25 °C, until equilibrium was reached. After that, to remove excess monoterpene, the respective solutions were filtered through a nylon filter of 0.45 μm. Liquid complexes were used for phase solubility diagrams, determining the concentration of entrapped monoterpene by GC/MS, and posterior spray drying process to obtain powdered dehydrated complexes.

From the phase diagrams of carvacrol or thymol, complexed with β-CDs, the parameters efficiency of complexation (CE) and the molar ratio (MR) were determined. CE is the ratio between the dissolved complex and free cyclodextrin (CD) concentration. It is independent of S_0 (aqueous solubility), and was calculated from the slope of the phase solubility profiles by using Equation (1).

$$CE\ (\%) = \frac{[dissolved - complex]}{[CD]_f} \tag{1}$$

The MR of β-CD–carvacrol and β-CD–thymol inclusion complexes was calculated using CE values with Equation (2).

$$MR = \frac{1}{\left(1 + \frac{1}{CE}\right)} \tag{2}$$

2.3. Atomization Process to Obtain Complexes in Solid State

To obtain complexes in solid state, the β-CD–carvacrol and β-CD–thymol solutions were subjected to an atomization process using a laboratory-scale atomizer and Büchi B290 Mini Spray Dryer (Flawil, Switzerland) working with air as the carrier gas at a flow rate 5 mL/min, pressure of 3.2 bar, and an inlet and outlet temperature of $170 \pm 2\,°C$ and $68 \pm 2\,°C$, respectively, using a 1.5 mm nozzle diameter. In each case, the entrapment efficiency (EE) was determined with respect to the theoretical number of monoterpenes present in the inclusion complex after atomization, using Equation (3).

$$EE = \frac{\text{Amount of active compound entrapped}}{(\text{Initial active compound amount})} \times 100 \tag{3}$$

Furthermore, the process performance (PP) was determined as follows:

$$PP = \frac{\text{Total weight obtained from solids after spray drying process (g)}}{(\text{Initial Initial weight } - CD \text{ in solution (g)})} \times 100 \tag{4}$$

Carvacrol and thymol concentrations in dehydrated complexes were quantified after spray drying. For that, β-CD–carvacrol and β-CD–thymol were diluted in ethanol (complex: ethanol, 20:80, v/v), to break the complexes formed. After that, β-CDs was removed from the solution, leaving only the active compound for further quantification in triplicate, by GC/MS analysis at Agilent Technologies 7890B (Palo Alto, CA, USA) coupled to a 5977A mass spectrometer, as previously described by Rodríguez-López et al. (2019) [18].

2.4. Fourier-Transform Infrared Spectroscopy (FTIR)

The FTIR spectra used to study changes of chemical structures of free carvacrol and thymol, and their respective complexes were acquired using a Varian FTIR 670 (Agilent Tech, Amstelveen, The Netherlands) spectrophotometer coupled with an accessory to analyze the attenuated total reflectance (ATR) with a wave number resolution of $0.10\ cm^{-1}$ in the range of 250–4000 cm^{-1}. A minimum of 32 scans were signal-averaged with a resolution of 4 cm^{-1} in the above ranges.

2.5. Boxes Production

The PLA samples were dried in an oven at 60 °C for 4 h to avoid bubbles in the molding process. After that, physical mixtures were performed using as ingredients PLA (100%, 98.5%, 97.5%, and 95% weight percentages, wt%), and dehydrated complexes of β-CD–carvacrol or β-CD–thymol at (0%, 1.5%, 2.5%, and 5% wt%), that were introduced in a pilot extruder to produce pellets. The extruder had a screw diameter (D) of 25.4 mm, screw length (L) of 406.4 mm (L/D ratio of 16), four heating zones, and a slot 1.75 mm matrix outlet. The barrel temperature profile was set at 150/170/180/180 °C with a screw speed of 30 rpm.

The pellets produced in the previous step were thermo-pressed in the Belken BLD-68 injector from AG Plastic (Querétaro, México), optimizing the parameters of heated mold (180 °C/100 bar), to ensure the adequate fluidity of the material to produce (12 × 10 × 3.0 cm) boxes (Figure 2). Once the material reached the cooling temperature, the boxes were then released from the molds.

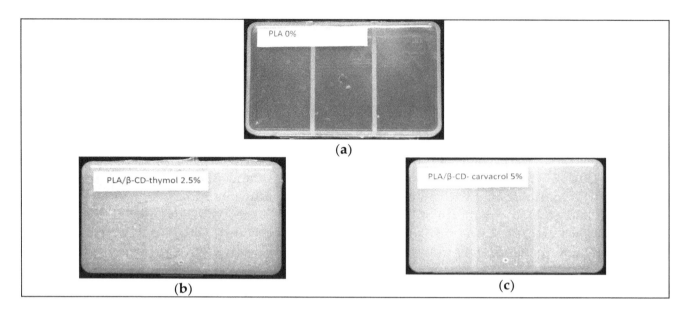

Figure 2. Boxes obtained by injection of the pellets. (**a**) (PLA); (**b**) (PLA/β-CD–thymol, 2.5%, wt%); (**c**) (PLA/β-CD–carvacrol, 5.0%, wt%).

2.6. Packaging Characterization

2.6.1. Mechanical Properties

Tensile proofs were carried out in the universal traction machine SFM 100 from United Testing Systems (Ontario, Canada). Previously the packages were manually cut to obtain assay pieces (ten of each formulation), according to dimensions established by the ASTM method D-638. The tests were conducted at room temperature, at 5 mm/min speed using an initial grip length of 25 mm. The parameters, namely, average of maximum stress (MPa), breaking point (%), and Young's modulus (MPa) were determined for the pieces of PLA and β-CD–carvacrol or β-CD–thymol, according to the aforementioned procedure [21].

2.6.2. Scanning Electron Microscopy (SEM)

The structure of the packaging material was determined by scanning electron microscopy MIRA3 model form TESCAN (Brno, Czech Republic). Packages were previously frozen at −80 °C, manually fractured, and later placed on the slide and gold coated during 90 s using a sputter coater. All the samples were evaluated using a voltage of 7.0 kV.

2.6.3. Thermal Characterization of the Developed Packaging

The thermal evaluation of the packaging material was done by differential scanning calorimetry (DSC) and thermogravimetric analysis (TGA). DSC assays were performed in DSC-Q100 (TA Instruments, New Castle, USA). Firstly, pieces of 5 mg were dried for 48 h in an oven at room temperature; after that, samples were placed in an aluminium capsule that was subjected to a temperature scan from 20–230 °C at a heating rate of 10 °C/min under inert nitrogen atmosphere. In addition, thermal stability of the materials was carried out by TGA using the gravimetric thermal analyzer TGA-550 (TA Instruments, New Castle, PA, USA). For that, samples of 10 mg were weighed and placed in platinum trays, which were subjected to a temperature scan of 20–600 °C at a heating rate of 20 °C/min under a nitrogen atmosphere.

2.7. Antimicrobial Activity

The antifungal activity of the packages with β-CD–thymol and β-CD–carvacrol was evaluated by vapor phase diffusion, in triplicates, according to Du et al. [22] using a strain of *A. alternata* (ATCC

42761). Pure fungal cultures in potato-dextrose agar medium plates with 14 days of incubation (23 °C) were suspended in 10 mL sterile distilled water containing 0.05% of Tween 20, and collected by gently scraping the surface of the agar with a sterile L-shaped glass rod. Next, the arthrospores concentration was adjusted to 10^6 spores/mL using the McFarland scale (Shumadzu, UVmini-1240), and the inoculum was used for in vitro bioassays.

For the bioassay, 3.0 µL of spore suspensions were placed in the centre of Petri dishes previously filled with inoculated potato dextrose agar (PDA). Subsequently, these boxes were incubated at 25 °C for 5 and 10 days, inverted and covered with parafilm, and were used as controls.

Packages containing different concentration of active compounds were aseptically cut into 50 mm rectangles and placed on top of the Petri dishes. Parafilm M (Bemis) was used to hermetically seal the Petri dishes, which were incubated at 25 ± 1 °C in an incubator (Binder ED), for 120–240 h. After the incubation period, the inhibition zone diameter created by the vapor and active compound (thymol or carvacrol complexed with CDs) released from the packaging into the culture medium was measured and related to the package antimicrobial activity.

The growth of fungal cultures as well as controls were daily evaluated by measuring the diameter of the colony or surface area (diameter at right angles to each other) of the plates occupied by the colony during incubation time. The measurements were carried out with a gauge on the agar surface reporting growth at 5 and 10 days. Due to the transparency of the materials used, these measurements were conducted without opening the box. Every assay was tested in triplicate and the results were statistically analyzed.

2.8. Statistical Analysis

The data corresponding to mechanical properties and the diameter of the colony in the antifungal activity were subjected to statistical analysis. Analysis of variance (ANOVA) and Tukey's multiple comparison test were performed using MINITAB 18 statistical software (Paris, France), at a 5% significance level.

3. Results and Discussion

3.1. Assessment of the Obtained Complexes

As described previously by Rodríguez-López et al., 2019 [18], phase solubility diagrams of carvacrol and thymol with β-CDs were carried out at pH 7.0 (25 °C), since the pH of the medium could condition its dissociation degree and consequently its solubility, thus determining the stability of the complexes. By using linear regression analysis of the phase solubility diagrams and considering the formation of β-CD–carvacrol and β-CD–thymol 1:1 complexes when the concentration of β-cyclodextrin was 11 mM, it was possible to determine the complexation constant (Kc), the complexation efficiency (CE), and molar ratio (MR) by applying Equations (1) and (2).

As can be seen in Table 1, β-CD–thymol and β-CD–carvacrol complexes show the same molar ratio (1:2), indicating that almost one of every two β-CDs molecules in solution is forming soluble complexes with carvacrol or thymol [18]. However, the efficiency of complexation obtained for carvacrol (105.6) is significantly higher than that obtained for thymol (69.3).

Table 1. Carvacrol and thymol aqueous solubility (S_0), complexation constant (Kc) with β-CDs, complexation efficiency (CE), and molar ratio (MR) at pH of 7.0.

Complexes	S_0 (mmol L^{-1})	K_C (L mol^{-1})	CE (%)	Molar Ratio
Carvacrol/β-CDs	5.64 ± 0.12	1871 ± 143	105.6 ± 10.3	1:2
Thymol/β-CDs	5.77 ± 0.15 *	1198 ± 115	69.3 ± 9.2	1:2

* SD, standard deviation of triplicate determinations.

For further packaging formulations with PLA, soluble complexes of β-CD–carvacrol and β-CDs–thymol were subjected to a spray drying process to obtain complexes in a solid state to improve their management.

After the dehydration process, entrapment efficiency (Equation (3)) and process performance (Equation (4)) parameters were determined (see Table 2), showing similar values for both the parameters, but slightly higher for thymol.

Table 2. Entrapment efficiency (EE) and process performance (PP) of β-CD–carvacrol and β-CD–thymol complexes in solid state.

Monoterpene	β-CD	EE (%)	PP (%)
Carvacrol	11 mM	45 ± 2.5 *	84 ± 3.2
Thymol	11 mM	47 ± 1.8	86 ± 3.7

* SD, standard deviation of triplicate determinations.

FTIR is a suitable technique for evidencing the formation of the β-CD–carvacrol and β-CD–thymol inclusion complexes (Figure 3), due the shift or vanishing of the stretching and bending vibrations of the functional groups of guest molecule once complexed.

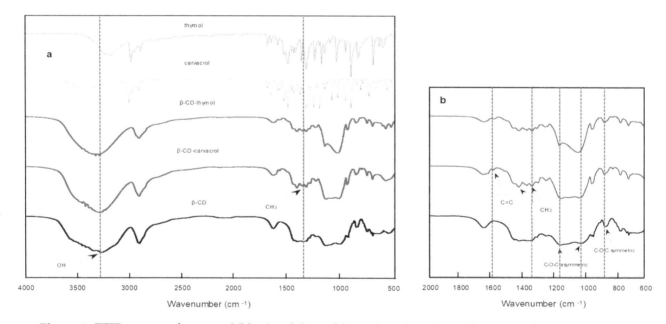

Figure 3. FTIR spectra of carvacrol (blue) and thymol (green); β-CD–thymol (green) and β-CD–carvacrol (blue) complexes, and β-CD (black) in normal (**a**) and broad view (**b**).

As can be seen in Figure 3a, the IR spectrum of thymol (structural isomer of carvacrol), shows several characteristics peaks: 3164 cm^{-1}, O-H stretching and bending vibrations (3164 cm^{-1} and 1453 cm^{-1}, respectively); C-H symmetric and asymmetric stretching bands at 2858 cm^{-1} and 2897 cm^{-1}, respectively; and three C=C stretching vibrations of weak intensity at 1624 cm^{-1}, 1592 cm^{-1}, and 1506 cm^{-1}, revealing the tri-substitution of the aromatic ring. With respect to the substituents of the aromatic ring, methyl (-CH$_3$) appears at 1344 cm^{-1} and a typical doubled signal (like a tooth) at 1410 cm^{-1} characteristic of isopropyl group was observed. The IR spectrum of β-CDs (Figure 3a) showed characteristic bands corresponding to stretching vibrations of O-H and C-H links, around 3268 cm^{-1} and 2875 cm^{-1}, respectively and O-H bending vibrations at 1623 cm^{-1}.

In addition, Figure 3b shows an approach of the IR spectrum of β-CDs revealing C–O–C symmetric and asymmetric vibrations at 890 cm^{-1}, 1170 cm^{-1}, and 1021 cm^{-1}; respectively. With respect to free β-cyclodextrin, the spectra of β-CD–carvacrol and β-CD–thymol inclusion complexes (Figure 3b) highlighted the presence of characteristics C=C peaks corresponding to carvacrol and thymol aromatic ring close to 1590 cm^{-1} and vibrations of their respective methyl (-CH$_3$) groups appear at 1430 cm^{-1} (asymmetric) and 1360 cm^{-1} (symmetric). These shifts relative to those of respective free compounds provide a clear evidence of host-guest interactions.

3.2. Mechanical Properties of PLA Packaging Loaded with β-CD–Carvacrol or β-CD–Thymol Inclusion Complexes

In order to prevent breakages during the packaging process, polymeric materials to be used in food packaging require sufficient flexibility [23]. In this sense, the mechanical properties of the PLA boxes with different concentrations (1.5%; 2.5%, and 5%, wt%) of β-CD–carvacrol (Table 3) and β-CDs–thymol (Table 4) inclusion complexes were evaluated, using PLA boxes without solid complexes as control.

Table 3. Mechanical properties of the enriched or not PLA trays with β-CD–carvacrol complexes.

PLA Boxes with Different % of β-CD–Carvacrol				
Parameter	0%	1.50%	2.50%	5%
Young's modulus (Mpa)	2873 ± 176	2327 ± 170 *	2259 ± 53 *	1960 ± 110 *
Maximum stress (MPa)	63.6 ± 4.5	49.9 ± 6.5 *	51.3 ± 4.9 *	47.5 ± 5.1 *
Breaking point (%)	2.4 ± 0.4	2.7 ± 0.3 *	2.9 ± 0.2 *	3.2 ± 0.4 *

Results expressed in mean ± standard deviation of ten determinations; symbol (*) in the same file indicates significant differences ($p < 0.05$) according to Tukey's test.

Table 4. Mechanical properties of the enriched or not PLA trays with β-CD–thymol complexes.

PLA Boxes with Different % of β-CD–Thymol				
Parameter	0%	1.50%	2.50%	5%
Young's modulus (Mpa)	2873 ± 176	2667 ± 161 *	2382 ± 69 *	2394 ± 118 *
Maximum stress (MPa)	63.6 ± 4.5	57.9 ± 6.8 *	53.2 ± 2.3 *	55.1 ± 5.2 *
Breaking point (%)	2.4 ± 0.4	2.8 ± 0.3 *	2.9 ± 0.2 *	3.1 ± 0.3 *

Results expressed in (mean ± standard deviation) of ten determinations; Symbol (*) in the same file indicate significant differences ($p < 0.05$) according to Tukey's test.

As can be seen in Table 3, Young's modulus ranged from 2873 to 1960 MPa for β-CD–carvacrol complexes and from 2873 to 2394 MPa for β-CD–thymol complexes (see Table 4), showing lower values than the control trays (only PLA). In fact, Young's modulus gradually decreases as the concentration (weight percentage, w%) of the dehydrated complexes increases, obtaining the lowest value of Young's modulus in the sample fortified with 5% of carvacrol (see Table 3), with a significant difference respect to the average value ($p < 0.05$). The same trend was observed when evaluating the maximum stress, with the lowest value being observed in the PLA package enriched with dehydrated complexes of β-CD–carvacrol (5%), 14% lower than the value obtained for PLA fortified with thymol complexes at the same concentration (w%). The different mechanical values observed when both the complexes were added to the PLA polymer could be due to the higher CE value obtained for carvacrol–β-CDs (105.6%), 65% higher than the value obtained for thymol–β-CDs (69.3%), revealing that CE values above 100% indicate that at pH 7.0, there are more β-CDs complexing carvacrol than free in solution. In the case of thymol, the number of β-CDs complexing thymol is lower, since CE is less than 100%, and in consequence, the decrease in mechanical properties is less pronounced.

Regarding the breaking point, a significant increment of this parameter was observed as the concentration (wt%) of the dehydrated complexes increased, improving 25% and 23% of the elongation capacity of the polymeric material (control), when 5% of β-CD–carvacrol or 5% of β-CD–thymol, respectively, were added to PLA. This behavior may be attributable to a plasticizing effect triggered by the addition of β-CD complexes to the polymer matrix disrupting the crystalline structure of PLA and increasing its ductile properties [24].

As a result, the relative high elongations achieved were beneficial since the boxes presented better flexibility. These results are consistent with those obtained by Ramos and López-Rubio, wherein an increase in elongation and breaking point in plastic films composed of polypropylene/carvacrol/thymol were evidenced [24,25].

3.3. Scanning Electron Microscopy

The fracture micrographs of the samples are shown in Figure 4.

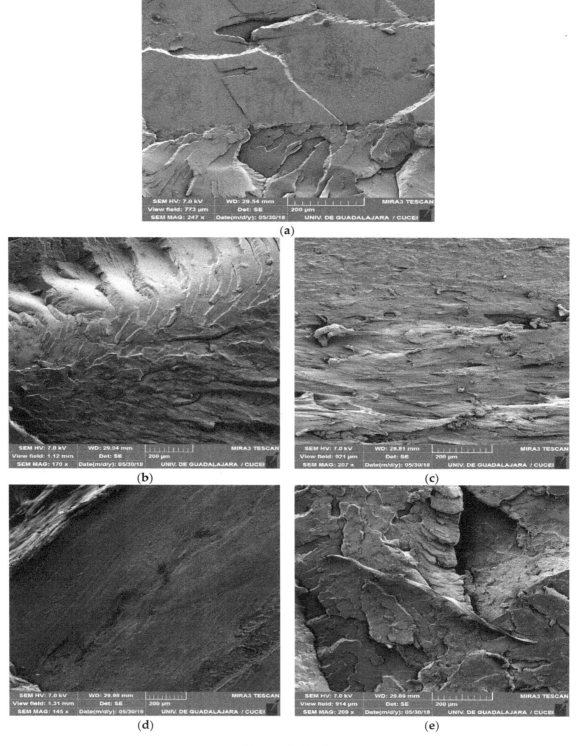

Figure 4. *Cont.*

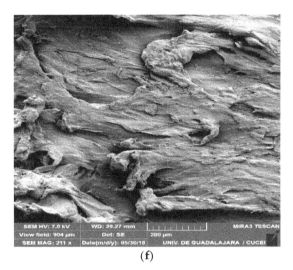 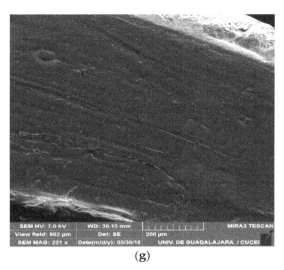

(f)   (g)

Figure 4. SEM micrographs of fracture samples: (**a**) 100% PLA; (**b**) 98.5% PLA with 1.5% β-CD–thymol; (**c**) 97.5% PLA with 2.5% β-CD–thymol; (**d**) 95.0% PLA with 5.0% β-CD–thymol; (**e**) 98.5% PLA with 1.5% β-CD–carvacrol; (**f**) 97.5% PLA with 2.5% β-CD–carvacrol; (**g**) 95.0% PLA with 5.0% β-CD–carvacrol.

As can be seen (Figure 4a), while control sample (only PLA) had an irregular surface, the PLA samples enriched with β-CD–carvacrol (Figure 4e–g) or β-CD–thymol (Figure 4b–d) complexes exhibited a more uniform surface as the concentration (wt%) increased. In this sense, the decrease in the number of cuts and folds of polymeric material was directly proportional to the concentration of added complex.

These results are in agreement with the values reported in mechanical tests (see Tables 3 and 4), evidencing that the increase of the concentration of complexes in the formulation of the plastic material favors obtaining more flexible packaging (decrease in Young's modulus), providing the formation of a smoother and continuous surface.

This fact could be due to encapsulation which helps incorporate the active compound (carvacrol or thymol) into the polymeric matrix, since different results have been described in the literature when raw essential oils (without encapsulation) were added to polymeric materials to produce heterogeneous structures with oil droplets trapped into the polymer [26,27].

3.4. Differential Scanning Calorimetry

To investigate the thermal transitions of the films studied, DSC measurements were accomplished. As can be seen in Table 5, the packages containing β-CD–carvacrol or β-CD–thymol complexes showed similar thermal properties, regardless of their concentration. The glass transition temperature (Tg) of the PLA-enriched materials was analogous to that obtained for PLA control, and similar to Tg values described in the literature [28], indicating that the amorphous phase of the PLA does not undergo any change.

On the other hand, the packaging with additives shows a significant variation in the cold crystallization temperature with respect to the control packaging (Tc = 102.7 °C; PLA 0%), increasing up to 3 °C and 5 °C for concentrations of 2.5% (wt%) of β-CD–thymol and β-CD–carvacrol, respectively, modifying the cold crystallization behavior of the PLA, and in consequence, the formation of the ordered structure of polymer matrix [29].

As can be seen in Figure 5, an endothermic peak was observed for all samples at a melting temperature, Tm, close to 168.5 °C, with slight temperature variations (lower than 1 °C), for PLA containing β-CD–carvacrol complexes. The little variations of cold crystallization and melting temperatures observed, when increasing β-CD–carvacrol and β-CD–thymol added to the PLA matrix, could be due to the increase in the chain mobility of the polymer matrix.

Table 5. Parametric values of DSC obtained from pure PLA and added β-CD–carvacrol or β-CD–thymol at 1.5%, 2.5%, and 5%, wt%.

Parameter	Control * 0%	PLA–Thymol–β-CDs (wt%)			PLA–Carvacrol–β-CDs (wt%)		
		1.5%	2.5%	5.0%	1.5%	2.5%	5.0%
Tg (°C)	59	61	59	61	60	60	60
Tcc (°C)	102.7	103.8	105.4	105.0	106.5	107.7	105.9
Tm (°C)	168.5	168.5	168.5	168.8	167.9	168.8	169.1
ΔHc Energy (J/g)	36.09	33.08	30.61	29.03	36.42	32.09	36.83
ΔHm Energy (J/g)	45.63	44.09	37.44	35.65	44.72	37.54	42.95

* Control, pure PLA without β-CD–carvacrol or β-CD–thymol; Tg, glass transition temperature; Tcc, cold crystallization transition temperature; Tm, melting transition temperature.

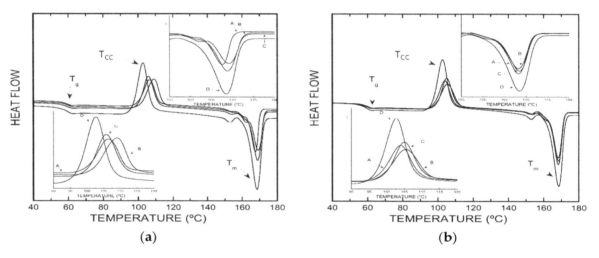

Figure 5. (**a**) DSC curves for: (**A**) PLA–(β-CD–carvacrol 1.5%, wt%); (**B**) PLA–(β-CD–carvacrol 2.5%, wt%); (**C**) PLA–(β-CD–carvacrol 5%, wt%); (**D**) PLA. (**b**) DSC curves for (**A**) PLA–(β-CD–thymol 1.5%, wt%); (**B**) PLA–(β-CD–thymol 2.5%, wt%); (**C**) PLA–(β-CD–thymol 5%, wt%); (**D**) PLA.

3.5. Thermogravimetry (TGA)

The thermal stability of PLA trays fortified with β-CD–carvacrol and β-CD–thymol complexes and non-fortified trays was measured by TGA, and all the samples had two weight loss steps (Figure 6).

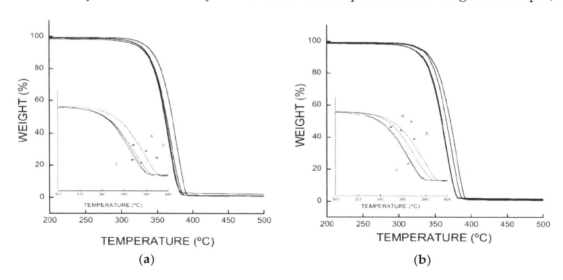

Figure 6. (**a**): Thermogravimetric analysis curves for (**A**) PLA–(β-CD–carvacrol 1.5%, wt%); (**B**) PLA–(β-CD–carvacrol 2.5%, wt%); (**C**) PLA–(β-CD–carvacrol 5, wt%); (**D**) PLA. (**b**): Thermogravimetric analysis curves for (**A**) PLA–(β-CD–thymol 1.5%, wt%); (**B**) PLA–(β-CD–thymol 2.5%, wt%); (**C**) PLA–(β-CD–thymol 5%, wt%); (**D**) PLA.

A significant mass loss between 320 °C and 390 °C could be observed, which is in agreement with PLA decomposition, following which the thermal analysis curves slow down from 390 °C up to 500 °C till the complexes achieve a constant mass. In addition, pure PLA has a slightly higher stability (Figure 6) than PLA–β-CD–carvacrol and PLA–β-CD–thymol and the thermal stability of the polymeric matrices diminishes with increasing concentrations of β-CD–carvacrol or β-CD–thymol. These results indicate that although all polymer samples are, in essence, thermally stable below 300 °C, the mixtures containing β-CD–carvacrol and β-CD–thymol have a faster weight loss rate than pure PLA at the same temperature.

In practice, the PLA-based packages used in the food industry will be at room temperature or lower; therefore, the thermal stability of the PLA–β-CD–carvacrol and PLA–β-CD–thymol materials developed herein, will not be compromised.

3.6. Antifungal Assays

The antifungal properties of PLA packages containing monoterpene–cyclodextrin complexes was monitored for 10 days of incubation to determine their prospective potential uses in the agrofood industry, and the results are shown in Table 6.

Table 6. Antifungal activity over incubation time of developed packaging materials against *Alternaria alternate*.

Type of Packaging	Encapsulation Concentration (% w/w)	Incubation Time	
		5 Days	10 Days
PLA-control	0.0%	29.7 [a]	69.0
PLA–β-CD–thymol	1.5%	28.3	71.6
	2.5%	30.0	60.0
	5.0%	3.3 *	0.0 *
PLA–β-CD–carvacrol	1.5%	29.7	65.3
	2.5%	0.0 *	0.0 *
	5.0%	0.0 *	0.0 *

[a] Diameter of the colony or surface area in mm. For each test, * values are statistically significant ($p < 0.05$).

As can be seen in Table 6, the results showed that PLA packages containing 2.5% and 5% β-CD–carvacrol or 5% β-CD–thymol (wt%) completely inhibited *Alternaria alternata* after 10 days of incubation (see Figure 7).

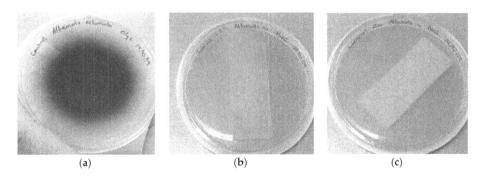

| (a) | (b) | (c) |

Figure 7. Photographs of the antifungal test: (**a**) Control *Alternaria alternata*; (**b**) PLA–β-CD–carvacrol packaging at 2.5%, wt%; (**c**) PLA–β-CD–carvacrol packaging at 5%, wt% with inoculation of *A. alternata* after 10 days of incubation.

These results are in agreement with those obtained by Llana-Ruiz-Cabello et al. (2016), revealing the antimicrobial properties, against yeasts and moulds, of PLA films containing 5% and 10% of oregano essential oil in ready-to-eat salads [30].

The addition of carvacrol and thymol encapsulates to the polymeric matrix of PLA, as well as the optimized injection temperature (180–190 °C) to produce the packaging material, allow to play an active role against the growth of moulds. The inhibitory effect of these vapor phase assets can be attributed to the accumulation of volatile substances in the medium, followed by interaction with the hydrophobic portion of the cell membrane [31].

Other investigations associate the inhibitory effect of the active compounds of essential oils such as carvacrol and thymol with changes in the morphology of hyphae due to penetration of active compounds into the plasma membrane [32]. The antifungal activity of both the compounds was preserved following the inclusion and injection process, due to encapsulation with β-CD. Previous investigations with similar encapsulation processes such as spray drying [33], freeze drying [34], and lyophilization [35] have shown that encapsulation helps to preserve the antimicrobial and antioxidant properties, mainly of essential oils, and that they are advantageous because they improve water solubility by forming inclusion complexes.

Carvacrol and thymol are volatile compounds; therefore, they could be highly effective in removing bacteria from packaging [36]. The antimicrobial action of carvacrol and thymol released by the PLA matrix against a wide range of phytopathogens constitutes an interesting topic for further studies. Indeed, studies are being conducted regarding the measurement of the effectiveness of packaging against other microorganisms in different food products at various storage temperatures by this research group.

4. Conclusions

In this work, PLA packages filled with thymol or carvacrol complexed in β-cyclodextrins (β-CDs) were prepared and characterized to evaluate their potential use as antibacterial materials. The results obtained by FTIR confirm that the inclusion of carvacrol and thymol in the apolar cavity of β -CDs yielded a significantly higher efficiency of complexation for carvacrol (105.6) than for thymol (69.3). Different proportions of β-CD–thymol or β-CD–carvacrol (0.0%, 1.5%, 2.5%, and 5.0%, wt%) complexes were mixed with PLA for packaging development by injection process, selecting 180–190 °C as the optimal temperature. The presence of β-CD–carvacrol or β-CDs–thymol complexes confer to polymer material plasticizers features that diminish intermolecular forces of the polymer chains, thereby reducing packaging stiffness. In TGA experiments for thermal behavior analysis, the presence of thymol– or carvacrol–β-cyclodextrins solid complexes in PLA formulations slightly decreased the thermal degradation temperature of the polymer, when compared with pure PLA. The performance of the developed polymer materials against *Alternaria alternata* inhibition after 10 days of incubation provided evidence for their potential use in agrofood industry, since packages containing 2.5% and 5% β-CD–carvacrol, or 5% β-CD–thymol, completely inhibited fungal growth. Additional studies are required to evaluate the diffusion and release kinetics of carvacrol and thymol complexes in the PLA polymer matrix during food storage.

Author Contributions: V.-C.F., experimental development and writing of the article; A.-P.H., FTIR and SEM analysis interpretation, and writing and revision of the article; N.-D.S.M., mechanical, structural, and thermal evaluation of polymer materials; N.-D.E., design of antifungal tests and revision of the article; G.J.A., experimental design and revision of the article.

References

1. Francis, G.A.; Gallone, A.; Nychas, G.J.; Sofos, J.N.; Colelli, G.; Amodio, M.L.; Spagno, G. Factors affecting quality and safety of fresh-cut produce. *Crit. Rev. Food Sci. Nutr.* **2012**, *52*, 595–610. [CrossRef] [PubMed]
2. De Oliveira, M.A.; De Souza, V.M.; Bergamini, A.M.; De Martinis, E.C. Microbiological quality of ready-to-eat minimally processed vegetables consumed in Brazil. *Food Control* **2011**, *22*, 1400–1403. [CrossRef]

3. Abadias, M.; Usall, J.; Anguera, M.; Solsona, C.; Viñas, I. Microbiological quality of fresh, minimally-processed fruit and vegetables, and sprouts from retail establishments. *Int. J. Food Microbiol.* **2008**, *123*, 121–129. [CrossRef] [PubMed]

4. França, K.R.S.; Silva, T.L.; Cardoso, T.A.L.; Ugulino, A.L.N.; Rodrigues, A.P.M.; De Mendonça Júnior, A.F. In vitro Effect of Essential Oil of Peppermint (Mentha x piperita L.) on the Mycelial Growth of Alternaria alternata. *J. Exp. Agric.* **2018**, *26*, 1–7. [CrossRef]

5. Muller, J.; González-Martínez, C.; Chiralt, A. Combination of poly (lactic) acid and starch for biodegradable food packaging. *Materials* **2017**, *10*, 952. [CrossRef] [PubMed]

6. Murariu, M.; Dubois, P. PLA composites: From production to properties. *Adv. Drug Deliv. Rev.* **2016**, *107*, 17–46. [CrossRef] [PubMed]

7. Zhou, L.; Zhao, G.; Jiang, W. Mechanical properties of biodegradable polylactide/poly (ether-block-amide)/thermoplastic starch blends: Effect of the crosslinking of starch. *J. Appl. Polym. Sci.* **2016**, *133*, 7. [CrossRef]

8. Qi, X.; Ren, Y.; Wang, X. New advances in the biodegradation of Poly(lactic) acid. *Int. Biodeterior. Biodegrad.* **2017**, *117*, 215–223. [CrossRef]

9. Huang, T.; Qian, Y.; Wei, J.; Zhou, C. Polymeric antimicrobial food packaging and its applications. *Polymers* **2019**, *11*, 560. [CrossRef]

10. Ahmed, J.; Hiremath, N.; Jacob, H. Antimicrobial, rheological, and thermal properties of plasticized polylactide films incorporated with essential oils to inhibit Staphylococcus aureus and Campylobacter jejuni. *J. Food Sci.* **2016**, *81*, 419–429. [CrossRef]

11. Suriyatem, R.; Auras, R.; Rachtanapun, C.; Rachtanapun, P. Biodegradable rice Strach/carboxymethyl chitosan films with added propolis extract for potential use as active food packaging. *Polymers* **2018**, *10*, 954. [CrossRef] [PubMed]

12. Ribeiro-Santos, R.; Andrade, M.; de Melo, N.R.; Sanches-Silva, A. Use of essential oils in active food packaging: Recent advances and future trends. *Trends Food Sci. Technol.* **2017**, *61*, 132–140. [CrossRef]

13. Wang, L.H.; Zhang, Z.H.; Zeng, X.A.; Gong, D.M.; Wang, M.S. Combination of microbiological, spectroscopic and molecular docking techniques to study the antibacterial mechanism of thymol against Staphylococcus aureus: Membrane damage and genomic DNA binding. *Anal. Bioanal. Chem.* **2017**, *409*, 1615–1625. [CrossRef] [PubMed]

14. Ramos, M.; Jiménez, A.; Garrigós, M.C. Carvacrol-Based Films: Usage and Potential in Antimicrobial Packaging. In *Antimicrobial Food Packaging*; Academic Press: Cambridge, MA, USA, 2016; pp. 329–338. [CrossRef]

15. Raybaudi-Massilia, R.M.; Mosqueda-Melgar, J.; Soliva-Fortuny, R.; Martín-Belloso, O. Control of pathogenic and spoilage microorganisms in fresh-cut fruits and fruit juices by traditional and alternative natural antimicrobials. *Compr. Rev. Food Sci. Food Saf.* **2009**, *8*, 157–180. [CrossRef]

16. Saad, N.Y.; Muller, C.D.; Lobstein, A. Major bioactivities and mechanism of action of essential oils and their components. *Flavour Fragr. J.* **2013**, *28*, 269–279. [CrossRef]

17. Ayala-Zavala, J.F.; Soto-Valdez, H.; Gonzalez-Leon, A.; Alvarez-Parrilla, E.; Martin-Belloso, O.; Gonzalez-Aguilar, G.A. Microencapsulation of cinnamon leaf (Cinnamomum zeylanicum) and garlic (Allium sativum) oils in beta-cyclodextrin. *J. Incl. Phenom. Macrocycl. Chem.* **2008**, *60*, 359–368. [CrossRef]

18. Rodríguez-López, M.I.; Mercader-Ros, M.T.; López-Miranda, S.; Pellicer, J.A.; Pérez-Garrido, A.; Pérez-Sánchez, H.; Gabaldón, J.A. Thorough characterization and stability of HP-β-cyclodextrin thymol inclusion complexes prepared by–microwave technology: A required approach to a successful application in food industry. *J. Sci. Food Agric.* **2019**, *99*, 1322–1333. [CrossRef]

19. Rodríguez-López, M.I.; Mercader-Ros, M.T.; Pellicer, J.A.; Gómez-López, V.M.; Martínez-Romero, D.; Núñez-Delicado, E.; Gabaldón, J.A. Evaluation of monoterpene-cyclodextrin complexes as bacterial growth effective hurdles. *Food Control* **2020**, *108*, 106814. [CrossRef]

20. Serna-Escolano, V.; Serrano, M.; Valero, D.; Rodríguez-López, M.I.; Gabaldón, J.A.; Castillo, S.; Guillén, F.; Zapata, P.J.; Martínez-Romero, D. Effect of Thymol and Carvacrol Encapsulated in Hp-B-Cyclodextrin by Two Inclusion Methods against Geotrichum citri-aurantii. *J. Food Sci.* **2019**, *84*, 1513–1521. [CrossRef]

21. American Society for Testing and Materials (ASTM). *Standard Test Method for Tensile Properties of Plastics*; ASTM: West Conshohocken, PA, USA, 2014.

22. Du, W.X.; Olsen, C.W.; Avena-Bustillos, R.J.; McHugh, T.H.; Levin, C.E.; Mandrell, R.; Friedman, M. Antibacterial effects of allspice, garlic, and oregano essential oils in tomato films determined by overlay and vapour-phase methods. *J. Food Sci.* **2009**, *74*, M390–M397. [CrossRef]

23. Arrieta, M.P.; López, J.; Ferrándiz, S.; Peltzer, M.A. Characterization of PLA-limonene blends for food packaging applications. *Polym. Test.* **2013**, *32*, 760–768. [CrossRef]

24. Ramos, M.; Jiménez, A.; Peltzer, M.; Garrigós, M.C. Characterization and antimicrobial activity studies of polypropylene films with carvacrol and thymol for active packaging. *J. Food Eng.* **2012**, *109*, 513–519. [CrossRef]

25. López-Rubio, A.; Lagaron, J.M. Improvement of UV stability and mechanical properties of biopolyesters through the addition of β-carotene. *Polym. Degrad. Stab.* **2010**, *95*, 2162–2168. [CrossRef]

26. Liu, D.; Li, H.; Jiang, L.; Chuan, Y.; Yuan, M.; Chen, H. Characterization of active packaging films made from poly (lactic acid)/poly (trimethylene carbonate) incorporated with oregano essential oil. *Molecules* **2016**, *21*, 695. [CrossRef]

27. Kumari, A.; Kumar, V.; Yadav, S.K. Plant extract synthesized PLA nanoparticles for controlled and sustained release of quercetin: A green approach. *PLoS ONE* **2012**, *7*, e41230. [CrossRef]

28. Carrasco, F.; Pagés, P.; Gámez-Pérez, J.; Santana, O.O.; Maspoch, M.L. Processing of poly(lactic acid): Characterization of chemical structure, thermal stability and mechanical properties. *Polym. Degrad. Stab.* **2010**, *95*, 116–125. [CrossRef]

29. Hwang, S.W.; Shim, J.K.; Selke, S.E.; Soto-Valdez, H.; Matuana, L.; Rubino, M.; Auras, R. Poly (L-lactic acid) with added α-tocopherol and resveratrol: Optical, physical, thermal and mechanical properties. *Polym. Int.* **2012**, *61*, 418–425. [CrossRef]

30. Llana-Ruiz-Cabello, M.; Pichardo, S.; Bermudez, J.M.; Banos, A.; Nunez, C.; Guillamon, E.; Aucejo, S.; Camean, A.M. Development of PLA films containing oregano essential oil (Origanum vulgare L. virens) intended for use in food packaging. *Food Addit. Contam. Part A* **2016**, *33*, 1374–1386.

31. Laird, K.; Phillips, C. Vapour phase: A potential future use for essential oils as antimicrobials. *Lett. Appl. Microbiol.* **2012**, *54*, 169–174. [CrossRef]

32. Soylu, E.M.; Kurt, S.; Soylu, S. In vitro and in vivo antifungal activities of the essential oils of various plants against tomato grey mould disease agent Botrytis cinerea. *Int. J. Food Microbiol.* **2010**, *143*, 183–189. [CrossRef]

33. Arana-Sánchez, A.; Estarrón-Espinosa, M.; Obledo-Vázquez, E.N.; Padilla-Camberos, E.; Silva-Vázquez, R.; Lugo-Cervantes, E. Antimicrobial and antioxidant activities of Mexican oregano essential oils (Lippia graveolens H. B. K.) with different composition when microencapsulated in beta-ciclodextrin. *Lett. Appl. Microbiol.* **2010**, *50*, 585–590. [CrossRef] [PubMed]

34. Santos, E.H.; Kamimura, J.A.; Hill, L.E.; Gomes, C.L. Characterization of carvacrol beta-cyclodextrin inclusion complexes as delivery systems of antibacterial and antioxidant applications. *LWT Food Sci. Technol.* **2015**, *60*, 583–592. [CrossRef]

35. Wang, T.; Li, B.; Si, H.; Lin, I.; Chen, L. Release characteristics and antibacterial activity of solid state eugenol/β-cyclodextrin inclusion complex. *J. Incl. Phenom. Macrocycl. Chem.* **2011**, *71*, 207–213. [CrossRef]

36. Goñi, P.; López, P.; Sánchez, C.; Gómez-Lus, R.; Becerril, R.; Nerín, C. Antimicrobial activity in the vapour phase of a combination of cinnamon and clove essential oils. *Food Chem.* **2009**, *116*, 982–989. [CrossRef]

4

The Impact of Cross-linking Mode on the Physical and Antimicrobial Properties of a Chitosan/Bacterial Cellulose Composite

Jun Liang [1,2,*]**, Rui Wang** [2] **and Ruipeng Chen** [1]

[1]   State Key Laboratory of Food Nutrition and Safety, Tianjin University of Science & Technology, Tianjin 300222, China; 13642198672@163.com
[2]   College of Packaging and Printing Engineering, Tianjin University of Science & Technology, Tianjin 300222, China; chenruipeng2016@163.com
*   Correspondence: jliang1118@tust.edu.cn

Abstract: The bacteriostatic performance of a chitosan film is closely related to its ionic and physical properties, which are significantly influenced by the mode of cross-linking. In the current work, chitosan with or without bacterial cellulose was cross-linked with borate, tripolyphosphate, or the mixture of borate and tripolyphosphate, and the composite films were obtained by a casting of dispersion. Mechanical measurements indicated that different modes of cross-linking led to varying degrees of film strength and elongation increases, while the films treated with the borate and tripolyphosphate mixture showed the best performance. Meanwhile, changes in the fractured sectional images showed a densified texture induced by cross-linkers, especially for the borate and tripolyphosphate mixture. Measurements of Fourier transform infrared showed the enhanced interaction between the matrix polymers treated by borate, confirmed by a slight increase in the glass transitional temperature and a higher surface hydrophobicity. However, the reduced antimicrobial efficiency of composite films against *E. coli*, *B. cinerea*, and *S. cerevisiae* was obtained in cross-linked films compared with chitosan/bacterial cellulose films, indicating that the impact on the antimicrobial function of chitosan is a noteworthy issue for cross-linking.

Keywords: chitosan; composite films; cross-linking; physical properties; bacteriostasis properties

1. Introduction

An increased consumer demand for fresh, high quality foodstuff has given rise to an intense interest in the characteristics of active packaging materials that alter the conditions of the packaged food for extending shelf life, improving sensory qualities, or inhibiting the propagation of spoilage and pathogenic microorganisms [1]. In light of their availability, unique qualities, and eco-friendliness, the use of biopolymers within multiple food-packaging applications is quite beneficial [2,3]. Among the natural polymers used for bio-degradable packaging development, chitosan stands out on account of its intrinsic anti-microbe, solid mechanical strength, excellent barrier capacity, and biodegradability, as well as superior film-forming properties [4]. These unique properties have allowed for an exploration of its potential usage in the food industry as active edible food coatings or films in terms of improving food conservation by resorting to its antifungal and antibacterial ability [5].

In spite of the numerous benefits and original properties of chitosan, the existence of multiple amino groups and hydroxyl in the framework enhances its strong adhesion to water, and therefore chitosan films possess a high water swelling degree [6]. After over absorption of water, chitosan films become brittle, and are thus not applicable for packaging [7,8]. As a result, several techniques can be used to obtain the improved barrier and mechanical properties of chitosan films, including blending with polyvinyl alcohol, poly N-vinyl pyrrolidone, polyethylene glycol etc. [9–11].

Bacterial cellulose that is approximately less than 100 nm in diameter possesses unusual physical and mechanical properties [12,13]. Besides, bacterial cellulose and chitosan share similar structures and mutually complementary qualities, resulting in reinforced molecular interactions among polysaccharide chains. Compared with films of pure chitosan, chitosan/bacterial cellulose composite films exhibit more advantageous mechanical qualities, such as a rational thermal stability and a low O_2 permeability [14]. Thereby, bacterial cellulose possesses great potential for being utilized as an agent of reinforcement in chitosan-based films, in terms of promising mechanical qualities and antimicrobial activity properties [15].

Another effective method for the improvement of films with desirable properties has emerged, namely, cross-linking treatment [16]. A polymer with an integrated network can be obtained by the cross-linking process, in which the polymer chains are interconnected by covalent or non-covalent linking [17]. In general, chitosan-based biomaterials are cross-linked by way of verified approaches, such as making use of cross-linking agents, the heat curing process, and ultraviolet irradiation or electron-beam [18,19]. There have been many investigations where chemically cross-linked chitosan-based biomaterials were treated with cross-linking agents, including glutaraldehyde, borate, formaldehyde, or 1,5- pentane-dial. In addition, researchers have investigated the physical property changes of films in relation to the degree of cross-linking [20]. A cross-linker is appropriate for biopolymer materials, in particular those obtained from proteins or carbohydrates, supplying reduced gas and water vapor permeability in food packaging materials [21,22]. Cross-linkers may make up for the intrinsic deficiencies in the barrier and mechanical properties of biopolymers, rendering them more applicable in comparison with petroleum-based counterparts [23]. Generally, remarkable mechanical properties, heat stability, and water resistance are obtained and the qualities of composite films might be controlled by means of adjusting the mode or the extent of cross-linking. However, a more detailed intermolecular force in the texture-property relationship cannot be achieved all the time.

Therefore, in this study, a selective cross-linking method was used in the preparation of cross-linked chitosan/bacterial cellulose films to explore the impact of cross-linking mode on the macroscopic physical characterization along with the microphysical characterization of films simultaneously (Scheme 1). In view of the molecular characteristics of chitosan and bacterial cellulose, borate is able to cross-link chitosan and bacterial cellulose by forming hydroxyl groups with hydrogen bonds in matrix polymers [24]. Whereas by means of electrostatic interaction with the amino group, tripolyphosphate can produce a cross-linked structure with chitosan [25]. Natural cationic chitosan (CS) has been widely applied in constructing bactericidal coatings due to its contact-active disruption of the pathogen cytoplasmic membrane [26–28]. Thus, a noteworthy issue is how the reduction of hydroxyl or amino groups induced by cross-linking influences the antimicrobial property of chitosan, since iconicity is the origin of its contact-active function, despite the fact that the effects of some inserted nano or micron particles have been wildly explored [29–31]. It has been reported that when chitosan is immobilized onto a substrate, its antibacterial activities might be significantly reduced, proposed for the impeded ability of chitosan diffusion onto the cell membranes of microbes [32–34]. Nevertheless, chitosan is generally considered to be unable to enter the cell interior and the inhibitory effect is only exhibited on the cell surface due to its molecular weight. As a result, it is worth paying much more attention to the influence of the ionic property of chitosan, essential for its interaction with microbes, on the bacteriostasis abilities.

Hence, the present study aims to fabricate a chitosan/bacterial cellulose film treated by varied ways of cross-linking, verify the transformation of intermolecular forces in accordance with film mechanical properties, and clarify the antimicrobial ability changes induced by cross-linking. The composite films are also investigated using an electronic universal material test machine, electron scanning microscope (SEM), contact angle meter, water absorption and surface hydrophobicity tests, differential scanning calorimeter (DSC), and IR spectroscopy, as well as inhibition tests of bacteria, fungi, and yeast.

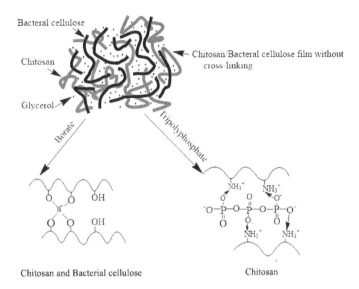

Scheme 1. The rationale of cross-linking mode using borate and tripolyphosphate.

2. Materials and Methods

2.1. Materials

Chitosan (viscosity: above 400 mPa·s; average MW: 50–100 KDa; deacetylation: 85%) was purchased from Shanghai Macklin Biochemical Co., Ltd. (Shanghai, China). Bacterial cellulose (produced with *Acetobacter xylinum*; mean diameter: 50–100 nm; mean length: 10,000–20,000 nm; end group: free hydroxyl), was generously provided by Qihong Sci. and Tech. Co., Ltd. (Guilin, China). Glycerol, borate, and tripolyphosphate (Analytical grade purity >99%) were obtained from Damao Chemical Reagent Beijing CO., Ltd. (Beijing, China). *E. coli.* ATCC 25922, *B. cinerea* ATCC 30387, and *S. cerevisiae* BY 4743 yeast were obtained from the School of Biotechnology at Tianjin University of Science and Technology (Tianjin, China).

2.2. Sample Preparation

Chitosan/bacterial cellulose films of different weight ratios were prepared by means of solution casting in 10 cm×10 cm petri dishes according to the published method, but with slight modifications [35]. A total of 2 g of chitosan was dissolved in 100 mL 1% acetic acid solution at room temperature to form 2.0 w/v% polymer solutions. Then, 0.5 wt% bacterial cellulose solution was obtained by dissolving 0.5 g of bacterial cellulose powder in 100 mL DI water at room temperature and afterwards, the solution was sheared for 30 min with a Kesun JLL350-B2 blender. Following this, 100 mL mixture solutions made from chitosan and bacterial cellulose stock solutions of different volume ratios were added by a corresponding weight of plasticizer glycerol to obtain final bacterial cellulose/chitosan 0, 1/64, 1/32, 1/16, 1/8, and 1/4 with 40 wt% glycerol in dry films. The film-forming conditions were set at 50 °C and relative ambient humidity in a drier for 48 h. As for the cross-linked films, cross-linkers were added in the bacterial cellulose/chitosan 1/32 matrix with final concentrations of 4% borate (CB-b), 4% tripolyphosphate (CB-t) or 2% borate, and 2% tripolyphosphate mixture (CB-bt) in the dried films, respectively. The dry films (d CB-b, d CB-t, d CB-bt) were peeled off the dishes, and kept in storage at room temperature against the desiccants DrieRite, so as to maintain zero relative humility (RH) for at least one week before measurements. The average thickness of the film was 0.10 ± 0.01 mm. To explore the impact of water absorption on the mechanical characteristics of films, the wet films (w CB-b, w CB-t, w CB-bt) were obtained by placing dry films in a man-made temperature humidity chamber at 25 °C and 80% relative humidity in a MEA15004-014 (Jufu, Beijing, China) for 12 h.

2.3. Cross Sectional Structure

The fractured sectional image of sample films was examined by scanning electron microscopy (SEM). For fractured section structure images, films were frozen in liquid nitrogen and were broken later. The film fracture section coating with a gauzy film of gold (Au) was then put on double-sided Scotch tape installed on a Luminal specimen holder, and imaged under a Geol scanning electron microscope (SU-1510, Hitachi, Tokyo, Japan).

2.4. Mechanical Qualities

The elongation and tensile strength at the break of films was tracked on a 3396 electronic universal material test machine (Intron, Boston, MA, USA) according to China National Standard GB/T 1040-2006. Rectangular dry or wet specimens (1 cm × 10 cm) were incised using a precision double-blade cutter. Initial grip separation was fixed at 70 mm and the cross-head speed was set at 100 mm/min. A total of five measurements were calculated and averaged for each sample.

2.5. Thermal Studies

As a function of cross-linkers, the glass transition temperature was measured on an 8000 differential scanning calorimeter (DSC) (PerkinElmer, Boston, MA, USA). Films were put in pressure-tight DSC cells (about 8.5 mg of matter per cell). Highly purified nitrogen was channeled into the sample compartment and flushed for at least 30 min before measurement. The first scanning was performed ranging from -10 to $50\,°C$ at the speed of $10\,°C$ per minutes and the samples were kept at $50\,°C$ for 30 min to expel the water residues in the films. After rapid cooling ($10\,°C/min$ to $-10\,°C$), measurements were carried out during a second scanning at $10\,°C/min$ from -10 to $220\,°C$. Glass transition temperature was determined in accordance with a half-variation in calorific capacity during phase transition.

2.6. Infrared Spectroscopy

FTIR measurements of the dry and wet films were carried out by way of a Nicolet 6700 spectrometer (Thermo, Waltham, MA, USA) at ambient temperature. Data were collected over 16 scans at a $16\,cm^{-1}$ resolution and analyzed using Origin 8.6.

2.7. Contact Angle

The measurements of static water contact angle were performed by means of a VCA option Contact Angle Analyzer (Jike, Shanghai, China) under optimal conditions. The image was captured by a CCD camera right after a 10 μL water drop was deposited onto the film surface and from which a contact angle analysis could be performed. A total of five measurements were averaged for each sample.

2.8. Film Swelling Degree, Water Vapor Absorption, and Water Activity

The swelling degree (SD) of a given sample film was calculated as follows: firstly, a thoroughly dried sample was immersed in distilled water at $25\,°C$ for 12 h; then, the samples were picked out and weighed after gentle wiping of the surface using absorbent paper. SD was measured as the percentage of initial film weight increase that occurred after swelling in water.

The water vapor absorption (WVA) of a given sample was measured as follows: sample films (10 cm × 10 cm) dried in desiccators were weighed (±1 mg) in glass dishes; then, they were put in a temperature humidity chamber at $25\,°C$ and 80% relative humidity for 12 h. Water vapor absorption (WVA) was measured as the percentage of initial film weight growth that occurred during moistening and was determined on a wet basis.

The water activity (WA) of each dry and wet film was determined by a water activity detector (Dongxi Yike, Beijing, China).

Triplicate measurements of SD, WVA, and WA were determined individually, prepared as replicated experimental units.

2.9. Antimicrobial Testing

As far as the antimicrobial testing is concerned, the PE film as a blank (1 cm × 1 cm) was immersed in 10 mL liquid culture in a tube with a calibrated suspension of *E. coli* (1×10^6 CFU), *B. cinerea* (5×10^5 CFU), and *S. cerevisiae* (5×10^5 CFU), respectively. After incubation at 37 °C with a constant shaking of 250 rpm for 12 h, 1 mL of the 100 times diluted microbe suspension was inoculated on an agar medium plate. Right after incubation at 37 °C for 24 h, the images of the agar plates were captured and the antimicrobial effects of the samples were evaluated on the basis of the colony count using the following equation:

$$E(\%) = [(A - B)/A] \times 100\% \qquad (1)$$

where A = the number of viable microbe colony for the PE film (blank) in the plate and B = the number of viable microbe colonies for chitosan-based samples in the plate. A total of three measurements were averaged for each sample.

3. Results and Discussion

3.1. Mechanical Properties

The maximum stress, tensile strengths, and yield at break for all films in dry or wet conditions are summarized in Figure 1 using the average statistical data produced from the five analyzed specimens. Obviously, dry films possess a higher film strength, i.e., maximum stress and tensile strength, in contrast with wet films. Since the weight and shape of the films used are the same, the measured tensile force can truly reflect the mechanical qualities of the films. Regarding the films from chitosan reinforced with bacterial cellulose, it has been shown that the tensile strength firstly enhanced, and then declined with the concentration increase of bacterial cellulose (Figure 1b). This phenomenon might be related to the formation of intermolecular hydrogen bonds between the chemical groups in bacterial cellulose (–OH) and chitosan (–OH and –NH_2), therefore restricting the motion of the matrix while promoting rigidity [36]. Consistently, it is also reported by Ciechańska that bacterial cellulose fibers form a three-dimensional network, which is hydrogen bonded to the glucan chain and staggered in chitosan to enhance the strength of the chitosan matrix [37]. The maximum reached the ratio of bacterial cellulose/chitosan 1/32. Accordingly, the bacterial cellulose/chitosan 1/32 film was selected for the subsequent experiments. As for CB-t and CB-bt films, however, the wet films manifested a better performance in yield at break (Figure 1f), indicating that the absorption of water induced an increase of elongation in matrices processed by tripolyphosphate and the borate/tripolyphosphate mixture. Studies by Salari et al. (2018) have shown that after the blending with 4% bacterial cellulose, tensile strength enhancement of chitosan film was close to 100%, but the performance of bacterial cellulose in our work is moderate, which might be due to the fact that we added 40 percent of glycerol and the films were plasticized and softened [38].

The impact of glycerol on the physical properties of polymer matrices has been extensively studied. Low-molecular weight compounds or diluents, acting as external plasticizers, are an integral part of polymeric systems. They serve to increase the flexibility and workability of the otherwise rigid neat polymers [39,40]. In the current work, the measured maximum strength and yield at the break for dry chitosan/bacterial cellulose with 40% glycerol are 20.4 ± 3.4 (MPa) and 64.8 ± 4.9 (%), respectively. Dhanavel and his staff have studied the chitosan film containing 40% glycerol. They obtained 9.4 MPa for the maximum strength and 66% for yield at the break [41]. Herein, the bacterial cellulose was proved to be able to effectively enhance the strength of chitosan films, improving its applicability in food packaging.

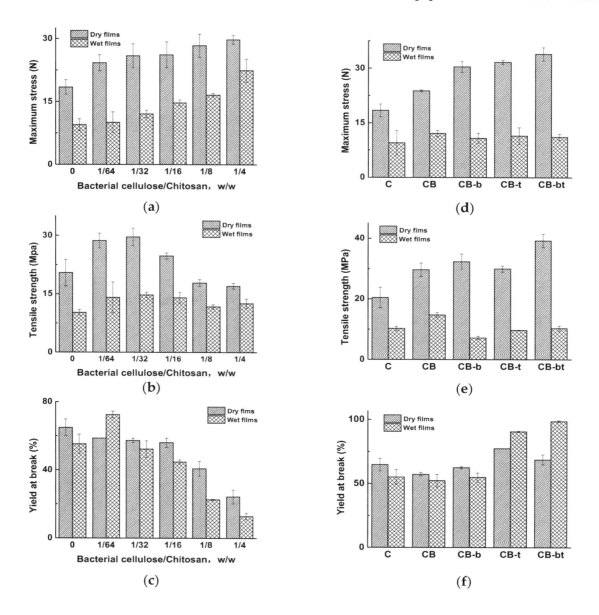

Figure 1. The impact of cross-linking treatment on the maximum strength, tensile strength, and yield at the break of dry and wet films of various ratios of bacterial cellulose/chitosan, chitosan (C), and 1/32 bacterial cellulose/chitosan (CB) treated with 4% borate (CB-b), 4% tripolyphosphate (CB-t), or a mixture of 2% borate and 2% tripolyphosphate (CB-bt). (**a**) maximum strength of Bacterial cellulose/Chitosan composite with varying weight ratio; (**b**) tensile strength of Bacterial cellulose/Chitosan composite with varying weight ratio; (**c**) yield at the break of Bacterial cellulose/Chitosan composite with varying weight ratio; (**d**) maximum strength 1/32 Bacterial cellulose/Chitosan composite with varying mode of cross linking; (**e**) tensile strength of 1/32 Bacterial cellulose/Chitosan composite with varying mode of cross linking; (**f**) yield at the break of 1/32 Bacterial cellulose/Chitosan composite with varying mode of cross linking.

Concerning dry 1/32 bacterial cellulose/chitosan films with or without cross-linkers, the values of maximum strength (Figure 1d) and tensile strength (Figure 1e) range from 18.4 to 34.0 N and 20.4 to 39.0 MPa, respectively. Meanwhile, regarding film elongation (Figure 1f), namely yield at break, the value ranges from 57.3 to 77.4%. The introduction of cross-linkers caused an increase in both film strength and elongation, which could be attributed to the improvement of the interaction among matrix polymers [42]. Additionally, the cross-linker mixture performed best in terms of improving the film strength. A reasonable explanation for this is that cross-linker mixture induced a stronger intermolecular force network in the matrices, which was further studied by DSC and IR in the following section.

Chitosan and bacterial cellulose were highly absorbent. After being kept in an 80% humidity environment, a certain amount of water was absorbed in the films, which were described as wet films. In general, an increase in moisture uptake weakened the tensile strength of chitosan films [43]. Concerning the wet films, the values for maximum strength and tensile strength ranged from 9.5 to 11.4 N and 7.2 to 14.7 MPa, respectively. For yield at break, the value ranged from 52.2 to 98.6%. Unexpectedly, the introduction of moisture in wet films leads to a conspicuous decrease in film strength and an increase in film elongation. For films treated with tripolyphosphate or a mixture of borate and tripolyphosphate, the film elongation of wet films showed an obvious increase compared with that of dry films, and the water clearly displayed a plasticizer effect in the composite films, i.e., softening the chitosan/bacterial cellulose film. For films treated with borate, the film elongation of wet films showed an obvious decrease compared with that of dry films, and the water showed an antiplasticizer effect, i.e., stiffening the chitosan/bacterial cellulose films. The phenomenon of a plasticizer being able to exert a role of an antiplasticizer is well-known and has received increasing attention from food scientists and technologists in recent years [44–46]. It has been reported that small molecules like glycerol, water, etc. can exert different effects, depending on their concentration. In most cases, there is a critical element that marks the onset of a change in functionality from antiplasticizer to plasticizer, depending on the polymer matrices [47]. In this work, the various cross-linking modes lead to different roles of water in the chitosan/bacterial cellulose film.

3.2. Cross Sectional Fracture Structure

Cross sectional fracture structure of C, CB, CB-b, CB-t, CB-bt, and bacterial cellulose is shown as Figure 2A–F, respectively. Glycerol is known to enter polyhydroxylated polymers' chain interior, disrupting inter-and intra-molecular interactions, rendering the polymer plasticized, as well as forming a continuous phase of plasticized film [48]. The homogenous structure of various films demonstrates the conspicuous compatibility of the two polymers and a tight structure deficient in phase separation. In the current work, it has been found that cross-linkers could further change the interactions of matrix molecules within the films. In addition to bacterial cellulose, the composite films present a clearer fiber structure. With the addition of borate and tripolyphosphate, the films were characterized by a compact, uniform, and homogeneous structure, indicating a closer interaction of matrix molecule induced by cross-linkers. A similar phenomenon has also been observed in previous studies [23].

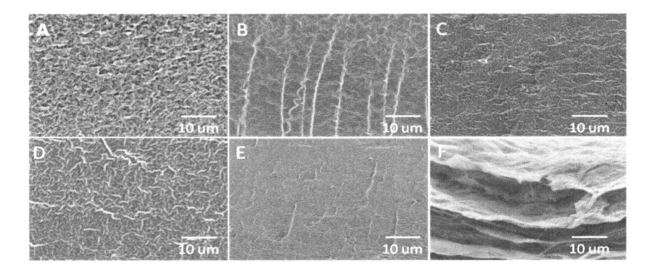

Figure 2. Fractured cross-sectional image of C (**A**), CB (**B**), CB-b (**C**), CB-t (**D**), CB-bt (**E**), and bacterial cellulose film imaged by SEM.

3.3. Thermal Studies

The thermal stability of biomaterials was determined using DSC. The thermograms for bacterial cellulose/chitosan 1/32-based films are shown in Figure 3. Glass transitions were observed at 91.03, 91.33, 110.47, 108.23, and 108.95 for C, CB, CB-b, CB-t, and CB-bt, respectively. Kadam et al. (2018) have reported that the mixing of polyphenolic cross-linkers could effectively increase the glass transitional temperature of chitosan films [49], which was in agreement with the results in the current work. We inferred that cross-linking could enhance the molecular interaction among the matrix polymers, and thus improve the thermal stability of bacterial cellulose/chitosan films. The hypothesis is to be further studied in the following section using FTIR. Only one T_g characteristic curve is detected for all samples, implying the entire compatibility of chitosan and bacterial cellulose. Inconsistent with the results of other studies [50], the broad endothermic peak could not be observed at approximately 110–120 °C in all samples, probably caused by the dissociation process of interchain hydrogen bonding of chitosan. Since a rapid drying method at a higher temperature was used to prepare the sample films, most of the polymers in the matrices were in an amorphous state [51], resulting in a different molecular aggregation state and configuration of chitosan in the films.

In addition, the sequence of T_g for all samples ranked as CB-b > CB-bt > CB-t > CB > C, indicating the degrees of difference in terms of the impact on the matrix molecular interaction by cross-linkers. T_g of the mixture is influenced by multiple factors, such as molecular weight [52], intermolecular interaction [53], compound compatibility [54], etc. Based on the complicated microcosmic matrix situation induced by different cross-linking methods, much more work is still needed to clarify such a thermo dynamic phenomenon. Nevertheless, it was interesting to find that the features of chitosan-based films, i.e., T_g, hydrogen bond strength, surface hydrophobicity, and antimicrobial capacities, were clearly interrelated, as shown in the following section.

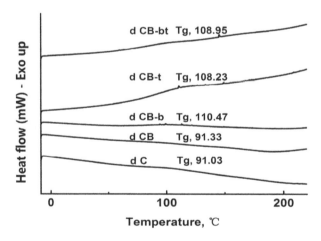

Figure 3. DSC thermograms for cast films C, CB, CB-b, CB-t, and CB-bt.

3.4. Infrared Spectroscopy

FTIR spectra were used for a more profound exploration of the molecular interactions among the chitosan/bacterial cellulose matrix polymers. The IR absorbance of C, CB, CB-b, CB-t, and CB-bt is shown in Figure 4. The peak above 3000 cm^{-1} corresponds to the combined stretching of hydroxyl [55]. Hydroxyl groups in a carbohydrate matrix can exist in the form of hydrogen, bonded or free, without being implicated in any hydrogen bonds [56]. The vibration frequency of both can be attributed to the range 3000–3600 cm^{-1}, while the characteristic IR band of free hydroxyls distributes in the higher frequency region [57]. Hence, as the hydrogen boned hydroxyls turn into free hydroxyls, the peak above 3000–3600 cm^{-1} shifts to a higher frequency range.

A strong switch of the characteristic absorption band of the hydrogen bond is measured as 3265, 3281, 3243, 3281, and 3264 cm^{-1} for C, CB, CB-b, CB-t, and CB-bt, respectively, indicating that the electrostatic interaction between hydroxyl and borate induces a stronger hydrogen bond network in chitosan/bacterial cellulose films. Furthermore, a band measured at 1557 cm^{-1} (CB) characteristic as amide (II) [8] shifts to the lower wavenumber at 1552 cm^{-1} (CB-b), conforming to the intensified intermolecular interaction for the hydroxyl group induced by borate. Inconsistent with other results by Benucci et al. [58], a decrease of the amide (II) signal was not detected, and it was speculated that the non-covalent cross-linking, caused by tripolyphosphate in the current work, did not lead to the reduction of hydrogen in the group. Meanwhile, it has been found that the rank of hydrogen bond strength followed the same order with that of T_g, i.e., CB-b > CB-bt ≥ CB-t ≥ CB, indicating that the intermolecular force was one of the main forces that modulated the matrix thermodynamic properties.

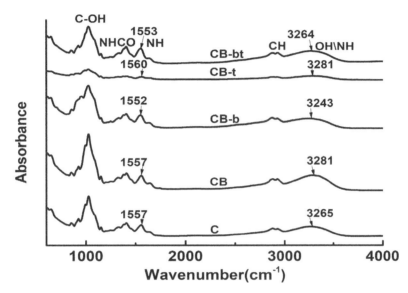

Figure 4. IR spectra of C, CB, CB-b, CB-t, and CB-bt.

3.5. Contact Angle

The contact angle with the film surface was measured to monitor the hydrophobic properties of produced matrices. Observed from the spatially varied curves of drips on the films, the contact angles were measured and the average was tabulated (Table 1). The contact value followed a rank of CB-b > CB-bt > CB-t > C> CB. Meanwhile, moisture absorption was found to decrease the surface contact angle for all tested films, according to the comparison of results of dry and wet films. The addition of cross-linkers, particularly borate, induced a higher contact angle for all tested samples, implying that the cross-linkers used in the current work might promote the hydrophobic properties of the film surface.

Table 1. The impact of cross-linking treatment on the contact angle of dry and wet films from chitosan (C), 1/32 bacterial cellulose/chitosan (CB), and 1/32 bacterial cellulose/chitosan treated with borate (CB-b), tripolyphosphate (CB-t), or a mixture of borate and tripolyphosphate (CB-bt).

Sample	Contact Angle/°	
	DRY FILM	Wet Films
C	93.4 ± 0.8	88.8 ± 0.5
CB	91.2 ± 0.2	85.2 ± 1.1
CB-b	103.8 ± 1.3	101.3 ± 0.5
CB-t	94.7 ± 0.9	85.5 ± 0.4
CB-bt	99.4 ± 0.4	94.5 ± 0.1

It is intriguing to note that the values of the film contact angle are in agreement with the reverse order sequence of the chitosan/bacterial cellulose-based matrix hydrogen bond network detected by IR spectroscopy, i.e., CB ≥ CB-t > CB-bt >CB- b. The strength of the matrix hydrogen bond network seemed to have a direct relationship with the surface hydrophobic properties. A reasonable explanation for this is that the cross-linkers led to a decrease of the free polar chemical group in the film surface [59]. Thus, the water/matrix polymer interactions were reduced while the surface hydrophobicity was heightened [60].

3.6. Water Vapor Absorption, Swelling Degree and Water Activity

In spite of possessing quite a few merits of chitosan films, reduced mechanical properties is a main shortcoming, especially after wet absorption [61]. Thus, hygroscopicity is considered as one of the essential qualities of a chitosan film or coating [62]. In this respect, water vapor absorption (WVA), swelling degree (SD), and water activity (WA) values of all chitosan/bacterial cellulose films were measured and tabulated in Table 2. On the basis of a statistical analysis, it was indicated that the mode of cross-linking significantly affected the SD of films, yet the changes for WVA and SD were minor. Interestingly, it has been found that it was CB-b, the film with the strongest hydrogen bond network according to the results of IR, that possessed the lowest swelling degree, implying that the resistance to water absorption was due to the reduced existence of polar chemical groups and the decreased free volumes of polymer molecule relaxation. The water absorption of polymeric films was closely related to the hydrophobic or hydrophilic properties, along with the existence of free molecule volumes in the matrices [63]. It is speculated that cross-linking might reduce the interaction of water with the matrix molecule, and further influence the retention of water inside the films. The result was in good accordance with the values achieved by Vartiainen et al., who investigated the water absorption of chitosan composite films reinforced with nanoclay [64].

Table 2. WVA, SD, and WA of C, CB, CB-b, CB-t, and CB-bt.

Samples	Water Absorption ($\times\%$)	Swelling Degree ($\times\%$)	Water Activity (Dry Films)
C	31.9 ± 0.5	71.7 ± 2.4	0.13 ± 0.01
CB	31.1 ± 0.3	42.0 ± 2.0	0.13 ± 0.01
CB-b	28.0 ± 1.7	20.7 ± 1.1	0.14 ± 0.01
CB-t	30.9 ± 1.5	29.5 ± 4.3	0.16 ± 0.01
CB-bt	29.7 ± 0.5	$32.4 + 1.5$	$0.14 + 0.01$

3.7. Antimicrobial Testing

In order to assess the antimicrobial properties of hybrid films, the *E. coli* (bacterial), *B. cinerea* (mold), and *S. cerevisiae* (yeast) were chosen as the representative microorganisms. Table 3 presents the antimicrobial results of C, CB, CB-b, CB-t, and CB-bt films. For the sake of reliability, the experiments were repeated three times. The pH of the bacterial culture medium was measured before and after the sample films were immersed, while no obvious change was found, probably due to the fact that most of the acetic acid volatilized in the sample preparation process. Apparently, the successive order of antimicrobial activity was shown as C $\geq$ CB > CB-b $\geq$ CB-t > CB-bt. As was mentioned in the literature, chitosan is antimicrobial against a wide variety of microbes, which originates from its polycationic character [65]. In this work, almost all *E. coli*, *B. cinerea*, and *S. cerevisiae* treated with C and CB were killed (Figure 5 shows the results of anti *S. cerevisiae* as representative). Unexpectedly, our results showed that the addition of cross-linkers remarkably weakened the antimicrobial ability of CB films.

Table 3. A comparison of antimicrobial activity of C, CB, CB-b, CB-t, and CB-bt against *E. coli*, *B. cinerea*, and *S. cerevisiae*.

Samples	Bacteria Inhibitory Rate ($\times\%$)	Mold Inhibitory Rate ($\times\%$)	Yeast Inhibitory Rate ($\times\%$)
C	92.4 ± 2.9	72.8 ± 10.7	100.0 ± 0.0
CB	90.8 ± 6.4	77.8 ± 11.4	95.5 ± 6.4
CB-b	85.5 ± 4.7	51.0 ± 8.9	62.9 ± 2.3
CB-t	85.0 ± 3.7	51.5 ± 18.6	65.9 ± 0.3
CB-bt	73.1 ± 2.7	36.9 ± 14.7	52.8 ± 19.6

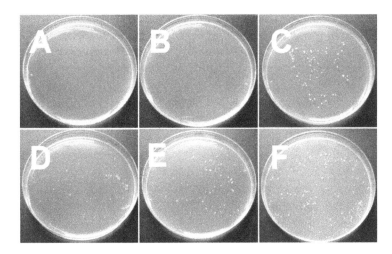

Figure 5. Images of antibacterial efficacy against *S. cerevisiae* for (**A**) C, (**B**) CB, (**C**) CB-b, (**D**) CB-t, (**E**) CB-bt, and (**F**) Blank (PE film).

The antimicrobial mechanism of chitosan is correlated with the cell death generated from the disruption of cell functioning or the destruction of the cell wall caused by the electrostatic interactions among cationic amino groups (NH_3^+) in chitosan molecules and anionic teichoic acids in the cell wall [66]. In our study, it was speculated that borate and tripolyphosphate could effectively disturb the interaction of chitosan with microbes by changing the ionic properties of amino groups, thus decreasing the efficacy of chitosan against tested *E. coli*, *B. cinerea*, and *S. cerevisiae*. Though cross-linkers have been widely used in enhancing the mechanical qualities of chitosan films [20], to the best of our knowledge, their impact on the antimicrobial effect has rarely been reported. In the current work, a critical issue has been put forward and investigated, i.e., certain cross-linkers actually significantly decrease the antimicrobial activity of chitosan.

4. Conclusions

This study focused on expounding the impact of cross-linking mode on the physical and antibacterial properties of chitosan/bacterial cellulose composite films. It has been found that the cross-linking mode microscopically affects the molecular interaction and microstructure, and macroscopically leads to changes in the films' mechanical and thermodynamic properties, hydrophobicity, and water absorption. It is also worth noting that cross-linking can reduce the bacteriostatic function of chitosan/bacterial cellulose composite films. The results of this work prove that due to the varied intermolecular interactions between the cross-linkers and matrix polymers in the matrices, cross-linking mode has a significant impact on the practicability of the composite packaging film.

Author Contributions: Experimental operation and data acquisition, R.W.; Writing—Original Draft Preparation, J.L.; Proofreading, R.C.

Acknowledgments: This work was supported by the Tianjin Science and Technology Planning Project (No. 18PTSYJC00140).

References

1. Werner, B.G.; Koontz, J.L.; Goddard, J.M. Hurdles to commercial translation of next generation active food packaging technologies. *Curr. Opin. Food Sci.* **2017**, *16*, 40–48. [CrossRef]

2.   Shahabi-Ghahfarrokhi, I.; Khodaiyan, F.; Mousavi, M.; Yousefi, H. Effect of γ-irradiation on the physical and mechanical properties of kefiran biopolymer film. *Int. J. Biol. Macromol.* **2015**, *74*, 343–350. [CrossRef]

3.   Cacciotti, I.; Mori, S.; Cherubini, V.; Nanni, F. Eco-sustainable systems based on poly(lactic acid), diatomite and coffee grounds extract for food packaging. *Int. J. Biol. Macromol.* **2018**, *112*, 567–575. [CrossRef] [PubMed]

4.   Vilela, C.; Pinto, R.J.B.; Coelho, J.; Domingues, M.R.M.; Daina, S.; Sadocco, P.; Santos, S.A.O.; Freire, C.S.R. Bioactive chitosan/ellagic acid films with uv-light protection for active food packaging. *Food Hydrocolloids* **2017**, *73*, 120–128. [CrossRef]

5.   Al-Naamani, L.; Dobretsov, S.; Dutta, J. Chitosan-zinc oxide nanoparticle composite coating for active food packaging applications. *Innov. Food Sci. Emerg. Technol.* **2016**, *38*, 231–237. [CrossRef]

6.   Wang, X.; Lou, T.; Zhao, W.; Song, G. Preparation of pure chitosan film using ternary solvents and its super absorbency. *Carbohydr. Polym.* **2016**, *153*, 253–257. [CrossRef]

7.   Rubentheren, V.; Ward, T.A.; Chee, C.Y.; Nair, P.; Salami, E.; Fearday, C. Effects of heat treatment on chitosan nanocomposite film reinforced with nanocrystalline cellulose and tannic acid. *Carbohydr. Polym.* **2016**, *140*, 202–208. [CrossRef]

8.   Chen, H.; Hu, X.; Chen, E.; Wu, S.; McClements, D.J.; Liu, S.; Li, B.; Li, Y. Preparation, characterization, and properties of chitosan films with cinnamaldehyde nanoemulsions. *Food Hydrocolloids* **2016**, *61*, 662–671. [CrossRef]

9.   Li, H.; Gao, X.; Wang, Y.; Zhang, X.; Tong, Z. Comparison of chitosan/starch composite film properties before and after cross-linking. *Int. J. Biol. Macromol.* **2013**, *52*, 275–279. [CrossRef]

10.   Tanuma, H.; Saito, T.; Nishikawa, K.; Dong, T.; Yazawa, K.; Inoue, Y. Preparation and characterization of peg-cross-linked chitosan hydrogel films with controllable swelling and enzymatic degradation behavior. *Carbohydr. Polym.* **2010**, *80*, 260–265. [CrossRef]

11.   Tripathi, S.; Mehrotra, G.K.; Dutta, P.K. Physicochemical and bioactivity of cross-linked chitosan-pva film for food packaging applications. *Int. J. Biol. Macromol.* **2009**, *45*, 372–376. [CrossRef] [PubMed]

12.   Qiu, Y.; Qiu, L.; Cui, J.; Wei, Q. Bacterial cellulose and bacterial cellulose-vaccarin membranes for wound healing. *Mater. Sci. Eng. C* **2016**, *59*, 303–309. [CrossRef] [PubMed]

13.   Cacicedo, M.L.; Castro, M.C.; Servetas, I.; Bosnea, L.; Boura, K.; Tsafrakidou, P.; Dima, A.; Terpou, A.; Koutinas, A.; Castro, G.R. Progress in bacterial cellulose matrices for biotechnological applications. *Bioresour. Technol.* **2016**, *213*, 172–180. [CrossRef] [PubMed]

14.   Picheth, G.F.; Pirich, C.L.; Sierakowski, M.R.; Woehl, M.A.; Sakakibara, C.N.; de Souza, C.F.; Martin, A.A.; da Silva, R.; de Freitas, R.A. Bacterial cellulose in biomedical applications: A review. *Int. J. Biol. Macromol.* **2017**, *104*, 97–106. [CrossRef] [PubMed]

15.   Abdul Khalil, H.P.S.; Saurabh, C.K.; Adnan, A.S.; Nurul Fazita, M.R.; Syakir, M.I.; Davoudpour, Y.; Rafatullah, M.; Abdullah, C.K.; Haafiz, M.K.M.; Dungani, R. A review on chitosan-cellulose blends and nanocellulose reinforced chitosan biocomposites: Properties and their applications. *Carbohydr. Polym.* **2016**, *150*, 216–226.

16.   Reddy, N.; Yang, Y.Q. Citric acid cross-linking of starch films. *Food Chem.* **2010**, *118*, 702–711. [CrossRef]

17.   Garavand, F.; Rouhi, M.; Razavi, S.H.; Cacciotti, I.; Mohammadi, R. Improving the integrity of natural biopolymer films used in food packaging by crosslinking approach: A review. *Int. J. Biol. Macromol.* **2017**, *104*, 687–707. [CrossRef]

18.   Li, S.; Donner, E.; Thompson, M.; Zhang, Y.; Rempel, C.; Liu, Q. Preparation and characterization of cross-linked canola protein isolate films. *Eur. Polym. J.* **2017**, *89*, 419–430. [CrossRef]

19.   Salmazo, L.O.; Lopez-Gil, A.; Ariff, Z.M.; Job, A.E.; Rodriguez-Perez, M.A. Influence of the irradiation dose in the cellular structure of natural rubber foams cross-linked by electron beam irradiation. *Ind. Crops Prod.* **2016**, *89*, 339–349. [CrossRef]

20.   Aryaei, A.; Jayatissa, A.H.; Jayasuriya, A.C. Nano and micro mechanical properties of uncross-linked and cross-linked chitosan films. *J. Mech. Behav. Biomed. Mater.* **2012**, *5*, 82–89. [CrossRef]

21.   Lopez de Dicastillo, C.; Rodriguez, F.; Guarda, A.; Galotto, M.J. Antioxidant films based on cross-linked methyl cellulose and native Chilean berry for food packaging applications. *Carbohydr. Polym.* **2016**, *136*, 1052–1060. [CrossRef]

22.   Uranga, J.; Leceta, I.; Etxabide, A.; Guerrero, P.; de la Caba, K. Cross-linking of fish gelatins to develop sustainable films with enhanced properties. *Eur. Polym. J.* **2016**, *78*, 82–90. [CrossRef]

23. Sun, S.; Liu, P.; Ji, N.; Hou, H.; Dong, H. Effects of various cross-linking agents on the physicochemical properties of starch/pha composite films produced by extrusion blowing. *Food Hydrocolloids* **2018**, *77*, 964–975. [CrossRef]

24. Shang, K.; Ye, D.-D.; Kang, A.H.; Wang, Y.-T.; Liao, W.; Xu, S.; Wang, Y.-Z. Robust and fire retardant borate-crosslinked poly (vinyl alcohol)/montmorillonite aerogel via melt-crosslink. *Polymer* **2017**, *131*, 111–119. [CrossRef]

25. Sacco, P.; Paoletti, S.; Cok, M.; Asaro, F.; Abrami, M.; Grassi, M.; Donati, I. Insight into the ionotropic gelation of chitosan using tripolyphosphate and pyrophosphate as cross-linkers. *Int. J. Biol. Macromol.* **2016**, *92*, 476–483. [CrossRef]

26. Antunes, B.P.; Moreira, A.F.; Gaspar, V.M.; Correia, I.J. Chitosan/arginine-chitosan polymer blends for assembly of nanofibrous membranes for wound regeneration. *Carbohydr. Polym.* **2015**, *130*, 104–112. [CrossRef]

27. Avcu, E.; Baştan, F.E.; Abdullah, H.Z.; Rehman, M.A.U.; Avcu, Y.Y.; Boccaccini, A.R. Electrophoretic deposition of chitosan-based composite coatings for biomedical applications: A review. *Prog. Mater. Sci.* **2019**, *103*, 69–108. [CrossRef]

28. Kalantari, K.; Afifi, A.M.; Jahangirian, H.; Webster, T.J. Biomedical applications of chitosan electrospun nanofibers as a green polymer—Review. *Carbohydr. Polym.* **2019**, *207*, 588–600. [CrossRef]

29. Hosseinnejad, M.; Jafari, S.M. Evaluation of different factors affecting antimicrobial properties of chitosan. *Int. J. Biol. Macromol.* **2016**, *85*, 467–475. [CrossRef] [PubMed]

30. Khan, A.; Gallah, H.; Riedl, B.; Bouchard, J.; Safrany, A.; Lacroix, M. Genipin cross-linked antimicrobial nanocomposite films and gamma irradiation to prevent the surface growth of bacteria in fresh meats. *Innov. Food Sci. Emerg. Technol.* **2016**, *35*, 96–102. [CrossRef]

31. Chung, Y. Effect of abiotic factors on the antibacterial activity of chitosan against waterborne pathogens. *Bioresour. Technol.* **2003**, *88*, 179–184. [CrossRef]

32. Cheah, W.Y.; Show, P.L.; Ng, I.S.; Lin, G.Y.; Chiu, C.Y.; Chang, Y.K. Antibacterial activity of quaternized chitosan modified nanofiber membrane. *Int. J. Biol. Macromol.* **2018**, *126*, 569–577. [CrossRef] [PubMed]

33. Wahid, F.; Hu, X.H.; Chu, L.Q.; Jia, S.R.; Xie, Y.Y.; Zhong, C. Development of bacterial cellulose/chitosan based semi-interpenetrating hydrogels with improved mechanical and antibacterial properties. *Int. J. Biol. Macromol.* **2019**, *122*, 380–387. [CrossRef] [PubMed]

34. Schuerer, N.; Stein, E.; Inic-Kanada, A.; Ghasemian, E.; Stojanovic, M.; Montanaro, J.; Bintner, N.; Hohenadl, C.; Sachsenhofer, R.; Barisani-Asenbauer, T. Effects of chitosan and chitosan n-acetylcysteine solutions on conjunctival epithelial cells. *J. EuCornea* **2018**, *1*, 12–18. [CrossRef]

35. Gao, P.; Wang, F.; Gu, F.; Ning, J.; Liang, J.; Li, N.; Ludescher, R.D. Preparation and characterization of zein thermo-modified starch films. *Carbohydr. Polym.* **2017**, *157*, 1254–1260. [CrossRef] [PubMed]

36. Urbina, L.; Guaresti, O.; Requies, J.; Gabilondo, N.; Eceiza, A.; Corcuera, M.A.; Retegi, A. Design of reusable novel membranes based on bacterial cellulose and chitosan for the filtration of copper in wastewaters. *Carbohydr. Polym.* **2018**, *193*, 362–372. [CrossRef] [PubMed]

37. Ciechańska, D. Multifunctional bacterial cellulose/chitosan composite materials for medical applications. *Fibres Text. East. Eur.* **2004**, *12*, 69–72.

38. Salari, M.; Sowti Khiabani, M.; Rezaei Mokarram, R.; Ghanbarzadeh, B.; Samadi Kafil, H. Development and evaluation of chitosan based active nanocomposite films containing bacterial cellulose nanocrystals and silver nanoparticles. *Food Hydrocolloids* **2018**, *84*, 414–423. [CrossRef]

39. Wang, S.; Jing, Y. Study on the barrier properties of glycerol to chitosan coating layer. *Mater. Lett.* **2017**, *209*, 345–348. [CrossRef]

40. Li, X.; Zhang, H.; He, L.; Chen, Z.; Tan, Z.; You, R.; Wang, D. Flexible nanofibers-reinforced silk fibroin films plasticized by glycerol. *Compos. Part. B Eng.* **2018**, *152*, 305–310. [CrossRef]

41. Zappino, M.; Cacciotti, I.; Benucci, I.; Nanni, F.; Liburdi, K.; Valentini, F.; Esti, M. Bromelain immobilization on microbial and animal source chitosan films, plasticized with glycerol, for application in wine-like medium: Microstructural, mechanical and catalytic characterisations. *Food Hydrocolloids* **2015**, *45*, 41–47. [CrossRef]

42. Ubaid, M.; Murtaza, G. Fabrication and characterization of genipin cross-linked chitosan/gelatin hydrogel for ph-sensitive, oral delivery of metformin with an application of response surface methodology. *Int. J. Biol. Macromol.* **2018**, *114*, 1174–1185. [CrossRef]

43. Zhang, S.; Kim, N.; Yokoyama, W.; Kim, Y. Effects of moisture content on mechanical properties, transparency, and thermal stability of yuba film. *Food Chem.* **2018**, *243*, 202–207. [CrossRef] [PubMed]

44. Michaelis, M.; Brummer, R.; Leopold, C.S. Plasticization and antiplasticization of an acrylic pressure sensitive adhesive by ibuprofen and their effect on the adhesion properties. *Eur. J. Pharm. Biopharm.* **2014**, *86*, 234–243. [CrossRef] [PubMed]

45. Chang, Y.P.; Abd Karim, A.; Seow, C.C. Interactive plasticizing–antiplasticizing effects of water and glycerol on the tensile properties of tapioca starch films. *Food Hydrocolloids* **2006**, *20*, 1–8. [CrossRef]

46. Aguirre, A.; Borneo, R.; León, A.E. Properties of triticale protein films and their relation to plasticizing–antiplasticizing effects of glycerol and sorbitol. *Ind. Crops Prod.* **2013**, *50*, 297–303. [CrossRef]

47. Liang, J.; Xia, Q.; Wang, S.; Li, J.; Huang, Q.; Ludescher, R.D. Influence of glycerol on the molecular mobility, oxygen permeability and microstructure of amorphous zein films. *Food Hydrocolloids* **2015**, *44*, 94–100. [CrossRef]

48. Ren, L.; Yan, X.; Zhou, J.; Tong, J.; Su, X. Influence of chitosan concentration on mechanical and barrier properties of corn starch/chitosan films. *Int. J. Biol. Macromol.* **2017**, *105*, 1636–1643. [CrossRef]

49. Kadam, D.; Lele, S.S. Cross-linking effect of polyphenolic extracts of lepidium sativum seedcake on physicochemical properties of chitosan films. *Int. J. Biol. Macromol.* **2018**, *114*, 1240–1247. [CrossRef]

50. Benucci, I.; Liburdi, K.; Cacciotti, I.; Lombardelli, C.; Zappino, M.; Nanni, F.; Esti, M. Chitosan/clay nanocomposite films as supports for enzyme immobilization: An innovative green approach for winemaking applications. *Food Hydrocolloids* **2018**, *74*, 124–131. [CrossRef]

51. Liang, J.; Wang, S.; Ludescher, R.D. Effect of additives on physicochemical properties in amorphous starch matrices. *Food Chem.* **2015**, *171*, 298–305. [CrossRef] [PubMed]

52. Seo, J.A.; Kim, S.J.; Kwon, H.J.; Yang, Y.S.; Kim, H.K.; Hwang, Y.H. The glass transition temperatures of sugar mixtures. *Carbohydr. Res.* **2006**, *341*, 2516–2520. [CrossRef] [PubMed]

53. Sterzyński, T.; Tomaszewska, J.; Andrzejewski, J.; Skórczewska, K. Evaluation of glass transition temperature of pvc/poss nanocomposites. *Compos. Sci. Technol.* **2015**, *117*, 398–403. [CrossRef]

54. Ou, Y.; Sun, Y.; Guo, X.; Jiao, Q. Investigation on the thermal decomposition of hydroxyl terminated polyether based polyurethanes with inert and energetic plasticizers by dsc-tg-ms-ftir. *J. Anal. Appl. Pyrolysis* **2018**, *132*, 94–101. [CrossRef]

55. Akyuz, L.; Kaya, M.; Koc, B.; Mujtaba, M.; Ilk, S.; Labidi, J.; Salaberria, A.M.; Cakmak, Y.S.; Yildiz, A. Diatomite as a novel composite ingredient for chitosan film with enhanced physicochemical properties. *Int. J. Biol. Macromol.* **2017**, *105*, 1401–1411. [CrossRef] [PubMed]

56. Branca, C.; D'Angelo, G.; Crupi, C.; Khouzami, K.; Rifici, S.; Ruello, G.; Wanderlingh, U. Role of the oh and nh vibrational groups in polysaccharide-nanocomposite interactions: A ftir-atr study on chitosan and chitosan/clay films. *Polymer* **2016**, *99*, 614–622. [CrossRef]

57. Hadjiivanov, K. Identification and characterization of surface hydroxyl groups by infrared spectroscopy. *Adv. Catal.* **2014**, *57*, 99–318.

58. Benucci, I.; Lombardelli, C.; Cacciotti, I.; Liburdi, K.; Nanni, F.; Esti, M. Chitosan beads from microbial and animal sources as enzyme supports for wine application. *Food Hydrocolloids* **2016**, *61*, 191–200. [CrossRef]

59. Peresin, M.S.; Kammiovirta, K.; Heikkinen, H.; Johansson, L.S.; Vartiainen, J.; Setala, H.; Osterberg, M.; Tammelin, T. Understanding the mechanisms of oxygen diffusion through surface functionalized nanocellulose films. *Carbohydr. Polym.* **2017**, *174*, 309–317. [CrossRef]

60. Soares, F.C.; Yamashita, F.; Müller, C.M.O.; Pires, A.T.N. Thermoplastic starch/poly(lactic acid) sheets coated with cross-linked chitosan. *Polym. Test.* **2013**, *32*, 94–98. [CrossRef]

61. Hejazi, M.; Behzad, T.; Heidarian, P.; Nasri-Nasrabadi, B. A study of the effects of acid, plasticizer, cross-linker, and extracted chitin nanofibers on the properties of chitosan biofilm. *Compos. Part. A Appl. Sci. Manuf.* **2018**, *109*, 221–231. [CrossRef]

62. Narayanan, A.; Kartik, R.; Sangeetha, E.; Dhamodharan, R. Super water absorbing polymeric gel from chitosan, citric acid and urea: Synthesis and mechanism of water absorption. *Carbohydr. Polym.* **2018**, *191*, 152–160. [CrossRef] [PubMed]

63. Aguirre-Loredo, R.Y.; Rodriguez-Hernandez, A.I.; Morales-Sanchez, E.; Gomez-Aldapa, C.A.; Velazquez, G. Effect of equilibrium moisture content on barrier, mechanical and thermal properties of chitosan films. *Food Chem.* **2016**, *196*, 560–566. [CrossRef] [PubMed]

64. Vartiainen, J.; Harlin, A. Crosslinking as an efficient tool for decreasing moisture sensitivity of biobased nanocomposite films. *Mater. Sci. Appl.* **2011**, *02*, 346–354. [CrossRef]

65. Fardioui, M.; Meftah Kadmiri, I.; Qaiss, A.E.K.; Bouhfid, R. Bio-active nanocomposite films based on nanocrystalline cellulose reinforced styrylquinoxalin-grafted-chitosan: Antibacterial and mechanical properties. *Int. J. Biol. Macromol.* **2018**, *114*, 733–740. [CrossRef] [PubMed]

66. Ngo, D.-H.; Vo, T.-S.; Ngo, D.-N.; Kang, K.-H.; Je, J.-Y.; Pham, H.N.-D.; Byun, H.-G.; Kim, S.-K. Biological effects of chitosan and its derivatives. *Food Hydrocolloids* **2015**, *51*, 200–216. [CrossRef]

5

Effect of Cinnamon Extraction Oil (CEO) for Algae Biofilm Shelf-Life Prolongation

Maizatulnisa Othman [1,*], Haziq Rashid [1], Nur Ayuni Jamal [1],
Sharifah Imihezri Syed Shaharuddin [1], Sarina Sulaiman [1], H. Saffiyah Hairil [2],
Khalisanni Khalid [3] and Mohd Nazarudin Zakaria [4]

[1]   Department of Manufacturing and Materials Engineering, Faculty of Engineering, International Islamic University Malaysia, Gombak 50728, Selangor, Malaysia; Mrhaziqhanif@gmail.com (H.R.); ayuni_jamal@iium.edu.my (N.A.J.); shaimihezri@iium.edu.my (S.I.S.S.); sarina@iium.edu.my (S.S.)

[2]   PERMATApintar College, National University Malaysia, Bangi 43600, Selangor, Malaysia; Hanasaffiyah2006@gmail.com

[3]   Agri-Nanotechnology Program, Biotechnology and Nanotechnology Research Center, Malaysian Agricultural Research and Development Institute (Mardi), MARDI Headquarters, Persiaran MARDI-UPM, Serdang 43400, Selangor, Malaysia; typhloids@gmail.com

[4]   Department of Biocomposite Technology, Faculty of Applied Sciences, Universiti Teknologi MARA, Shah Alam 40450, Selangor, Malaysia; nazarudin@salam.uitm.edu.my

*   Correspondence: maizatulnisa@iium.edu.my.

Abstract: This study was conducted to improve the life-span of the biofilm produced from algae by evaluating the decomposition rate with the effect of cinnamon extraction oil (CEO). The biofilm was fabricated using the solution casting technique. The soil burying analysis demonstrated low moisture absorption of the biofilm, thus decelerating the degradation due to low swelling rate and micro-organism activity, prolonging the shelf-life of the biofilm. Hence, the addition of CEO also affects the strength properties of the biofilm. The maximum tensile strength was achieved with the addition of 5% CEO, which indicated a good intermolecular interaction between the biopolymer (algae) and cinnamon molecules. The tensile strength, which was measured at 4.80 MPa, correlated with the morphological structure. The latter was performed using SEM, where the surface showed the absence of a separating phase between the biofilm and cinnamon blend. This was evidenced by FTIR analysis, which confirmed the occurrence of no chemical reaction between the biofilm and CEO during processing. The prolongation shelf-life rate of biofilm with good tensile properties are achievable with the addition of 5% of CEO.

Keywords: cinnamon extraction oil; algae; biodegradation; shelf-life; food packaging

1. Introduction

Quotidian plastic materials for food packaging in the market is proffered using a synthetic polymer base like polyproline and polyethylene, which is strenuous to decompose. Based on every day activities of Malaysian households and industries, high amounts of solid waste materials seem to have polluted and harmed the landfill with poison. In the interest of conserving the landfill, while overcoming the solid waste pollution issues, a few groups of individuals. such as researchers, the government, and industrial players [1], came up with several solutions. Starting off, the government promotes the routine of recycling, reducing, and reusing on a daily basis to the community [2]. Industrial players and researchers further focus on recycling plastics to be used as secondary materials and attempts to recycle and convert those materials into synthetic fiber threads and yarns to produce jerseys, shoes, and other textile products [3]. Nonetheless, the main problems are nowhere near solved at this point. The solid waste keeps on continuing its revolution in our daily life. The development of the first environmentally friendly materials consist partly of conventional plastic polyethylene (PE) or polypropylene (PP) and partly of nature's own material—chalk (40% by weight). Until early 2004, the research growth on producing the bio-based product was expanding throughout the entire globe [4]. PLA (polylactic acid) films, produced from lactic acid, have shown the highest commercial potential and are now produced on a comparatively large scale. Most bioplastics are produced based on natural biomaterials, such as corn, starch, and soybean [5]. Moreover, other food harvests, for example, cassava, wheat, potato, and sago, have also been transformed into plastic to supplant oil-based plastic [6]. Since those products are food assets for human beings, continuous transformation of those yields into plastic will soon interfere with human sustenance supply by reducing the world's sustenance assets. In an effort to prevent interference with food assets, other biomaterials should be assessed. Contemporaneous edible films have the potential to substantially reduce the environmental burden due to food packaging and limit moisturization, aroma, and lipid migration between food components [7]. Dealing with the biopolymer materials, it is difficult to avoid fugacious shelf-life issues. Biopolymer, which derives from natural sources, is commonly known for having transitory shelf-life as compared to synthetic biopolymer. Problems and errors may arise during the storage time either for logistic purposes or during transportation of the biopolymer to the industrial consumer. Humidity, temperature, micro-organism, and fungi attack may affect the quality and deteriorate the strength of the biopolymer. Accordingly, this study focuses on producing biofilm with sustainable shelf-life, high quality, low toxicity, and cheap costs, which could provide efficient food chain supply. Therefore, algae are chosen to be used in biopolymer film fabrication. Algae, or seaweed, is an environmental asset that exists in boundless amounts, which can be cultivated naturally. The Agarose chemical structure provides a good support for films. In addition, it is reported that the films, which are made of algae, are transparent, strong, and flexible [2]. In order to enhance the shelf-life and improve the strength of the algae film, cinnamon extract was used to act as a co-primer and anti-microbial in the film. Based on a previous study, anti-microbial polymer film was able to restrain microbial development, hence, broadening the time span of usability of sustenance [6]. Cinnamaldehyde is an organic compound with the formula of $C_6H_5CH=CHCHO$ and occurs naturally as a predominant trans (E) isomer, giving cinnamon its flavor and odor [1]. It is a type of flavonoid that is naturally synthesized by the shikimate pathway [2]. This pale yellow, viscous liquid occurs in the bark of cinnamon trees and other species of the genus *Cinnamomum* sp. The essential oil of cinnamon bark contains about 50% cinnamaldehyde [3]. Cinnamaldehyde is also used as a fungicide [8]. Proven effective on over 40 different crops, cinnamaldehyde is typically applied to the root systems of plants [5]. Its low toxicity and well-known properties makes it ideal for agricultural activities. Thus, "cinnamaldehyde" is an effective insecticide, and its scent is also known to repel animals, such as cats and dogs [8]. It has also been tested as a safe and effective insecticide against mosquito larvae [9]. At a concentration of 29 ppm, cinnamaldehyde can kill half of Aedes aegypti mosquito larvae within 24 h [10]. The "trans-cinnamaldehyde" also works as a potent fumigant and practical repellent for adult mosquitos [11]. By adding cinnamon active chemicals into the packaging system [5], the growth rate of microorganisms in food can be inhibited or reduced. Among other antimicrobials, cinnamaldehyde, which is a major component of cinnamon, also possesses antimicrobial activity and has been utilized in the processing of milk, chicken, and meat [5,6]. The objectives of this study were (1) to assess the suitable percentage loading of cinnamon extract with algae film, and (2) to characterize the effect of cinnamon with the biofilm based on the soil bury test, tensile test, FTIR, and SEM.

2. Materials and Methods

Raw algae and cinnamon were purchased from a local store, located in Gombak, Malaysia, while glycerol and acetic acid were then purchased from Sigma Aldrich (Selangor, Malaysia). For the preparation of raw materials of the biofilm, algae were processed though cleaning, drying, and shredding into powder form (ranging from 50 μ–100 μ). The cinnamon oil was collected using the microwave essential oil extraction method. Temperature was set at 60 °C for 6 min as the cinnamon started to steam up and condensation took place for the production of the extraction oil. Next, we let the extraction oil cool down for another 20 min before collecting it. Using a separator, we collected the oil and applied low heat (33–35 °C) to separate the oil and water. After 20 min of heating process, the oil particle moved on top of the water surface. The oil was collected using the pipet and placed in the vial.

2.1. Solution Casting Method

The algae biofilm was prepared using the solution casting method. The ingredients to prepare biofilm with CEO is as follows: 2% algae powder, three different percentages of CEO at 1%, 3%, and 5%; 1.5 mL glycerol solution; 1%, 3%, 5%, 7%, and 9% acetic acid with 0.2% molarity and distilled water were weighed individually using an electronic mass balance. Acetic acid was obtained in liquid form. It was diluted to 0.2% (w/v) using distilled water. The algae, glycerol, acetic acid, and distilled water were mixed in a beaker, which was then heated up to 90 °C on a hot plate and held at that temperature for 25 min. The stirring speed of the magnetic stirrer was set at a constant speed of 250 rpm, to avoid the formation of bubbles and maintain the homogeneity of the solution. Then, the mixed solution was cooled down to 65 °C for 35 min. During cooling, stirring was continued to prevent the formation of bubbles and solidification of the solution. The second batch of the biofilm was repeated with the addition of the CEO. Pure biofilms (0% CEO) were set up as a control sample. The solution was cast into a square form (18 × 27 cm) of the acrylic plate. Upon casting, the drying process took place in an oven at a temperature of 50 °C for 24 h. The biofilm thickness was measured using an electronic gauge (Digitronic Caliper, Gombak, Selangor, Malaysia), with accuracy ranging between 0.1% and 1% as a function of thickness value (0–100 μm or 0–1000 μm). Seven replicates were made for each type of biofilm formulation.

2.2. Soil Burial Test

The compostability of the biofilms and CEO additions were performed according to soil bury test ISO/DIS 17088. The biofilm dimensions of 20 mm × 20 mm were cut and weighed and five replicates were made for each formulation. The bury test area was plotted at a cool and shaded corner of the garden. The soil temperature was based on the normal climate change, which is from 33 to 35 °C, while soil type was black garden soil. Each sample was buried in a convenient depth of 50 mm to allow for aerobic soil bury composting, as the compost has to be turned at regular intervals in this process. The area was plotted with granite or brick to prevent interruption or error during the investigation. Each time the specimen was retrieved from the ground, the plotted area was covered with layers of dried leaves or thin layers of soil to allow air to permeate the hole and accelerate the growth and expansion of fungi or bacteria.

2.3. Tensile Test

The Instron tensile test ASTM D882-02 machine (Gombak, Selangor, Malaysia) was used for this test. The load of the machine was set at 5 kN with the speed at 10 mm/min. Seven replicates of strips for each composition were cut at dimensions of 70 mm × 10 mm. The result of the tensile strength and elongation at break were assessed through the graph of the stress-strain curve.

2.4. Fourier Transform Infrared Spectrometer (FTIR)

An FTIR Spectrometer (Perkin Elmer System spectrum 100; PerkinElmer, Gombak, Selangor, Malaysia) is an analytical technique used to identify organic, polymeric, and, in some cases, inorganic materials. The FTIR analysis method uses infrared light to scan test samples and observe chemical properties. The resolution was set up at $4 cm^{-1}$ in a spectral range of 4000 to 600 cm^{-1} and 32 scans per sample. Different peaks (various functional groups of chemical elements) of the IR spectrum were observed along the selected initial angle to the final angle.

2.5. Scanning Electron Microscopy (SEM)

The surface morphology of the films was studied using a Scanning Electron Microscope (SEM) JSM 5600 (Gombak, Selangor, Malaysia) with magnifications up to 1000×. Prior to carrying out the observation, the samples were subjected to sputter coating with a layer of carbon using a Polaron SC515 (Gombak, Selangor, Malaysia). This procedure was performed to ensure the sample morphology could be clearly observed under SEM and to prevent any electrostatic charging during observation.

3. Results and Discussions

3.1. Biofilm Thickness

In a polymer film packaging application, thickness is a crucial aspect, which requires specific attention from the material design. The thickness of the biofilm will highly influence other important properties, such as the strength, elasticity, and moisture content. Researchers [5] found that the main purpose of effective biofilm for food packaging is to secure the food from food pathogens, thus extending the shelf-life of the food, and will ensure the quality of the food and its nutrients to be intact. The general thickness of biofilms for packaging is ±0.3 mm [12]. Table 1 shows the thickness of the algae-based film by varying algae sample.

Table 1. Thickness of algae-based film with and without CEO.

Sample (% Algae)	Biofilm without CEO Thickness (mm)	Biofilm with CEO Thickness (mm)
0	0.2 ± 0.01	0.2 ± 0.01
1	0.2 ± 0.01	0.2 ± 0.01
2	0.2 ± 0.01	-
3	0.2 ± 0.01	0.2 ± 0.02
4	0.2 ± 0.02	-
5	0.2 ± 0.01	0.2 ± 0.03

The second batch formulation was focused on biofilm with the addition of several percentages of CEO. Based on this observation, it could be highlighted the importance of using CEO compared to the cinnamon powder. By using CEO, it was easier to control the thickness of the biofilm, as the resulting film thickness was not significantly different compared to the thickness of the control biofilm (0%, as depicted in Table 1). This is because the CEO used in the solution form mixed well within the blends. In comparison, using the cinnamon powder, the course cinnamon particle will not dissolve in the biofilm solution during processing, hence affecting the thickness of the bioplastic. In industry, this parameter is important for food packaging. This trend was similar to the previous study [13], where whey and pectin protein powder were incorporated into cinnamon and researchers had difficulty to control the thickness of the biofilm.

3.2. Algae-Based Biofilm with Acetic Acid

3.2.1. Tensile Data

(A) Tensile strength, (B) modulus elasticity and (C) elongation at break of algae-based biofilms were affected by the different percentages of acetic acid, as demonstrated in Figure 1. The first attempt to produce algae-based biofilm failed because the sample was too fragile, as the algae is rich with starch content and becomes hydrophilic in nature. The glycerol was chosen to alter the delicateness of the biofilm. Once again, the biofilm produced was low in strength, as the particles of glycerol may have leaked during processing and the biofilm produced was found to tear easily. Next, the flexibility of the biofilm was improved with the help of the acetic acid content, to aid the glycerol to be fully efficient in the algae-based biofilm. Acetic acid is a weak acid that has one carboxylic acid group and is usually use in food additive [5]. Although there has been no specific research on the usage of acetic acid as crosslinking agents, it has reported that the presence of acetic acid increased the interfacial interaction in the properties of coconut shells filled with low a polyethylene composite [14]. Researchers [15] also indicated that the combination of acetic acid with CO_2 packaging can extend the shelf life from 12 to 20 days for chicken retail cuts without negatively affecting the quality and sensory properties of the broiler meat. The addition of acetic acid into the blend may help the infusion of glycerol into the algae molecular structure by accelerating the disintegration and suspension of algae sediment. Previous researchers also stated similar strength results in a PVA and chitosan blend with glycerol and acetic acid [16]. However, from the analysis done, the addition of 1%, 3%, 5%, and 7% acetic acid in the biofilm decreases the (A) tensile strength to half compared to the control biofilm, probably due to the different molecular structures of the acetic acid, even though it comes from the same carboxylic acid family [17]. Figure 1 also illustrated a reduction in (B) modulus of algae-based biofilms, similar to the tensile strength, with an increase in the concentration of acetic acid. Based on the figure, the lowest elastic property was recorded for 7% acetic acid, further than this percentage will continuously drop the modulus strength of the biofilm. This finding was similar to previous attempts using glycerol to increase the percentage of citric acid by up to 15% [3]. On the other hand, the (C) elongation at break (Eb) results were vice versa to the tensile and modulus strength results. Based on the Eb graph, it was found that the biofilms were capable of resisting changes in shape without crack formation with the addition of the acetic acid. Figure 1 demonstrated that the (C) elongation at break gradually increased with the increasing percentage of acetic acid by up to 7%. The highest elongation at break was measured with 5% acetic acid at 27.34% of Eb, while the lowest Eb was shown by 0% acetic acid at 20.14%. Physically, the biofilm with the addition of acetic acid was better in flexibility, less fragile, and good in modulus. The biofilm obtained was transparent and could not easily tear off when folded. However, the addition of acetic acid neither improves the tensile strength nor the elongation at break of the biofilm.

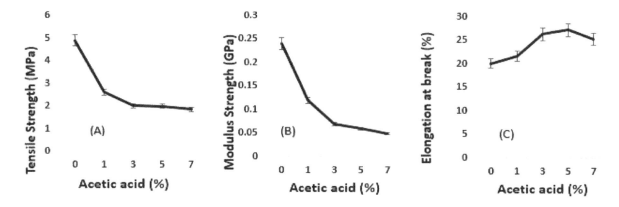

Figure 1. Effect of acetic acid on (**A**) tensile strength; (**B**) modulus strength; and (**C**) elongation at break of algae-based biofilm.

3.2.2. Soil Burial Test

The algae used in this research to form a biofilm is a green algae species known as Neochloris Oleoabundans or Ettlia Oleoabundans [18]. These unicellular green algae are freshwater based and rich in starch content [18,19]. Starch is hydrophilic in nature and easy to degrade due to microbial and moisture contact. The degradation of algae-based bioplastic film via starch by micro-organisms in the soil produced carbon dioxide, biomass formed by extraction of algae carbon, and soluble CEO compound. From the soil bury test analysis, gradual biodegradation was observed in the biofilm surface degradation as shown in Figure 2. Figure 2 demonstrates the physical appearance of the algae biofilm after soil burial test for 28 weeks. The samples for soil bury test were exposed to the actual weather. Under rainy conditions, excess water permeated through the soil and diffused into the biofilm samples causing swelling and softening of the biofilm. Based on the physical observation, the biofilm started to deteriorate after 14 weeks and onwards, most likely due to hydrophilic nature of the algae.

At the end of the 28 week period, the sample could barely retain the shape and began to wrinkle and tear apart. From the pictures, the sample showed high number of pores, and the number of pores continues to spread and increase in size as the length of soil bury test was prolonged. This confirmed that the biofilm sample had undergone biodegradation phases.

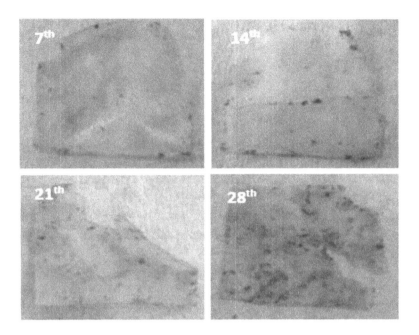

Figure 2. Physical appearance of algae film after soil burial at the **7th, 14th, 21th,** and **28th** weeks.

3.3. Algae-Based Biofilm with CEO

3.3.1. Tensile Data

Algae-based biofilm with 3% acetic acid was found to yield good (A) tensile strength and (B) modulus properties, and, therefore, was selected to be used with CEO. This percentage was selected to be used for further investigation alongside the addition of different ranges of CEO (phase two). The percentages of CEO tested were 1%, 3%, 5%, 7%, and 9%. The control sample is label as 0 in the tabulated figures. Figure 3 demonstrates the (A) tensile strength of algae film, which increases with increasing percentages of CEO. The control sample without acetic acid possessed the least tensile strength. Meanwhile, the maximum tensile strength was achieved with 5% CEO due to the good intermolecular interaction between algae and starch and cinnamon molecules. This finding was also supported by a previous study [20], which also recorded a similar pattern where tensile strength increased with the addition of cinnamon bark oil into the alginate film. Based on Figure 3, the (B) modulus elasticity of the algae-based biofilm was found to display the same trend as the tensile strength results. The addition of 5% CEO was found to increase the stiffness of the algae-based biofilm up to 0.323 GPa, compared to algae-based biofilm with 1%, 3%, 7%, and 9% of CEO, which recorded lower modulus strength. However, in this study, the 5% CEO loading did affect the (C) elongation at break compared to the other percentages. Figure 3 indicated that the 5% CEO loading has low elongation at break compared to the 3% and 7% of CEO loading. Based on the current findings, 5% CEO with 3% acetic acid yielded a good and accepted elongation at break of the algae-based biofilm. This was made possible with the right amount of acetic acid in strengthening and adhering to the intermolecular bonds between the algae and cinnamon molecules. Hence, the addition of acetic acid into algae-based biofilm clearly indicated that the acetic acid molecules affect the adjacent molecules by increasing the distance and reducing the internal force, resulting in a more flexible film. The interference with adjacent molecules affects the intermolecular and intramolecular linkage of the polymer, thus strengthening the structure of the algae-based biofilm [21].

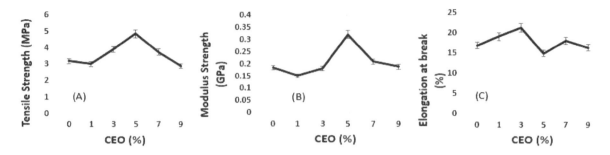

Figure 3. Effect of CEO on (**A**) tensile strength; (**B**) modulus strength; and (**C**) elongation at break of algae-based film.

3.3.2. Soil Burial Test

Figure 4 depicted the physical appearance and Figure 5 shows the SEM analysis of biofilm with 5% CEO after soil burial at the 7th, 14th, 21th, and 28th weeks, respectively. The analysis is similar to control algae-based biofilm in Figure 2, it was noticed that the color of the biofilms with CEO turned darker and the darkening of the biofilms is a sign of biodegradation [22]. Changes in the appearance of the biofilms are explainable through the high-moisture absorption property and low intensity of cinnamaldehyde in the CEO percentages [23]. Based on the physical appearance of the biofilm in Figure 4, the sample in this research would have behaved similarly to the findings by Zhang et al. [20], where the alginate films incorporated with cinnamon bark oil showed less biodegradation potential compared to the alginate film without cinnamon bark oil. Therefore, it can be postulated that the addition of 3% acetic acid into the recipe assists in reducing the decomposition rate of the biofilm with CEO, compared to the sample of 5% CEO without acetic acid content. From the SEM analysis in Figure 5, the agglomeration of biofilm became more obvious as the biofilm began to swell, which in turn caused slow degradation. The addition of 5% of CEO into the biofilm exhibited physical changes. Besides which, different volumes of CEO used in this study resulted in varying biodegradation rates and behaviors. The antimicrobial (cinnamaldehyde functional group) and repellent properties of cinnamon may also decelerate the degradation rate of the film. These findings are in accordance with the previous research [23], where the higher the cinnamon percentage, the slower the composability rate. The film with 5% CEO demonstrated lesser pore percentages. A higher amount of CEO tends to reduce the degradation rate because of the hydrophobicity of the acetic acid, and the strong aroma of the cinnamon itself may repel insects and micro-organisms from attacking the biofilm [24].

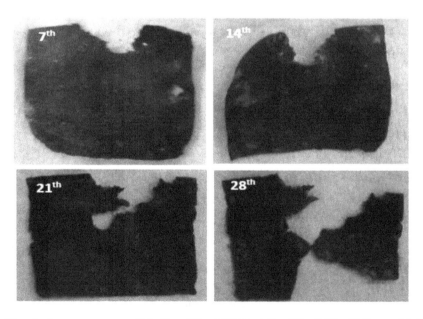

Figure 4. Physical appearance of biofilm 5% of CEO at the **7th, 14th, 21th,** and **28th** weeks.

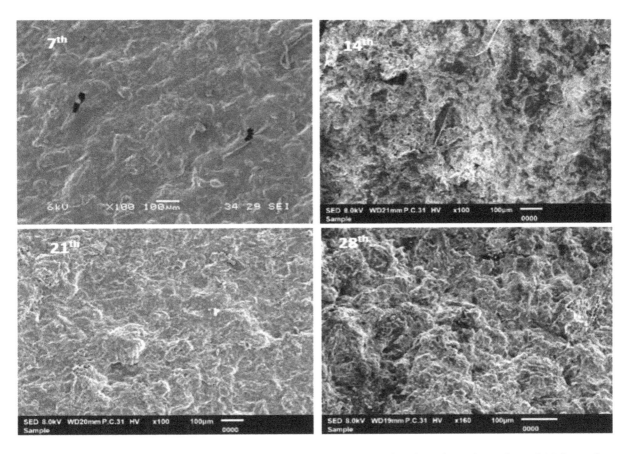

Figure 5. SEM analysis of algae film with 5% CEO after soil burial at the **7th, 14th, 21th,** and **28th** weeks.

3.4. Fourier Transform Infra-Red (FTIR) Spectroscopy Data

Figure 6A illustrated the FTIR spectra of biofilm with 5% CEO content and B biofilm with 3% of acetic acid, respectively, both displaying individual peaks within the range of 4000–500 cm^{-1}. Peak A presented a broad absorption band at about 3310 cm^{-1}, which represents the hydroxyl (OH) group. The peak at 2924 cm^{-1} was recognized due to the C–H stretching of methane. Besides this, the peak at 1606 cm^{-1} was formed due to the stretching vibration of the conjugated peptide bond formation by amine (NH$_2$) and acetone groups in the algae. The peak at 1441 cm^{-1} was due to an ester sulfate group. The characteristic peaks at 1013 cm^{-1} and 931 cm^{-1} indicated C–O stretching groups of 3,6-anhydrogalactose [25]. In addition, based on the FT-IR spectrum of CEO in A, the absorption band or frequency ranged from 3500 cm^{-1} to 3200 cm^{-1} broad, exhibiting the presence of O–H stretch. The specific absorbance band at 1635 cm^{-1} revealed the stretching vibration of the C=O bond for cinnamaldehyde [25]. Due to the influence of conjugation and an aromatic ring, the peak is wider than usual for aldehyde compounds. A strong absorption band between 900 cm^{-1} and 690 cm^{-1} indicated the presence of aromatic C=C bonds [26]. Cinnamaldehyde is the main active component in cinnamon, which can be used as a natural antimicrobial in food preservation to retard or inhibit the bacterial growth of pathogenic and spoilage bacteria, which in turn extends the shelf life of the food products [20]. Since there was no peak observed at 1700–1720 cm^{-1} in Figure 6B, which shows that there was no crosslinking between acetic acid and the algae-based blends due to the absence of chemical reaction. The sighting of a peak at wavenumber ranges between 1700–1720 cm^{-1} indicates the presence of cellulose-fatty acids ν(C=O), a stretching vibration of the esters. The slope was transmitted obviously in (A), however, slowly lowering down with the addition of acetic acid as showed in sample (B). The combination of CEO and acetic acid was significantly reduced the cellulose fatty acid presence in the algae as shown in sample (C) which is possibly occurs due to the formation of a physical reaction between CEO, acetic acid and algae [26]. Figure 6C represents the FT-IR spectrum of 3% acetic acid blends with 5% CEO biofilms. A peak at 1716 cm^{-1} was observed, indicating an association with C=O, which is attributed to the carboxyl and ester carbonyl bands. This confirmed the existence of acetic acid in the specimen. However, the peak at 3328 cm^{-1} became less intense when cinnamon was added into the formulation.

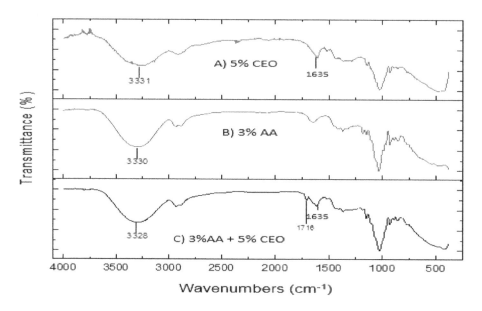

Figure 6. FTIR spectra of (**A**) Biofilm with 5% CEO; (**B**) biofilm with 3% acetic acid; and (**C**) biofilm with 5% CEO and 3% acetic acid.

4. Conclusions

The tensile test of biofilm demonstrated good enhancement upon the incorporation of 5% CEO. The biofilm achieved tensile strength at 4.8 MPa and elongation of 15%. Based on the SEM morphology, higher amounts of CEO in the presence of acidic acid leads to a reduction in the degradation rate of the biofilm. The biofilm demonstrated a continuous phase and exhibited a characteristic band at 1716 cm^{-1} in the FTIR analysis. Hence, in conclusion, 5% CEO and 3% acetic acid are the suitable blend that could tremendously affect the tensile behavior and the biodegradation rate of the biofilm.

Author Contributions: Conceptualization, M.O. and K.K.; methodology, H.R.; validation, S.I.S.S., S.S., and H.S.H.; formal analysis, H.R.; investigation, M.N.Z.; resources, N.A.J.; writing—original draft preparation, M.O.; writing—review and editing, H.R. and H.S.H.; visualization, H.R. and H.S.H.; supervision, M.O.; project administration, K.K.

Acknowledgments: Would like to acknowledge IIUM, MARDI, and UITM bio-composite department for support given for administrative and technical support.

References

1.   Webb, H.K.; Arnott, J.; Crawford, R.J.; Ivanova, E.P. Plastic degradation and its environmental implications with special reference to poly (ethylene terephthalate). *Polymers* **2013**, *5*, 1–18. [CrossRef]
2.   Ku, K.J.; Hong, Y.H.; Song, K.B. Mechanical properties of a Gelidium corneum edible film containing catechin and its application in sausages. *J. Food Sci.* **2008**, *73*, 217–221. [CrossRef] [PubMed]
3.   Reddy, N.; Yang, Y. Citric acid cross-linking of starch films. *Food Chem.* **2010**, *118*, 702–711. [CrossRef]
4.   Hosseini, M.H.; Razavi, S.H.; Mousavi, M.A. Antimicrobial, physical and mechanical properties of chitosan-based films incorporated with thyme, clove and cinnamon essential oils. *J. Food Process. Preserv.* **2009**, *33*, 727–743. [CrossRef]
5.   Sung, S.Y.; Sin, L.T.; Tee, T.T.; Bee, S.T.; Rahmat, A.R.; Rahman, W.A.; Vikhraman, M. Antimicrobial agents for food packaging applications. *Trends Food Sci. Technol.* **2013**, *33*, 110–123. [CrossRef]

6. Cui, H.Y.; Zhou, H.; Lin, L.; Zhao, C.T.; Zhang, X.J.; Xiao, Z.H.; Li, C.Z. Antibacterial activity and mechanism of cinnamon essential oil and its application in milk. *J. Anim. Plant Sci.* **2016**, *26*, 532–541.

7. Du, W.X.; Olsen, C.W.; Avena-Bustillos, R.J.; McHugh, T.H.; Levin, C.E.; Friedman, M. Effects of allspice, cinnamon, and clove bud essential oils in edible apple films on physical properties and antimicrobial activities. *J. Food Sci.* **2009**, *74*, 372–378. [CrossRef] [PubMed]

8. Gautam, N.; Kaur, I. Soil burial biodegradation studies of starch grafted polyethylene and identification of Rhizobium meliloti therefrom. *J. Environ. Chem. Ecotoxicol.* **2013**, *5*, 147–158. [CrossRef]

9. Herwig, G. *Plant Natural Products: Synthesis, Biological Functions and Practical Applications*; Wiley: Hoboken, NJ, USA, 2014; pp. 19–21.

10. Singh, G.; Maurya, S.; de Lampasona, M.P.; Catalan, C.A.N. A comparison of chemical, antioxidant and antimicrobial studies of cinnamon leaf and bark volatile oils, oleoresins and their constituents. *Food Chem. Toxicol.* **2007**, *45*, 1650–1661. [CrossRef]

11. Leja, K.; Lewandowicz, G. Polymer biodegradation and biodegradable polymers—A review. *Pol. J. Environ. Stud.* **2010**, *19*, 255–266.

12. Bahram, S.; Reaei, M.; Soltani, M.; Kamali, A.; Ojagh, S.M.; Abdollahi, M. Whey Protein Concentrate Edible Film Activated with Cinnamon Essential Oil. *J. Food Process. Preserv.* **2014**, *38*, 1251–1258. [CrossRef]

13. Faisal ZH, T.; Amri, F.; Salmah, H.; Tahir, I. The Effect of Acetic Acid on Properties of Coconut Shell Filled Low Density Polyethylene Composites. *Indones. J. Chem.* **2010**, *10*, 334–340.

14. Desai, M.A.; Kurve, V.; Smith, B.S.; Campano, S.G.; Soni, K.; Schilling, M.W. Utilization of buffered vinegar to increase the shelf life of chicken retail cuts packaged in carbon dioxide. *Poult. Sci.* **2014**, *93*, 1850–1854. [CrossRef] [PubMed]

15. Adinew, B. GC-MS and FT-IR analysis of constituents of essential oil from Cinnamon bark growing in South-west of Ethiopia. *Int. J. Herb. Med.* **2014**, *1*, 22–31.

16. Reddy, N.; Li, Y.; Yang, Y. Alkali-catalyzed low temperature wet crosslinking of plant proteins using carboxylic acids. *Biotechnol. Prog.* **2009**, *25*, 139–146. [CrossRef] [PubMed]

17. Seligra, P.G.; Jaramillo, C.M.; Famá, L.; Goyanes, S. Biodegradable and non-retrogradable eco-films based on starch-glycerol with citric acid as crosslinking agent. *Carbohydr. Polym.* **2016**, *138*, 66–74. [CrossRef] [PubMed]

18. Wu, J.; Sun, X.; Guo, X.; Ge, S.; Zhang, Q. Physicochemical properties, antimicrobial activity and oil release of fish gelatin films incorporated with cinnamon essential oil. *Aquac. Fish.* **2017**, 4–11. [CrossRef]

19. Maitra, J.; Shukla, V.K. Cross-linking in Hydrogels—A Review. *Am. J. Polym. Sci.* **2014**, *4*, 25–31. [CrossRef]

20. Zhang, Y.; Ma, Q.; Critzer, F.; Davidson, P.M.; Zhong, Q. Physical and antibacterial properties of alginate films containing cinnamon bark oil and soybean oil. *LWT Food Sci. Technol.* **2015**, *64*, 423–430. [CrossRef]

21. Awadhiya, A.; Kumar, D.; Verma, V. Crosslinking of agarose bioplastic using citric acid. *Carbohydr. Polym.* **2016**, *151*, 60–67. [CrossRef]

22. Fernando, I.P.S.; Sanjeewa, K.K.A.; Samarakoon, K.W.; Lee, W.W.; Kim, H.S.; Kim, E.A.; Jeon, Y.J. FTIR characterization and antioxidant activity of water soluble crude polysaccharides of Sri Lankan marine algae. *Algae* **2017**, *32*, 75–86. [CrossRef]

23. Kwak, Y.S.; Kim, S.J.; Kim, H.Y. The antibacterial effect of Cinnamomum verum extract. *Biomed. Res.* **2017**, *28*, 6667–6670.

24. Nowak, A.P.; Lisowska-Oleksiak, A. Red algae—An alternative source of carbon material for energy storage application. *Int. J. Electrochem. Sci.* **2014**, *9*, 3715–3724.

25. Bunghez, F.; Socaci, C.; Zăgrean, F.; Fetea, F. Application of FT-MIR Spectroscopy for Fingerprinting Bioactive Molecules of Plant Ingredients and a New Formula with Antimicrobial Effect. *Bull. UASVM Food Sci. Technol.* **2013**, *70*, 64–65. [CrossRef]

26. Dilek, Y.D.; Abel, U.U.; Tulay, B.O.; Aydin, A.; Ilkay, A.E.; Kazim, Y.; Deniz, G. Fourier transform infrared (FTIR) spectroscopy for identification of Chlorella vulgaris Beijerinck 1890 and Scenedesmus obliquus (Turpin) Kützing 1833. *Afr. J. Biotechnol.* **2012**, *11*, 3817–3824. [CrossRef]

6

Antibacterial Nanocomposites Based on Thermosetting Polymers Derived from Vegetable Oils and Metal Oxide Nanoparticles

Ana Maria Diez-Pascual

Department of Analytical Chemistry, Physical Chemistry and Chemical Engineering, Faculty of Sciences, Alcalá University, 28871 Madrid, Spain; am.diez@uah.es.

Abstract: Thermosetting polymers derived from vegetable oils (VOs) exhibit a wide range of outstanding properties that make them suitable for coatings, paints, adhesives, food packaging, and other industrial appliances. In addition, some of them show remarkable antimicrobial activity. Nonetheless, the antibacterial properties of these materials can be significantly improved via incorporation of very small amounts of metal oxide nanoparticles (MO-NPs) such as TiO_2, ZnO, CuO, or Fe_3O_4. The antimicrobial efficiency of these NPs correlates with their structural properties like size, shape, and mainly on their concentration and degree of functionalization. Owing to their nanoscale dimensions, high specific surface area and tailorable surface chemistry, MO-NPs can discriminate bacterial cells from mammalian ones, offering long-term antibacterial action. MO-NPs provoke bacterial toxicity through generation of reactive oxygen species (ROS) that can target physical structures, metabolic paths, as well as DNA synthesis, thereby leading to cell decease. Furthermore, other modes of action—including lipid peroxidation, cell membrane lysis, redox reactions at the NP–cell interface, bacterial phagocytosis, etc.—have been reported. In this work, a brief description of current literature on the antimicrobial effect of VO-based thermosetting polymers incorporating MO-NPs is provided. Specifically, the preparation of the nanocomposites, their morphology, and antibacterial properties are comparatively discussed. A critical analysis of the current state-of-art on these nanomaterials improves our understanding to overcome antibiotic resistance and offers alternatives to struggle bacterial infections in public places.

Keywords: metal oxide nanoparticles; vegetable oils; thermosetting polymers; antibacterial properties; nanocomposites; reactive oxygen species

1. Introduction

Over the past few decades, the manufacture and applications of synthetic polymeric materials exhibited an extraordinary boost. Nonetheless, concerns regarding diminution of fossil resources and the sharp rise in petroleum cost, health-related issues together with stringent environmental government protection policies have led to an incessantly increasing attention to the development of sustainable, safe, and environmentally friendly plastics from renewable resources. Despite the current growth of renewable polymers, their contribution to the profitable market is still very low mainly due to their expensiveness and inferior performance compared with synthetic polymers based on petroleum feedstock [1]. The key challenge for the biorenewable industry is to produce materials with properties that match or even go beyond those of petroleum-based ones. The most widely used renewable resources are plant oils, polysaccharides, and proteins [2].

1.1. Thermosetting Polymers Obtained from Vegetable Oils

Plant or vegetable oils (VOs) can be used as raw materials for polymers due to their abundance in nature, moderate cost, biodegradability, and very low toxicity [3]. Furthermore, their polar character bestows improved antibacterial properties to the resulting polymeric systems. The main elements of VOs are triglycerides, which are the result of the esterification of glycerol with three fatty acids (Scheme 1). Fatty acids represent 95% of the whole weight of triglycerides and their composition changes depending on the plant, harvest, season, growing conditions, and purification methods [4]. The fatty acids have different levels of saturation and unsaturation [5], being the unsaturated part the one that can be functionalized to yield the liquid resin to be used for the polymer synthesis. Many reactive points of the triglycerides—like double bonds, allylic carbons, and ester groups—can be used to introduce polymerizable groups. The chain length of fatty acids in triglycerides occurring in nature typically varies between 16 and 20 carbon atoms. Table 1 collects the most frequent fatty acids present in vegetable oils. Most of them exhibit a straight chain with an even number of carbons and double bonds in a cis configuration.

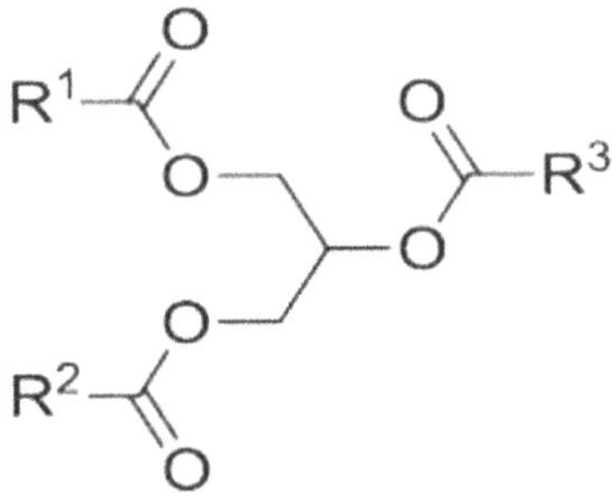

Scheme 1. General structure of a triglyceride from a plant oil (R^1, R^2, and R^3 represent fatty acid chains).

Table 1. Formulas and structures of the most common fatty acids in vegetable oils

Fatty Acid	Formula	Structure
Palmitic	$C_{16}H_{32}O_2$	 COOH
Palmitoleic	$C_{16}H_{30}O_2$	 COOH
Stearic	$C_{18}H_{36}O_2$	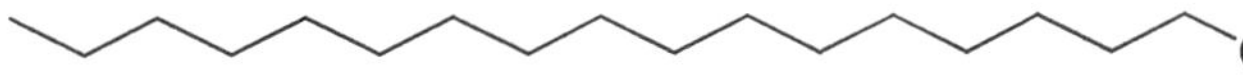 COOH
Oleic	$C_{18}H_{34}O_2$	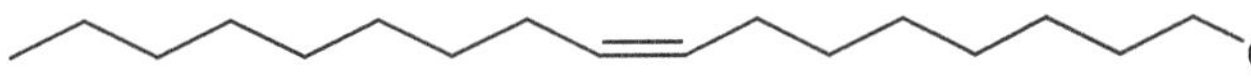 COOH
Linoleic	$C_{18}H_{32}O_2$	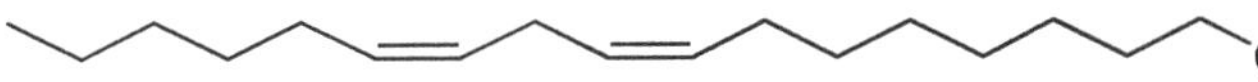 COOH
Linolenic	$C_{18}H_{30}O_2$	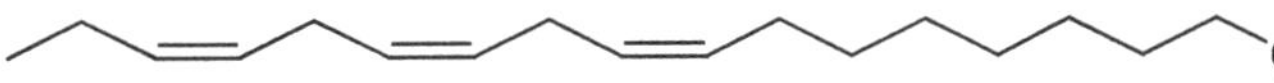 COOH
α-Eleostearic	$C_{18}H_{30}O_2$	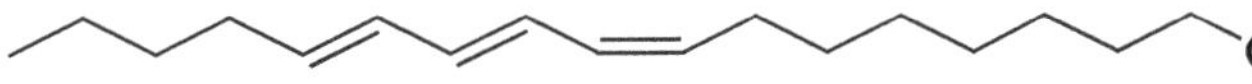 COOH

Table 2 summarizes the proportions of the different fatty acids in the most frequent VOs. The chemical and physical properties of VOs are strongly influenced by the level of unsaturation, which can be calculated by measuring the iodine value (IV), which indicates the amount of iodine (mg) that reacts with the C=C double bonds in 100 g of the VOs; the higher the IV value, the more number of C=C per triglyceride of the VOs. Thus, VOs can be divided into drying oils (IV > 130), semi-drying oils (100 < IV < 130), and non-drying oils (IV < 100). The IV values of the most widespread VOs are also collected in Table 2.

Table 2. Properties and fatty acid compositions of the most frequent vegetable oils.

Vegetable Oil	Double Bonds [a]	Iodine Value [b] (mg/100g)	Fatty Acids (%)				
			Palmitic	Stearic	Oleic	Linoleic	Linolenic
Palm	1.7	44–58	42.8	4.2	40.5	10.1	-
Olive	2.8	75–94	13.7	2.5	71.1	10.0	0.6
Groundnut	3.4	80–106	11.4	2.4	48.3	31.9	-
Rapeseed	3.8	94–120	4.0	2.0	56.0	26.0	10
Sesame	3.9	103–116	9.0	6.0	41.0	43.0	1.0
Cottonseed	3.9	90–119	21.6	2.6	18.6	54.4	0.7
Corn	4.5	102–130	10.9	2.0	25.4	59.6	1.2
Soybean	4.6	117–143	11.0	4.0	23.4	53.3	7.8
Sunflower	4.7	110–143	5.2	2.7	37.2	53.8	1.0
Castor [c]	4.8	83–88	1.3	1.2	4.0	5.2	0.3
Linseed	6.6	168–204	5.5	3.5	19.1	15.3	56

[a] Average number of double bonds per triglyceride. [b] The amount of iodine (mg) that reacts with the double bonds in 100 g of vegetable oil. [c] Contains 89% of unsaturated ricinoleic acid

Linseed, soybean, sunflower, palm, olive, cottonseed, castor, and rapeseed oils are the most frequently used for the synthesis of biopolymers [6,7], and can be used in a number of applications including lubricants, coatings, soaps, cosmetic products, biodiesel, and so forth. Different derivatives of VOs have been used as polymerizable monomers in radiation-curable systems to yield flexible but tough resins such as epoxy, urethane, and polyester [8]. Combinations of epoxidized soybean oil and traditional epoxy resins or their copolymers with styrene and divinylbenzene as well as polyurethane resins derived from castor oil have been used to fabricate composite materials with outstanding mechanical and barrier performance [9]. Furthermore, epoxidized vegetable oils (EVOs) such as soybean and palm have been used in UV-curable coating systems [7,8]. Despite the use of VOs in paints and coatings is quite old, emphasis is currently being made on these materials to attain novel properties, superior performance together with environmentally friendliness at reasonable costs [10].

Different VOs have been used for the preparation of cationic polyurethane (PU) coatings with antimicrobial activity to be applied in the biomedical or food packaging industries. For instance, Xia et al. [11] developed PU coatings based on five amino polyols derived from soybean oil and studied the influence of their structure and functionalities on the antimicrobial, mechanical, and thermal properties. Those comprising N-methyldiethanolamine (MDEA)- and N-ethyldiethanolamine (EDEA), with the shortest side chains, exhibited the best antibacterial action. The same authors [12] modified the crosslinking density of the materials using soybean polyols with different number of hydroxyl moieties as starting materials. Reduction of the crosslink density by lessening the hydroxyl number of the soybean polyol promoted the physical interaction with the target bacteria via increasing the molecular chain mobility, thus resulted in superior antibacterial properties. In another study, Liang et al. [13] employed castor oiland MDEA to prepared antimicrobial PUs, and improved action was also observed via reduction of the polyol functionality.

On the other hand, polymers derived from VOs display reduced mechanical and barrier properties, which restrict their realistic applications [1]. Thus, their brittleness, elevated gas and vapor permeability, and low heat distortion temperature have limited their commercial uses. To solve these issues, different strategies have been reported, including mixing with other polymers [14] or incorporating nanofillers [15–18]. The addition of nanofillers to thermosetting polymers based on VOs results in bionanocomposites with enhanced performance [15,16]. Bio-based PUs show poor antimicrobial activity, which can be significantly improved via incorporation of regularly dispersed metal or metal-oxide nanoparticles [19].

1.2. Antimicrobial Effect of Metal-Oxide Nanoparticles

Metallic elements combine with oxygen to form metal oxides (MOs), which have been demonstrated to interact with bacteria by means of electrostatic interactions that modify the prokaryotic cell wall and damage the DNA via reactive oxygen species (ROS) generation [20]. MOs have attracted a lot of attention as a nanomaterial that can be synthesized into a specific size and shape. Bearing these in mind, metal oxide nanoparticles (MO-NPs), in particular those based on ZnO, TiO_2, CuO, and MgO, are being used as antimicrobial agents [20]. The mechanism of MO-NPs as antibacterial agents is influenced not only by the chemical composition, but also by the shape, size, solubility, degree of agglomeration, and surface charge of the NPs [21]. In particular, the morphology and size of NPs has been demonstrated to be crucial in the antimicrobial efficacy: the smallest spherical NPs were the most effective in destroying the bacteria, since are more prone to go through the bacterial cell walls due to their increased surface area to volume ratio [22]. On the other hand, low NP solubility has been demonstrated to bring out a reduced cytotoxic response [22]. The NP solubility also influences its tendency to agglomerate, which controls the NP–cell interactions. Thus, highly agglomerated NPs are not able to enter a cell or generate a considerable amount of ROS [23]. This can be due to the decrease in the surface area. Nevertheless, when NPs are well dispersed, they increase the interactions with the cells and the ROS fabrication. In addition, the NP surface charge influences the bactericide action. NPs with positive zeta potentials allow for electrostatic interactions with bacteria comprising negatively charged surfaces. This promotes NP penetration into the cell membrane. High zeta potentials support a strong interaction, provoking membrane disruption, bacteria flocculation, and viability lessening. From all the above facts, it can be concluded that the antimicrobial effects of MO-NPs can be tailored by controlling their surface charge, size, and surface morphology.

1.2.1. Antimicrobial Effect of Zinc Oxide

Zinc is a fundamental nutrient that plays a key role in growth of mammals. At the nanoscale, ZnO nanostructures have involved a lot of interest within the scientific community due to their inexpensiveness, ease of use, possibility of functionalization and biocompatibility. This nanomaterial is a semiconductor with a wide bandgap (3.37 eV), hence widely used for UV-Vis optical devices [24]. Furthermore, its bandgap can be modified by mixing with MgO and CdO, which is interesting for optoelectronic applications. It possesses large specific surface area, radiation resistance, good mechanical properties [25] (i.e., a Young's modulus close to 100 GPa and hardness of about 5 GPa), low coefficient of thermal expansion and high thermal conductivity (around 1 W cm^{-1} K^{-1}) [26]. At ambient temperature and pressure, ZnO crystallizes in the wurtzite structure. It is a hexagonal lattice including two interconnecting sublattices of Zn^{2+} and O^{2-}, in which each Zn^{2+} is surrounded by a tetrahedra of O^{2-} and vice-versa [24]. This structure results in polar symmetry along the hexagonal axis, being accountable for the piezoelectricity and spontaneous polarization that has been observed in ZnO.

The main benefits of using these MO-NPs as antimicrobial agents are that they enclose mineral elements vital for humans and show strong action when provided in small quantities. The antimicrobial action of ZnO NPs has been endorsed to different mechanisms [27,28], as schematized in Figure 1: a) bringing on oxidative stress due to formation of reactive oxygen species (ROS). Upon light exposure, electron–hole pairs are formed on the ZnO surface; the holes split water molecules and cause ROS formation, explicitly •OH, H_2O_2, and •O^{2-}, which can subsequently interact with proteins, DNA, and lipids of the bacteria, thereby inducing cell death; b) cell membrane disruption due to accumulation of the NPs at the bacterial membrane followed by NP internalization within the cell and destruction of organelles; c) release of Zn^{2+} ions that join to the bacteria membrane; d) lipid peroxidation can occur on the bacterial membrane, weakening membrane integrity, and promoting cell lysis.

The attachment of the ZnO-NPs to the bacterial membrane is a crucial step in the three aforementioned mechanisms. The way the NPs anchor to the membrane is not fully clear yet, though it seems that arises from electrostatic forces. Subsequent to the attachment, "pitting" takes place at the membrane due to ROS generation, which lethally harms the cells [29].

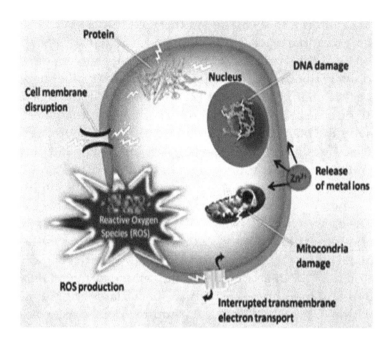

Figure 1. Mechanisms of antimicrobial activity of ZnO nanoparticles. Adapted from [27] permission from Elsevier.

ROS generation has been reported to increase with increasing ZnO surface area while decreasing level of crystallinity and particle size [30]. Even though ZnO has been used as an bactericide agent against both Gram-positive (*B. subtilis*, *S. aureus*) and Gram-negative bacteria (*P. aeruginosa*, *E. coli*), it displays higher activity towards Gram-positive ones, ascribed to different reasons, the most important being the difference in the membrane thickness [17,31]. Gram-positive bacteria contain a cell wall composed of peptidoglycans, teichoic, and lipoteichoic acids that it is simpler to be penetrated in comparison to the complex wall of Gram-negative ones, with an outer membrane of lipopolysaccharides and a peptidoglycan layer.

1.2.2. Antimicrobial Effect of Titanium Oxide

TiO_2 nanostructures are n-type semiconductors with exceptional properties for optics, photonics and electronic applications [32]. They are non-toxic, low reactive, rather inexpensive and environmentally friendly, and possess outstanding mechanical properties, low coefficient of thermal expansion, high thermal conductivity, and strong UV absorption.

In nature, three main polymorphs of TiO_2 can be found: anatase (tetragonal), brookite (orthorhombic), and rutile (tetragonal), the former being the most commonly employed in photocatalytic applications [33]. Furthermore, it is the most toxic form, thus used to obtain antimicrobial properties [34], being employed in medical devices, household cleaning products, air-conditioning surfaces, water treatment facilities, and so forth. Anatase is the least thermodynamically stable polymorph as a bulk phase, although it generally predominates when the grain size is below 100 nm. Nanostructured anatase-TiO_2 can be synthesized by a large number of techniques including liquid-phase sol–gel, microemulsion, hydro- and solvothermal, aerogel, sonochemical or surfactant-templated methods [35].

TiO_2 NPs have been used as bactericide agents for both Gram-positive and Gram-negative bacteria [35], and their properties are photodependent. Similarly to ZnO, they can produce different types of ROS, including hydroxyl radicals $\bullet OH$, superoxide radicals $\bullet O^{2-}$ and singlet oxygen ($^1O_2$) [36], as well as other species like H_2O_2 or O_2 that can disrupt the bacteria cell membrane. In particular, hydroxyl radicals seem to play the most important role on the antibacterial action of these NPs. Due to peroxidation, these free radicals affect the lipopolysaccharide, peptidoglycan, and phospholipid bilayers [37]. Owed to their photo-dependent properties and large effective surface area, TiO_2 NPs are frequently applied in surface coatings. Furthermore, the smaller the NPs, the more prone to enter the bacteria cell due to the increased surface area to volume ratio. In contrast, highly agglomerated NPs are not able to penetrate the cells or make a significant amount of ROS. Agglomeration is influenced by a number of factors, including the hydrophobicity of the material, the interactions with the dispersed medium (i.e., pH, protein content), and the surface charge [38].

1.2.3. Antimicrobial Effect of Copper Oxide

CuO is a p-type semiconductor with a narrow band gap of 1.2–1.9 eV that has attracted special attention due to its valuable physical properties including high temperature superconductivity, electron correlation effects, and spin dynamics [39]. It is quite economical, easily mixed with polar liquids (i.e., water) and polymers, and stable in terms of both chemical and physical properties. The crystal structure of CuO belongs to the monoclinic space group C2/c; each Cu^{2+} is bonded in a square co-planar geometry to four equivalent O^{2-} [40]. Elemental copper and its oxides have been renowned as antimicrobial agents by the US Environmental Protection agency (EPA). This makes them suitable for numerous applications in paints, fabrics, agriculture, and in hospitals both as powder and as coating.

CuO antibacterial action has been ascribed to a rapid lessening in cell membrane integrity and production of ROS, namely H_2O_2, hydroxyl radicals •OH, superoxide radicals •O^{2-} and singlet oxygen ($^1O_2$), being the hydroxyl radical the most common. Thus, CuO-NPs generate ROS through several mechanisms, likely including Fenton-like and Haber–Weiss reactions at the nanoparticle surface or in solution via release of Cu^{2+} dissolved from the nanoparticle surface [41]. Other proposed mechanisms include DNA damaging via formation of a DNA/Cu^{2+}/H_2O_2 complex or Cu^{2+}-bound•OH as the damaging species [42]. $^1O_2$ can also be formed in the presence of CuO-NPs under oxidative stress conditions, and it decomposes into •OH. Furthermore, in bacterial cells, Cu^{2+} ions are reduced by sulphydryl to Cu^+, and the reduced ions are responsible for causing oxidative stress via Cu^+-driven ROS. Nonetheless, the exact pathway of antimicrobial action of CuO is not clear in the literature [40].

1.2.4. Antimicrobial Effect of Iron Oxide

Fe_3O_4 occurs in nature as the mineral magnetite. It is a ferrimagnetic oxide, with a Curie temperature of 858 K, with semiconducting properties, with conductivities ranging from 10^2–10^3 Ω^{-1}cm [43]. Fe_3O_4 nanomaterials have attracted incredible interest since they are biocompatible and show exceptional electric, thermal, mechanical, and magnetic properties. Nanostructures with a variety of morphologies have been synthesized and applied in fields like lithium-ion batteries, wastewater treatment, magnetic resonance imaging contrast agents, therapeutics for cancer treatment, and radiation oncology. The crystal structure of magnetite corresponds to the cubic inverse spinel pattern in which the oxygen ions form a cubic face centered packing, and the iron ions locate at interstitial tetrahedral and octahedral sites [44].

The antimicrobial activity of magnetite NPs is believed to be the result of their interaction with bacterial membranes and their penetration into the bacterial cell, causing membrane damage and inactivation of bacteria. H_2O_2, which is produced from the metabolic activity of the bacteria, can react with the species present in magnetite: different oxido-reduction reactions take place involving both Fe^{3+} and Fe^{2+}, (Fenton-like and Haber–Weiss reactions), leading to the formation of ROS, in particular •OH and •HO_2 free radicals. The ROS produced at the microorganism surface depolarizes the bacterial membranes, causing oxidative stress and cell membrane damage [45].

2. Preparation of Vegetable Oil-Based Thermosetting Polymers Incorporating Metal Oxide Nanoparticles

2.1. Synthesis of Acrylated Epoxidized Linseed Oil (AELO)/TiO2 Nanocomposites

Firstly, the epoxidized linseed oil (ELO) was synthesized from the vegetable oil in a four-necked glass flask incorporating a mechanical stirrer, a thermometer, and a reflux condenser. The VO and formic acid were mixed and subsequently H_2O_2 was added dropwise. Then, the reaction continued for

4 h; when the product was cooled down, it was filtered and washed repetitively with distilled water, and finally dried in an oven overnight.

Acrylated epoxidized linseed oil (AELO) was prepared from ELO using acrylic acid as ring opening agent, triethylamine (TEA) as a catalyst and hydroquinone as a free radical inhibitor. In short, a mixture of ELO and acrylic acid was heated in the presence of TEA for about one day until an acid value of ~8 mg KOH/g was attained. Unreacted acrylic acid and TEA were eliminated by extraction. The esterification extent was determined as 88% measuring the acid values at the start and finish of the reaction. The schematic representation of the synthesis of AELO is presented in Scheme 2 [46].

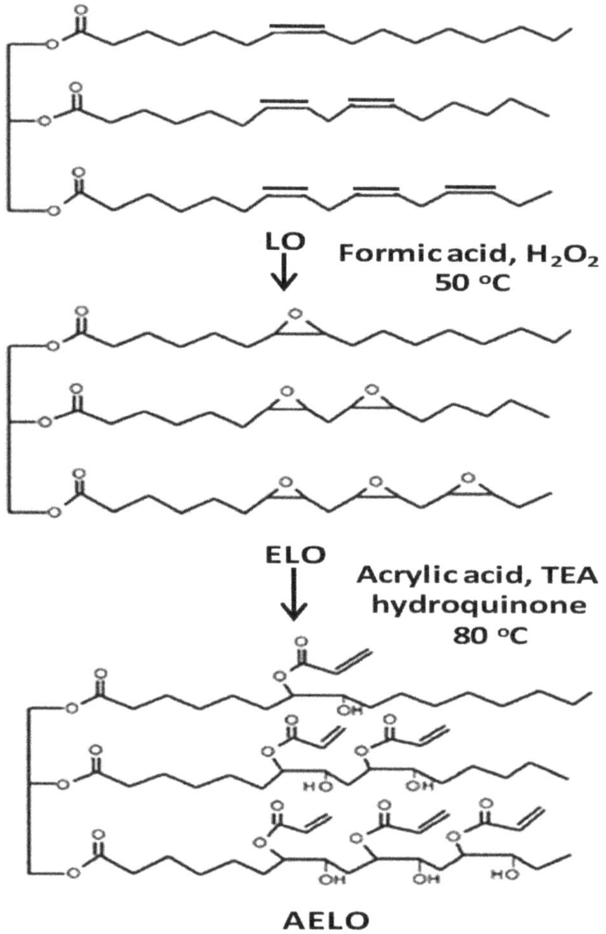

Scheme 2. Representation of the synthesis of AELO resin. Reprinted from [46], permission from the Royal Society of Chemistry.

AELO resin was then crosslinked with trimethylolpropane trimethacrylate monomer using benzophenone as photoinitiator, 2-dimethylaminoethanolas activator, and TiO_2 nanoparticles as fillers, and subjected to UV irradiation to obtain the cured nanocomposites, with nanoparticle loadings of 1.0, 2.5, 5.0, and 7.5 wt %.

2.2. Synthesis of Crosslinked Castor Oil (CO)/Chitosan-Modified ZnO Nanoparticles (CS-ZnO NPs)

Firstly, CS-modified ZnO NPs were synthesized by dissolving a small amount of ZnO nanopowder in an acetic acid solution. Then, the same amount of CS was added and the mixture was ultrasonicated, followed by drop-wise addition of NaOH until a pH of 10 was reached. The resulting product was heated, filtered, washed with distilled water, and dried in an oven.

The nanocomposites were then fabricated by solution mixing and casting process. Briefly, CO and the desired amounts of CS-ZnO NPs to obtain loadings of 1.0, 2.5, 5.0, and 7.5 wt %, were stirred, and then hexamethylene diisocyanate (HDI) and glutaraldehyde (GLA) were added as crosslinking agents to react with the CO in the presence of stannous octoate as catalyst. The mixture was heated and then poured onto Teflon plates. Curing was performed under vacuum in an oven. The schematic representation of the synthesis of CO/CS-ZnO NPs nanocomposites is shown in Scheme 3 [18].

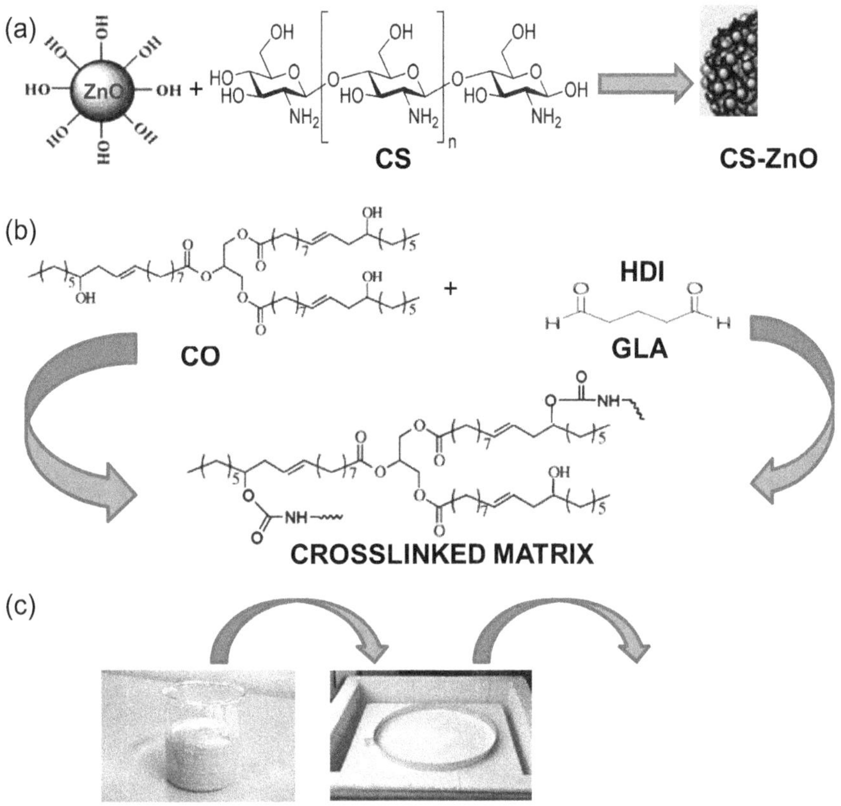

Scheme 3. Representation of the synthesis of CO/CS-ZnO NPs nanocomposites. (**a**) CS-ZnO NPs; (**b**) CO, HDI, GLA, and the crosslinked matrix; (**c**) illustration of the film casting process. Reprinted from [18] permission from the American Chemical Society.

2.3. Synthesis of Epoxidized Soybean Oil (ESO)/ZnO Nanocomposites

Initially, the VO and formic acid were mixed in a four-necked vessel, and H_2O_2 was added dropwise to start the epoxidation. The reaction continued for a few hours, and then the product, the epoxidized soybean oil (ESO) was cooled down, washed and dried overnight in an oven. ESO was mixed under stirring with different amounts of ZnO nanoparticles to yield nanocomposites containing 1.0, 3.0, 5.0, and 7.0 wt % loading in the presence of 4-dimethylaminopyridine as catalyst. The nanocomposites were degassed, decanted into silicone moulds, and cured in a vacuum oven [17].

2.4. Synthesis of Geranium-Derived Oil (Ge)/ZnO Nanocomposites

Nanocomposites were synthesized via plasma polymerization under vacuum in a cylindrical glass tube equipped with two external Cu rings that acted as electrodes connected to a generator. The glow discharge was created by a radio frequency generator. Firstly, the pressure was reduced and subsequently the monomer gas was brought into the tube until the pressure attained a stable value. The tube had an external ceramic fiber heater, where zinc acetylacetonate hydrate powder, the Zn precursor, was placed in a ceramic boat and heated. The vapor of the precursor entered the manufacture zone, ZnO nanoparticles were formed in the gas phase and then embedded into the geranium oil (Ge) matrix during the in situ polymerization. Two nanocomposites were prepared, with input radio frequency powers of 10 W and 50 W, which resulted in NP loadings of 0.79% and 1.57%, respectively [47].

2.5. Synthesis of Sunflower Oil Derived Hyperbranched Polyurethane (HBPU)/Fe₃O₄ Nanocomposites

The sunflower oil derived hyperbranched polyurethane (HBPU) was prepared in a two step process [48]: firstly, toluene diisocyanate, butanediol and polycaprolactone were mixed under an inert atmosphere. Secondly, pentaerythritol (a branching tetra-functional) was added with the monoglyceride of sunflower oil, and the reaction continued for a few hours. Separately, Fe_3O_4 NPs were prepared by a coprecipitation method: $FeCl_2$ and $FeCl_3$ were stirred in water under a nitrogen atmosphere followed by addition of NH_4 solution until a basic pH was attained.

Nanocomposites with different amounts of Fe_3O_4 (0, 5, 10, and 15 wt %) were prepared by solution casting. The NPs were dispersed in N,N-dimethyl acetamide by ultrasonication and added to the VO-derived polyurethane in the second stage of the polymerization reaction. The mixtures were stirred and then cooled down, followed by casting on inert substrates and curing.

A very similar approach was used to prepared nanocomposites based on the same matrix filled with Fe_3O_4 nanoparticles coated with multiwalled carbon nanotubes (MWCNTs), which were prepared in a single stage reaction (Scheme 4) [49].

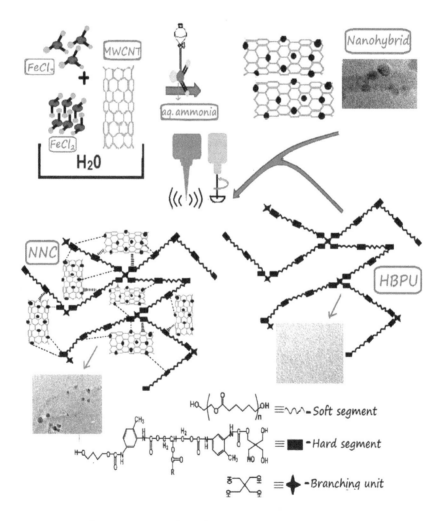

Scheme 4. Representation of preparation of Fe_3O_4–MWCNT hybrids and the corresponding sunflower oil derived polyurethane nanocomposites. Reprinted from [49] permission from the Royal Society of Chemistry.

Initially, $FeCl_2 \cdot 4H_2O$ and $FeCl_3$ were dispersed in water under an inert atmosphere. Separately, the required amount of MWCNTs was also dispersed in water followed by ultrasonication. The dispersed MWCNTs were then added to the first dispersion and the mixture was stirred again. Finally, aqueous NH_4 solution was added dropwise to the mixture until a basic pH was attained.

2.6. Synthesis of Linseed Oil (LO) Derived Polyol/CuO Nanocomposites

The linseed oil (LO) polyol was synthesized by mixing the LO, acetic acid, and H_2O_2 under mechanical stirring. Then, the temperature was raised for a few hours, during which epoxidation, and hydration reactions took place. The resulting polyol was successively washed with sodium bicarbonate aqueous solution, distilled water, and sodium chloride aqueous solution, and finally dried. The nanocomposites were prepared in situ via a 'solventless one-pot' chemical reaction. The polyol was then mixed with phthalic anhydride under stirring, followed by addition of different amounts of copper acetate. The reaction mixture was heated and refluxed. Three nanocomposites with CuO contents of 0.04, 0.05, and 0.06 mol were prepared [50].

3. Morphology of Vegetable Oil-Based Thermoset Polymers with Metal Oxide Nanoparticles

The morphology of the nanocomposites plays a key role in determining their antibacterial activity, as will be discussed in a following section. Hence, it is interesting to analyze the size, shape, etc., of the developed nanocomposites.

3.1. Linseed Oil-Based Nanocomposites

The surface of AELO/TiO_2 nanocomposites was investigated by scanning electron microscopy (SEM), and a typical illustration of the sample with 7.5 wt % loading is shown in Figure 2. The image shows a great number of spherical nanoparticles (bright spots) with an average diameter of 40 nm arbitrarily dispersed within the epoxidized VO, and also a few small clusters comprising several NPs. The interactions between the OH groups on the TiO_2 surface and the polar groups of the AELO avoid NP agglomeration, and no particle surface treatments or interfacial modifiers are required for the manufacturing of these biomaterials, making them cheaper and more environmentally friendly. Figure 2 also shows the energy dispersive X-ray (EDX) spectrum of the composite, which corroborates that the sample contains Ti, C, N, and O atoms.

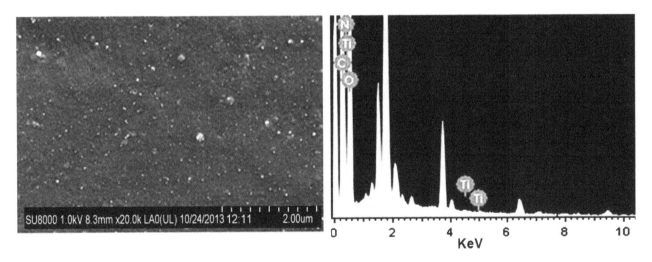

Figure 2. SEM micrograph (left) and EDX (right) of AELO/TiO_2 (7.5 wt %) nanocomposite. Reprinted from [46] permission from the Royal Society of Chemistry.

Transmission electron microscopy (TEM) micrographs of LO derived polyol/CuO nanocomposites reveal the presence of spherical nanosized NPs with diameters ranging between 50 and 60 nm Figure 3.

With increasing NP loading, the nanocomposites become denser, hence with higher refractive index and viscosity.

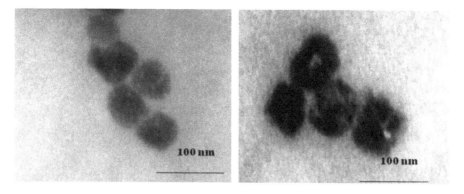

Figure 3. TEM micrographs of an epoxidized LO/CuO nanocomposite. Reprinted from [50] permission from Elsevier.

3.2. Castor Oil-Based Nanocomposites

SEM images of ZnO NPs (Figure 4a) revealed their quasi-spherical shape, with diameters ranging from 40 to 180 nm. Furthermore, due to the big surface area and high surface energy of the NPs, some agglomeration took place, and many small clusters can be observed. Upon functionalization with chitosan, their surface turned out to be more porous, and their average size raised to 180 nm (Figure 4b). Additionally, the agglomeration decreased, since the interaction between chitosan and the NPs reduced the attractive forces between nanoparticles [18].

Figure 4. SEM images of: (**a**) as synthesized ZnO nanoparticles; (**b**) chitosan-modified ZnO nanoparticles; (**c**) cured castor oil/chitosan-modified ZnO nanocomposite (7.5 wt %). Reprinted from [18] permission from the American Chemical Society.

On the other hand, the nanocomposites were transparent and flexible, as shown in Figure 4c. The CS-ZnO NPs can be observed as bright spheres homogeneously and randomly dispersed within the CO matrix. The H-bonding interactions between the urethane moieties of the cured CO and the OH moieties of the CS-NPs avoid aggregation and improve the matrix-NP compatibility. The porosity of the nanocomposites was found to increase with increasing NP loading, from 10% for neat CO to almost 100% at 7.5 wt % NP content. This augment in porosity is valuable for wound healing applications, since would enable the absorption of higher volumes of wound exudates and promote the distribution of nutrients to the cells.

3.3. Soybean Oil-Based Nanocomposites

The epoxidized SO showed a sponge-like morphology (Figure 5a), with many round pores of about 300 nm size [17]. However, in the nanocomposites filled with ZnO NPs (Figure 5b), the pore dimension was about half that of the matrix, while the surface roughness increased. The spherical NPs, with an average size of 80 nm, which increased slightly with increasing loading, become visible as light spots in the images, and are very well distributed. The interactions between the polar groups of the SO and the OH moieties of the ZnO NPs avert aggregation without the necessity of particle surface treatments or compatibilizers.

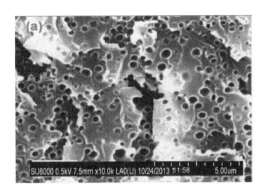 

Figure 5. SEM images of ESO (**a**) and ESO/ZnO (5.0 wt %) nanocomposite (**b**). Reprinted from [17] permission from the American Chemical Society.

3.4. Geranium Oil-Based Nanocomposites

SEM images were attained to assess the size and morphology of the synthesized ZnO NPs in the polymer matrix [47]. Ball-like NPs randomly and homogenously distributed were found in all the composites. Nonetheless, the NP size moderately increased with increasing power deposition, from about 60 nm at 10 W to around 80 nm at 50 W, indicating that a higher amount of NPs can be incorporated into the polymer matrix at higher deposition powers.

The surface morphology of Ge and the nanocomposites was also examined via atomic force microscopy (AFM). Due to the soft Ge surface, the tapping mode was used. The surface of neat Ge was flat, soft, and uniform (Figure 6a), with a mean particle roughness of 0.25 nm. The average roughness increased with the incorporation of ZnO (Figure 6b), and the nanocomposites with NP loadings of 0.79% and 1.57%, which correspond to frequency powers during the plasma polymerization of 10 and 50 W, showed average roughnesses of 33.7 and 37.2 nm, respectively. The nanocomposites have a more porous surface with a random distribution of the NPs, which are present in the form of protusions.

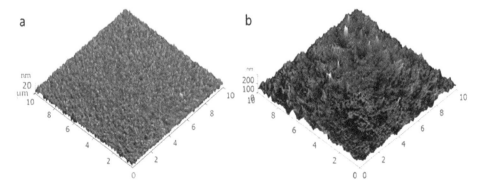

Figure 6. AFM images of neat Ge (**a**) and the corresponding nanocomposite with 1.57% ZnO (Zn/Ge)(**b**). Reprinted from [47] permission from Springer Nature.

3.5. Sunflower Oil-Based Nanocomposites

According to SEM and TEM images, a very homogeneous dispersion of the Fe_3O_4 NPs and the Fe_3O_4–MWCNT hybrids within the VO matrix was achieved during the in situ polymerization. The MWCNTs were relatively small and short (~10 nm diameter, length < 5 μm, Figure 7d,e). Fe_3O_4 NPs with diameters ranging from 9 to 12 nm decorated the wall of the MWCNTs (Figure 7f,g), which was attained via debundling of the MWCNTs upon sonication and mechanical shearing. Besides, the presence of dangling bonds on the MWCNT aid the NP nucleation on the nanotubes surface (Figure 7b,h,j). The Π–Π interactions between the aromatic rings of toluene diisocyanate and those of the MWCNTs and the interaction between the –C=O and –N–H of the polyol resin with the OH groups of the NPs led to strong interfacial adhesion between them, Figure 7c,d [49].

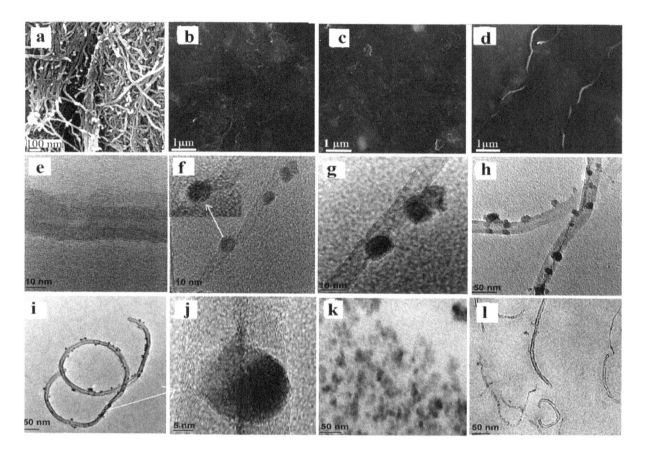

Figure 7. SEM images of Fe_3O_4–MWCNT nanohybrids (**a**), NNC (**b**), FNC (**c**), and MNC (**d**); TEM images of MWCNT (**e**), Fe_3O_4–MWCNT nanohybrid (**f** and **g**), NNC (**h–j**); FNC (**k**) and MNC (**l**). Nomenclature: carboxyl functionalized MWCNT-based nanocomposites (MNC); Fe_3O_4-based nanocomposites (FNC) and Fe_3O_4–MWCNT-based nanocomposites (NNC). Reprinted from [49] permission from the Royal Society of Chemistry.

4. Antimicrobial Activity of Vegetable Oil-Based Thermoset Polymers with Metal Oxide Nanoparticles

4.1. Antimicrobial Effect of Zinc Oxide-Reinforced Nanocomposites

The antibacterial activity of neat Ge and the corresponding nanocomposites with ZnO nanoparticles was investigated against two well-known pathogens in hospitals and implantable devices: Gram-negative *E. coli* and Gram-positive *S. aureus* bacteria, using the live/dead staining method [47]. Viability was estimated as the ratio of viable bacteria to the total number of bacteria attached to the sample surface, and the results are shown in Figure 8 [47]. About 80% of *S. aureus* were alive on the control, whilst the viability on the neat Ge prepared via plasma polymerization with radio frequency powers of 10 and 50 W were 53% and 50%, and those of the corresponding nanocomposites with ZnO were 31% and 42%, respectively (Figure 8a). Likewise, 81% of *E. coli* were viable on the control surface, whereas the viability on the neat oil were 60% and 76%, and for the composites were 33% and 44%, respectively.

The antibacterial action of VO-based polymers is related to their surface chemistry and nanoscale topography. The functional groups of the VO (ie. hydroxyl, carboxylic, methyl) are reported to disturb the microbial growth and avoid a good anchoring of microbial cells [51]. However, pristine VO showed low antibacterial activity versus both Gram-positive and Gram-negative bacteria. A plausible explanation could be that plasma treatment would have worsened the antimicrobial activity of the neat oil due to the crosslinking of many molecules, making them incapable of interacting with the cell membranes [52].

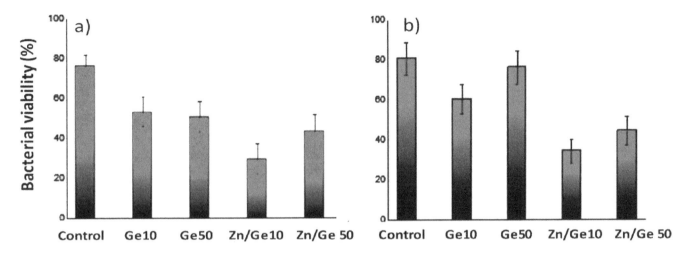

Figure 8. *S. aureus* (**a**) and *E. coli* (**b**) viability against neat geranium oil (Ge) and geranium oil/ZnO nanocomposites (Zn/Ge) prepared via plasma polymerization with radio frequency powers of 10 W (Ge 10 and Zn/Ge 10) and 50 W (Ge 50 and Zn/Ge 50). Adapted from [47] permission from Springer Nature.

In contrast, the Ge-based ZnO nanocomposites revealed enhanced bactericidal performance, corroborating that ZnO NPs were straightforwardly implicated in the inhibition of the pathogens. The antimicrobial mechanisms of MO-NPs, as described earlier, differ from those of the pristine polymers. Despite both being capable of anchoring to the outer microorganism surface—thus modifying the permeability of cell walls—NPs, due to their extremely small size, are also able to enter the cell, thus gathering in the cytoplasm, disrupting cellular activities and provoking membrane disruption. Besides, upon interaction with the bacteria cells, ZnO NPs can produce ROS, as detailed above; which can deteriorate the proteins, peptidoglycan, ribosomes, and DNA of the bacteria, causing inhibition of enzymatic activities and amino acid production, thus leading to cell lysis. This accounts for the improved performance observed for the nanocomposites compared to the neat Ge. In addition, it is possible the existence of synergistic effects of both composite components on inhibiting bacteria viability, as reported for nanocomposites reinforced with graphene and Ag NPs [53] or ZnO and Ag NPs [54], which were more efficient versus pathogens than the individual components alone.

On the other hand, the bactericide action of Zn/Ge 10 W was greater compared to Zn/Ge 50 W, probably related to the different NP size, since the smaller power during the plasma polymerization led to smaller particles, as discussed in Section 3.4, which are more likely to penetrate the bacteria cell. Furthermore, the lower resistance of *S. aureus* against ZnO NPs compared to *E. coli* is consistent with previous reports [55], ascribed to the outer membrane layer external to the peptidoglycan cell wall in the Gram-negative bacteria that can provide better resistance to ZnO penetration. Nonetheless, there is no agreement in the literature on this point, since ZnO NPs are synthesized with different dimensions, morphologies, surface modifications, surface defects, etc., that strongly condition their antibacterial action.

CO/CS-ZnO nanocomposites were tested against Gram-positive *S. aureus* and *M. luteus* and Gram-negative *E. coli* with and without UV illumination (Figure 9) [18]. Experiments were performed following the ISO 22196:2007 standard. Samples were sterilized in an autoclave and then submerged in a nutrient broth of ~2.0×106 colony forming units (CFU)/mL. After incubation at 37 °C for 24 h, the number of CFU per sample was calculated manually. The antibacterial activity was estimated as: log(viable cell count in the control/viable cell count in the composite), where the control was a flask containing bacteria and no sample. It has been stipulated that efficient antibacterial activity should be higher than 2.

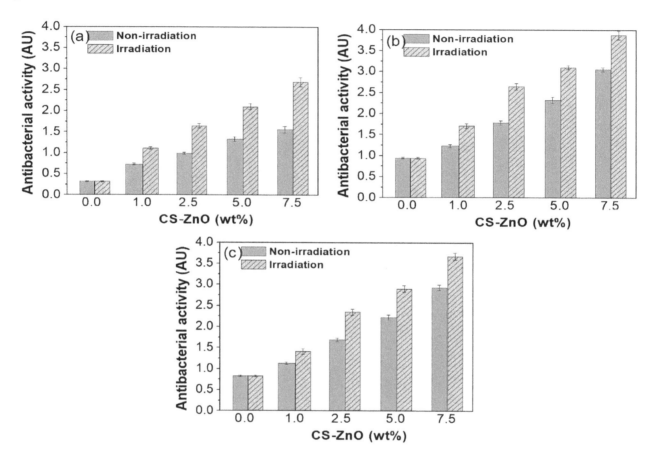

Figure 9. Antibacterial activity of CO/CS-ZnO nanocomposites versus (**a**) *E. coli*, (**b**) *S. aureus*, and (**c**) *M. luteus*. Solid and dashed bars correspond to experiments performed without and with UV light irradiation, respectively. Reprinted from [18] permission from the American Chemical Society.

Cured CO shows some antimicrobial action versus the three bacteria tested, being the activity stronger against Gram-positive ones both with and without UV light. A number of works have confirmed that fatty acids can restrain the bacterial growth [56], and even their mechanism is not well understood up to date, it appears to be connected with the electron transport disruption and oxidative phosphorylation at the cell membrane. Moreover, their bactericide action could arise from enzymatic activity inhibition and nutrient uptake or lysis of bacterial cells. More importantly, the antimicrobial activity increases as the CS-ZnO loading rises under both conditions tested. Nonetheless, while *E. coli* activity was restrained for NP contents ≥ 5.0 wt % under UV irradiation, *S. aureus* and *M. luteus* were inhibited by all the nanocomposites except for that with 1.0 wt % loading. This is consistent with the results previously discussed for Ge/Zn nanocomposites [47], corroborating that the antibacterial action is strongly influenced by the structural and chemical composition of the bacterial cell. Interestingly, CO/ZnO composites also show antibacterial activity without UV light, being only slightly toxic to *E. coli* while more effective against *S. aureus* and *M. luteus* (Figure 9b,c). Despite the generation of ROS is supposed to be the key reason for the antibacterial activity of ZnO NPs in the presence of UV light, the cause of their bactericidal effect without irradiation is still unknown; it is not likely related to ROS production albeit dependent on the ZnO attachment to bacterial cell walls, and increasing concentrations of Zn^{2+} ions in the bacterial cytoplasm due to local dissolution of the attached ZnO.

Noticeably, the biocide action found for these composites is significantly higher than that found for Ge/Zn nanocomposites prepared via plasma polymerization [47], indicating that both the functionalization of the nanoparticles and the nanocomposite synthesis process can play a key role in the biocide action. In particular, the biological activity of ZnO has been reported to be strongly dependent on the surface functional groups. Thus, Betancourt-Galindo et al. [57] modified the ZnO NPs by di-functional alcohol, and the resulting NPs showed improved performance, and were used in medical devices. In addition, it should be highlighted the exceptional antimicrobial properties of CS [58]. This polysaccharide is successful against both Gram-negative and Gram-positive microorganisms, while its efficiency is conditioned by factors such as its molecular weight, degree of deacetylation, and concentration. It is supposed to be able to go through the bacterial cell wall by means of pervasion and fabrication of a polymer membrane on the cell wall surface. Furthermore, its positively charged amino groups are prone to interact with the negatively charged bacterial membranes, resulting in protein leakage and affecting the phospholipid bilayer structure, thereby altering its permeability [59]. CS hybrid materials incorporating metal and MO-NPs have been prepared with outstanding properties derived from synergistic effects [60–64]. It has been reported that CS or modified CS combined with Ag leads to hydrogels with better antimicrobial activity [63]. Besides, mixtures of CS, polyvinylpyrrolidone and nanosized TiO_2 [61] or Ag_2O [62] exhibit superb wound healing and antimicrobial characteristics. ZnO–CS complexes with antimicrobial and antibiofilm properties have been also prepared by nano spray drying and precipitation methods [64].

The antibacterial activity of ESO/ZnO nanocomposites was also tested against *E. coli* and *S. aureus* with and without UV light (Figure 10) [17], following the same protocol as indicated earlier [18]. Neat ESO exhibits a little antimicrobial effect, ~0.1 and 0.4 versus *E. coli* and *S. aureus* respectively, under both conditions examined, the effect being smaller than that found for CO [18]. The differences are likely related to their chemical structure, since CO comprises a large number of reactive hydroxyl groups that improve the antibacterial activity. This is consistent with a previous study on ESO modified with quaternary ammonium salts and hydroxyl groups onto the backbone [65], which was reacted with different diisocyanate monomers to prepare polyurethane coatings that showed considerably enhanced antibacterial activity (about 95% bacterial reduction). The positive charges of the quaternary ammonium salts can interact with negatively charged surfaces of both Gram-positive and Gram-negative bacteria via electrostatic forces, which promote NP penetration Thus, the antimicrobial activity of VOs is confirmed to be strongly dependent on their surface functional groups, as mentioned earlier.

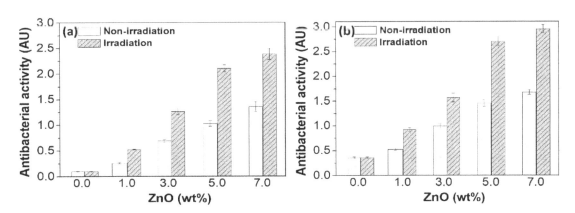

Figure 10. Antibacterial activity of ESO/ZnO nanocomposites against (**a**) *E. coli* and (**b**) *S. aureus*. Solid and dashed bars as indicated in Figure 9. Reprinted from [17] permission from the American Chemical Society.

On the other hand, as found for the other nanocomposites reinforced with ZnO NPs [18,47], the antibacterial action increases upon rising NP loading both in the presence and the absence of UV light. This trend arises from the uniform nanoparticle dispersion within the VO matrix. Thus, the higher the NP content, the higher the effective surface area available for interaction with the bacteria cell. Again, a more efficient bactericide action is found versus Gram-positive bacteria, related to the differences in structure and chemical composition of the cell walls. The thicker and multipart structure of *E. coli* should minimize the harm from oxidation radicals. It should be noted that for both bacteria tested, the antibacterial action of ESO/ZnO nanocomposites is weaker than that found for CO/CS-ZnO ones [18], likely due to the functionalization of the NPs, as discussed earlier and the synergistic effect of CS and ZnO. Improved bactericide action has also been reported upon functionalization of other nanomaterials like graphene oxide [66] or boron nitride nanotubes [67] with biocompatible polymers such as polyethylene glycol.

4.2. Antimicrobial Effect of Titanium Oxide-Reinforced Nanocomposites

Following the same protocol described above, the antibacterial action of AELO/ZnO nanocomposites was also investigated versus *S. aureus* and *E. coli* under both UV light and dark conditions (Figure 11) [46]. For both bacteria, AELO only exerts a minor antimicrobial action, about 0.15 and 0.3 respectively, the effect being alike under both conditions, and slight smaller than that found for ESO [17] and CO-based [18] nanocomposites. This is in agreement with studies on the bactericidal activity of VOs, which found a strong dependence on their content of phenolic compounds, together with their unsaturation degree and chain length: the longer the chain, the stronger the bacterial action is [68].

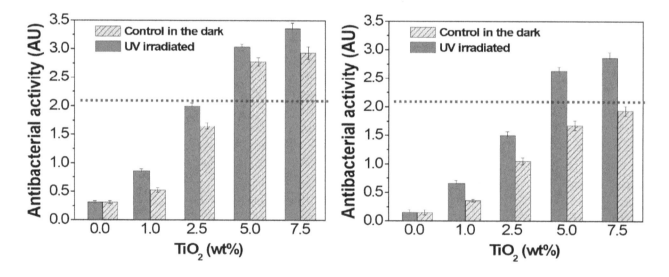

Figure 11. Antibacterial action of AELO/TiO$_2$ nanocomposites versus *S. aureus* (left) and *E. coli* (right). The dotted line shows the limit for effective antibacterial action. Reprinted from [46] permission from the Royal Society of Chemistry.

As observed for nanocomposites reinforced with ZnO NPs, the antibacterial action rises as the NP loading rises, both with and without UV light. It is expected that higher TiO$_2$ loadings will lead to more NPs in contact with the bacteria, and in consequence, produce more toxicity. Furthermore, the state of NP dispersion also conditions the bactericide action: the more homogenous the NP dispersion, the better the antibacterial property. Thus, the nanocomposite with 7.5 wt % TiO$_2$ presents similar action to that with 5.0 wt % due to the presence of small clusters as corroborated by SEM (Figure 2). Again, stronger bactericide effect is found against Gram-positive bacteria, as also reported for TiO$_2$/poly(lactic-co-glycolic acid) PLGA composites [69]. The antibacterial effect could be attributed to inactivation of cellular enzyme and DNA by the TiO$_2$ NPs causing small pores in the bacterial cell wall that result in increased permeability and cell death. Furthermore, oxidative damage provoked by ROS produced via redox reactions between adsorbed species (i.e., water, oxygen) and electrons and holes generated on irradiation of TiO_2 with UV light have also occurred. Reaction of the holes in the valence band with H_2O or hydroxide ions on the surface results in the formation of •OH radicals, whilst electrons in the conduction band react with O= to yield superoxide ions •O^{2-}. Both hydroxyl and superoxide ions can react with the phospholipid components of the bacteria cell membrane, thereby causing the deterioration of the membrane and finally cell lyses and bacteria inactivation [70].

Experimental data confirm that TiO_2-reinforced composites possess antibacterial activity under dark conditions, pointing towards other mechanism of bacterial inactivation besides the photocatalytic process. These could include NP-bacteria surface physicochemical interactions that injure the cells; redox reactions at the NP–cell interface concerning reduction of Ti(IV) to Ti(III) that provoke oxidative degradation of the cell membrane; internalization of the NPs in the cell and bacteria damage from the interior; bacterial phagocytosis by the NPs resulting in cell death [71]. Although the precise mechanism is not known up to date, the bacteria-TiO_2 contact appears to be critical for attaining strong bactericide effect.

4.3. Antimicrobial Effect of Copper Oxide-Reinforced Nanocomposites

To determine the bactericide effect of LO/CuO nanocomposites, E. coli and S. aureus bacteria were cultured for 24 h at 37 °C. At certain times aliquots were taken and growth was followed turbidometrically at 595 nm with a spectrophotometer [50]. Results are shown in Figure 12.

The growth curves indicated three regions: initial phase, active lag phase and stationary phase. The absorbance for the control (only bacteria) revealed that E. coli and S. aureus attained the stationary phase after 14 and 12 h, respectively. A similar trend is found for the nanocomposites, although the lag phase was extended by an average of 6 h in the case of E. coli, while for S. aureus it remained almost the same. More importantly, in the case of E. coli, about 17, 24, and 62% growth was inhibited compared to the control for nanocomposites with 0.04, 0.05, and 0.06% CuO, respectively, showing and increase in bactericide action with increasing NP loading. In contrast, for S. aureus, the growth inhibition was in the range of 60–55% for the three nanocomposites.

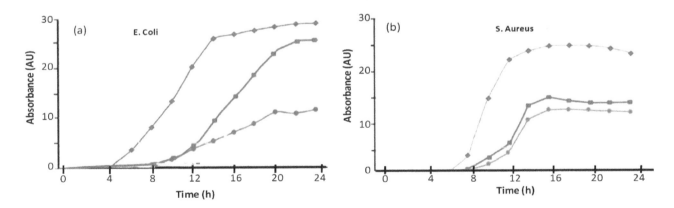

Figure 12. Absorbance versus time showing the effect of LO/CuO nanocomposites on the growth of E. coli (a) and S. aureus (b). – control; – 0.04 % CuO; – 0.05 % CuO; – and –0.06 % CuO; Adapted from [50] permission from Elsevier.

The antibacterial action of LO derived polyol-CuO nanocomposites seems to arise from the disruption of membrane integrity [72]. Electrostatic interactions can initially take place between positive residue in the composites and negative charges onto the bacteria surface, provoking the adhesion of the nanocomposites to the surface. Then, the polymeric chains could phagocytize the bacteria, and simultaneously the CuO would bind to the lipid and peptidoglycan layers, causing the cell death. Thus, as the NP content increased, the bactericide action improved, though in the case of S. aureus the improvement with increasing loading was slightly significant, likely because a high efficiency was already attained for the lowest NP content. Furthermore, this behavior could be the result of their tendency to form clusters like bunches of grapes, which impedes the whole penetration of the nanocomposites to the core of the cluster [50]. In addition, ROS formation can take place by released Cu^{2+}; in the presence of either superoxide or other reducing agents such as ascorbic acid, Cu^{2+} can be reduced to Cu^+ catalyzing the formation of •OH radicals from H_2O_2 via the Haber–Weiss reactions [41]. •OH is the most powerful oxidizing radical reacting with almost every biomolecule. It can start oxidative damage by abstracting the hydrogen both from an amino-bearing carbon to form a carbon centered protein radical and from an unsaturated fatty acid to form a lipid radical. Moreover, it is expected that in addition to the effect of CuO, the hydrophobic nature of LO and its fatty acid chains also boost the bactericide action.

4.4. Antimicrobial Effect of Iron Oxide-Reinforced Nanocomposites

The antibacterial action of sunflower oil-derived HBPU/Fe_3O_4 nanocomposites was tested against *S. aureus* and *K. pneumonia* bacteria via the agar well diffusion method [73]. The microorganisms were incubated for 24 h at 37 °C. Upon incubation, the inhibition zone for the different samples was calculated, using ampicillin as control, and results are shown in Figure 13 [48].

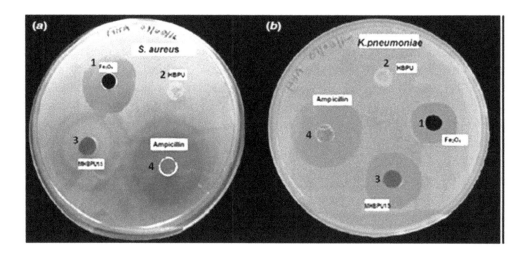

Figure 13. Inhibition zone showing the antibacterial activity against of *S. aureus* (**a**) and *K. pneumoniae* (**b**), respectively, of Fe_3O_4 (**1**), sunflower oil derived-polyurethane (**2**), the corresponding nanocomposite with 15 wt % loading (**3**), and the control (**4**). Adapted from [48] permission from IOP publishing.

The neat polyurethane derived from the VO did not have antibacterial activity while Fe_3O_4 NPs alone showed inhibition zones of 13 and 11 mm versus Gram-positive *S. aureus* and Gram-negative *K. pneumoniae*, respectively. Surprisingly, the nanocomposite with 15 wt % NP displayed a zone of inhibition of 15 and 13 mm for the indicated bacteria. This corresponds to normalized widths of the antimicrobial "halo" (nw_{halo}) of 0.43 and 0.42, respectively, calculated from the diameter of the inhibition zone (d_{iz}) and the disk diameter (d) as: $[(d_{iz} - d)/2]/d$ [74]. These results indicate that the NPs provided bactericide action to the VO-based polymer, and this activity increased with the rise in NP loading, as reported for other nanocomposites based on VOs discussed earlier. Nonetheless, the bactericide action found for this nanocomposite seems smaller compared to those based on semiconducting MO-NPs: ZnO [17,18,47], TiO_2 [46], or CuO [50], which could be related to the fact that in the semiconductors electron–hole pairs are formed, and the holes can split water molecules and cause ROS formation. In the case of magnetite NPs, the mechanism of bactericide action should be different (i.e., via Fenton-like and Haber–Weiss reactions, as discussed earlier), and albeit ROS could also be generated (mainly H_2O_2 when Fe^{2+} responded to oxygen that interacts with the outer bilayer of bacteria, thus entering the cell membrane and causing bacterial disruption), these would be less harmful or would be produced in a less efficient way. Furthermore, combined mechanism can also take place. Thus, the ferrous irons can subsequently react with the produced H_2O_2 through Fenton reaction, thus leading to hydroxyl radicals that damage the bacteria cell wall.

In contrast, a higher bactericide action was found for similar nanocomposites reinforced with Fe_3O_4–MWCNT nanohybrids (NNC in Figure 14 [49]), which showed an inhibition zone of 22 mm against *K. pneumoniae*, and close to 20 mm for *S. aureus* (nw_{halo} values of 0.39 and 0.4, respectively), likely arising from the synergistic effect of the NPs and the MWCNTs on the biocide effect. The antibacterial activity of MWCNTs has been ascribed to several mechanisms [75]: disruption of the membrane integrity by strong electrostatic forces between bacterial outer surface and MWCNTs, leading to oxidation of the membrane; ROS generation that induces destruction of the bacterial plasmid DNA; impurity components (i.e., metallic nanoparticles, catalysts, suspension) that are introduced into the MWCNT structure during the synthesis process; bactericidal oxidative stress. Furthermore, the toxicity of MWCNTs is highly influenced by several factors such as diameter, length, residual catalysts, electronic structure, surface functional groups, and surface chemistry [76]. In particular, the tube length is critical for the interactions with the cell membrane: the shorter the tube, the more chances there are for interaction between open ends of the nanotubes and the bacteria, leading to extra cell membrane damage, thus the bactericidal performance is stronger. The tube diameter also plays a key role in the bacterial inactivation process. Smaller diameters can damage the cell membrane via cell-surface interactions, while larger diameters (~15–30 nm) mostly interact with bacteria by their side walls [77].

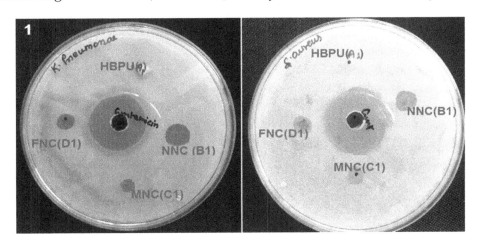

Figure 14. Inhibition zone showing the antibacterial activity against *K. pneumonia* and *S. aureus* for the neat polyurethane derived from the oil (HBPU), NNC, FNC, and MNC. The nomenclature is indicated in Figure 7. Reprinted from [49] from the Royal Society of Chemistry.

In the study by Das et al. [49], the MWCNTs were relatively small and short (~10 nm diameter, length < 5 μm, Figure 7d,e), which could account for their improved antibacterial activity. Furthermore, due to the MWCNT shape, they have less bactericidal action toward rod-shaped bacteria (*K. pneumonia*) when compared to spherical ones (*S. aureus*). Overall, the nanohybrids showed significantly improved performance compared to the nanocomposites reinforced only with Fe_3O_4 or MWCNTs.

The antibacterial activities of VO/MO-NPs nanocomposites are summarized in Table 3. Systematically, stronger bactericide effect is found against Gram-positive spherical bacteria compared to Gram-negative rod-shaped ones, ascribed to the structural and chemical compositional differences of their cell walls. For similar NP loadings, stronger antibacterial activity is obtained for smaller NPs, which are more prone to penetrate into the bacteria cell. Nonetheless, the two most important parameters determining the bactericide action appear to be the NP concentration and its level of surface functionalization. Systematically, the higher the NP loading, the more intense the NP-bacteria interactions are, hence the stronger the bactericide effect is. However, in some cases a stabilization of the antibacterial activity is envisaged for NP loadings >7.5 wt%, since NP agglomeration can occur, thus reducing the NP specific surface area, which results in lower NP-bacterial interfacial contact area.

NP surface functionalization, in particular with positively charged groups, improves the antibacterial efficiency, since cationic groups are susceptible to interact with the negatively charged bacterial membranes via electrostatic forces, resulting in protein leakage and damaging the phospholipid bilayer structure. Besides, synergistic effects typically take place between the MO-NPs and their surface functional moieties on improving antibacterial activity. Regarding the NP shape, slight influence is predicted, although most of the studies were carried out with spherical or quasi-spherical NPs, hence it is difficult to reach a clear conclusion on this issue. The composition of the VO matrix also influences the antibacterial activity of the nanocomposites: the higher its phenolic content, unsaturation degree, and chain length, the stronger the bacterial action is. Thus, composites based on CO, that comprises 89% of unsaturated ricinoleic acid (Table 2), display the highest antibacterial effectiveness. In contrast, those based on sunflower or geranium oil, with lower level of unsaturation, exert less bactericide action. This is consistent with the fact that neat CO showed bactericide action towards both Gram-positive and Gram-negative bacteria, while neat Ge or the HBPU derived from sunflower oil hardly showed bactericide action. Finally, the nanocomposite synthesis process can also play a key role in the biocide action. Thus, plasma polymerization methods can worsen the antimicrobial activity due to the high level of crosslinking between the molecules, which makes their interaction with the cell membranes more difficult.

Table 3. Antibacterial activity of VO/MO-NPs nanocomposites.

Nanocomposite Type (Processing)	Antibacterial Activity (AU)	Bacteria Strain	Average NP Size (nm)	NP Shape	NP Concentration (wt %)	REF
Zn/Ge 10 (PP)	31 [a]	S. aureus	60	Ball-like	0.79	47
Zn/Ge 10 (PP)	33 [a]	E. coli	60	Ball-like	0.79	47
Zn/Ge 50 (PP)	42 [a]	S. aureus	80	Ball-like	1.57	47
Zn/Ge 50 (PP)	44 [a]	E. coli	80	Ball-like	1.57	47
ESO/ZnO (SM + C)	0.30	E. coli	65	Spherical	1.00	17
ESO/ZnO (SM + C)	0.55	S. aureus	65	Spherical	1.00	17
ESO/ZnO (SM + C)	0.60	E. coli	73	Spherical	3.00	17
ESO/ZnO (SM + C)	0.98	S. aureus	73	Spherical	3.00	17
ESO/ZnO (SM + C)	1.12	E. coli	80	Spherical	5.00	17
ESO/ZnO (SM + C)	1.47	S. aureus	80	Spherical	5.00	17
ESO/ZnO (SM + C)	1.32	E. coli	92	Spherical	7.00	17
ESO/ZnO (SM + C)	1.68	S. aureus	92	Spherical	7.00	17
AELO/TiO$_2$ (IP + C)	0.41	E. coli	40	Spherical	1.00	46
AELO/TiO$_2$ (IP + C)	0.60	S. aureus	40	Spherical	1.00	46
AELO/TiO$_2$ (IP + C)	1.12	E. coli	42	Spherical	2.50	46
AELO/TiO$_2$ (IP + C)	1.62	S. aureus	42	Spherical	2.50	46
AELO/TiO$_2$ (IP + C)	1.73	E. coli	46	Spherical	5.00	46
AELO/TiO$_2$ (IP + C)	2.68	S. aureus	46	Spherical	5.00	46
AELO/TiO$_2$ (IP + C)	1.92	E. coli	64	Spherical	7.50	46
AELO/TiO$_2$ (IP + C)	2.81	S. aureus	64	Spherical	7.50	46
CO/CS-ZnO (SM + C)	0.75	E. coli	168	Quasi-spherical	1.00	18
CO/CS-ZnO (SM + C)	1.20	S. aureus	168	Quasi-spherical	1.00	18
CO/CS-ZnO (SM + C)	1.10	M. luteus	168	Quasi-spherical	1.00	18
CO/CS-ZnO (SM + C)	0.98	E. coli	177	Quasi-spherical	2.50	18
CO/CS-ZnO (SM + C)	1.80	S. aureus	177	Quasi-spherical	2.50	18
CO/CS-ZnO (SM + C)	1.71	M. luteus	177	Quasi-spherical	2.50	18
CO/CS-ZnO (SM + C)	1.32	E. coli	175	Quasi-spherical	5.00	18
CO/CS-ZnO (SM + C)	2.30	S. aureus	175	Quasi-spherical	5.00	18
CO/CS-ZnO (SM + C)	2.24	M. luteus	175	Quasi-spherical	5.00	18
CO/CS-ZnO (SM + C)	1.55	E. coli	180	Quasi-spherical	7.50	18
CO/CS-ZnO (SM + C)	3.05	S. aureus	180	Quasi-spherical	7.50	18
CO/CS-ZnO (SM + C)	2.98	M. luteus	180	Quasi-spherical	7.50	18
LO/CuO (SL)	17 [b]	E. coli	50	Quasi-spherical	0.04 [c]	50
LO/CuO (SL)	55 [b]	S. aureus	50	Quasi-spherical	0.04 [c]	50
LO/CuO (SL)	24 [b]	E. coli	56	Quasi-spherical	0.05 [c]	50
LO/CuO (SL)	57[b]	S. aureus	56	Quasi-spherical	0.05 [c]	50
LO/CuO (SL)	62 [b]	E. coli	59	Quasi-spherical	0.06 [c]	50
LO/CuO (SL)	60 [b]	S. aureus	59	Quasi-spherical	0.06 [c]	50
HBPU/Fe$_3$O$_4$ (SC + C)	0.43[d]	S. aureus	9	Spherical	15.0	48
HBPU/Fe$_3$O$_4$ (SC + C)	0.42 [d]	K. pneu-moniae	9	Spherical	15.0	48
HBPU/Fe$_3$O$_4$-CNT (SC + C)	0.40 [d]	S. aureus	11	Spherical	15.0	49
HBPU/Fe$_3$O$_4$-CNT (SC + C)	0.39 [d]	K. pneu-moniae	11	Spherical	15.0	49

[a] Percentage of cell viability; [b] Percentage of cell growth inhibition; [c] Concentration in mol; [d] Normalized width of the antimicrobial "halo"; PP: plasma polymerization; SM: solution mixing; IP: in situ polymerization; SC: solution casting; C: curing; SL: solventless one-pot reaction.

5. Conclusions

The incorporation of MO-NPs into VO-based thermosetting polymers is a versatile path to take advantage of their outstanding antimicrobial properties, leading to new bactericide nanocomposite materials with a wider range of applications. The antibacterial properties depend on a number of parameters, including the NP size, shape, level of functionalization and in particular, their concentration in the nanocomposite. It has been shown that the biocide action systematically increases with increasing NP loading, due to larger NP-bacteria interfacial contact area. Moreover, surface functionalization of the NPs or hybridization with other nanomaterials significantly improves their efficiency, in particular when this is carried out with substances that also exhibit antimicrobial activity (i.e., polymers like chitosan or MWCNTs). The main mechanism of action is likely the generation of ROS, which can disrupt cell membranes, damage DNA synthesis, and perturb metabolic paths, finally resulting in cell death. Furthermore, these NPs also exhibit biocide action under dark conditions, corroborating the existence of other modes of action like redox reactions at the NP–cell interface, bacterial phagocytosis, lipid peroxidation, etc. Although several examples related with MO-NPs embedded into VO-derived thermoset polymers have been reported, as discussed throughout this manuscript, additional research is required to support the progress of novel nanocomposites to be used in hospital equipment, in prostheses, or as antibacterial materials in public places. In this regard, novel synthesis approaches based either on commercial matrices or on VOs/MO-NP coatings that could be straightforwardly implemented at an industrial scale should be developed. Future research areas might include scaling up processes, optimization of the NP dispersion, use of novel MO-NPs with antimicrobial properties, different polymeric matrices derived from VOs and toxicity studies. Overall, more research and investment are needed to attain fully sustainable materials with antimicrobial activity as effective substitutes for the existing ones.

Acknowledgments: A.M. Diez-Pascual wishes to acknowledge the MINECO for a "Ramón y Cajal" Potsdoctoral Fellowship cofinanced by the EU. Financial support from the University of Alcalá via project reference CCG2018/EXP-011 is gratefully acknowledged. The aid of J.A. Luceño-Sanchez with the literature search is also acknowledged.

References

1.   Lligadas, G.; Ronda, J.C.; Galià, M.; Cádiz, V. Renewable polymeric materials from vegetable oils: A perspective. *Mater. Today* **2013**, *16*, 337–343. [CrossRef]

2.   Babu, R.P.; O'Connor, K.; Seeram, R. Current progress on bio-based polymers and their future trends. *Prog. Biomater.* **2013**, *2*, 8. [CrossRef] [PubMed]

3.   Raquez, J.-M.; Deléglise, M.; Lacrampe, M.-F.; Krawczak, P. Thermosetting (bio)materials derived from renewable resources: A critical review. *Prog. Polym. Sci.* **2010**, *35*, 487–509. [CrossRef]

4.   Gunstone, F. *Fatty Acid & Lipid Chemistry*; Blackie Academic & Professional: New York, NY, USA, 1996; pp. 1–252. ISBN 978-1-4613-6852-6.

5.   Belgacem, M.N.; Gandini, A. *Monomers, Polymers and Composites from Renewable Resources*; Elsevier: Amsterdam, The Netherlands, 2008; pp. 39–66. ISBN 9780080453163.

6.   Allen, R.R.; Formo, M.V.; Krishnamurthy, R.G.; McDermott, G.N.; Norris, F.A.; Sonntag, N.O.V. *Bailey's Industrial Oil and Fat Products*; Wiley: New York, NY, USA, 1982; ISBN 9780471839583.

7.   Mosiewicki, M.A.; Aranguren, M.I. A short review on novel biocomposites based on plant oil precursors. *Eur. Polym. J.* **2013**, *49*, 1243–1256. [CrossRef]

8.   Meier, M.A.R.; Metzger, J.; Schubert, U.S. Plant oil renewable resources as green alternatives in polymer science. *Chem. Soc. Rev.* **2007**, *36*, 1788–1802. [CrossRef] [PubMed]

9.   Wool, R.P. *Biobased Polymers and Composites*; Elsevier Academic Press: Burlington, VT, USA, 2005; ISBN 978-0-12-763952-9.

10. Alam, M.; Akram, D.; Sharmin, E.; Zafar, F.; Ahmad, S. Vegetable oil based eco-friendly coating materials: A review article. *Arab. J. Chem.* **2014**, *7*, 469–479. [CrossRef]

11. Xia, Y.; Zhang, Z.; Kessler, M.R.; Brehm-Stecher, B.; Larock, R.C. Antibacterial Soybean-Oil-Based Cationic Polyurethane Coatings Prepared from Different Amino Polyols. *Chem. Sus. Chem.* **2012**, *5*, 2221–2227. [CrossRef]

12. Garrison, T.F.; Zhang, Z.; Kim, H.; Mitra, D.; Xia, Y.; Pfister, D.P.; Brehm-Stecher, B.; Larock, R.C.; Kessler, M.R. Thermo-Mechanical and Antibacterial Properties of Soybean Oil-Based Cationic Polyurethane Coatings: Effects of Amine Ratio and Degree of Crosslinking. *Macromol. Mater. Eng.* **2014**, *299*, 1042–1051. [CrossRef]

13. Liang, H.; Liu, L.; Lu, J.; Chen, M.; Zhang, C. Castor oil-based cationic waterborne polyurethane dispersions: Storage stability, thermo-physical properties and antibacterial properties. *Ind. Crops. Prod.* **2018**, *117*, 169–178. [CrossRef]

14. Wang, R.; Schuman, T.P. Vegetable oil-derived epoxy monomers and polymer blends: A comparative study with review. *Express Polym. Lett.* **2013**, *7*, 272–292. [CrossRef]

15. Zhu, L.; Wool, R.P. Nanoclay reinforced bio-based elastomers: Synthesis and characterization. *Polymer* **2006**, *47*, 8106–8115. [CrossRef]

16. Lu, Y.; Larock, R.C. Novel Biobased Nanocomposites from Soybean Oil and Functionalized Organoclay. *Biomacromolecules* **2006**, *7*, 2692–2700. [CrossRef] [PubMed]

17. Díez-Pascual, A.M.; Díez-Vicente, A.L. Epoxidized Soybean Oil/ZnOBiocomposites for Soft Tissue Applications: Preparation and Characterization. *ACS Appl. Mater. Interfaces* **2014**, *6*, 17277–17288. [CrossRef] [PubMed]

18. Díez-Pascual, A.M.; Díez-Vicente, A.L. Wound Healing Bionanocomposites Based on Castor Oil Polymeric Films Reinforced with Chitosan-Modified ZnO Nanoparticles. *Biomacromolecules* **2015**, *16*, 2631–2644. [CrossRef] [PubMed]

19. Deka, H.; Karak, N.; Kalita, R.D.; Buragohain, A.K. Bio-based thermostable, biodegradable and biocompatible hyperbranched polyurethane/Ag nanocomposites with antimicrobial activity. *Polym. Degrad. Stab.* **2010**, *95*, 1509–1517. [CrossRef]

20. Gold, K.; Slay, B.; Knackstedt, M.; Gaharwar, A.K. Antimicrobial Activity of Metal and Metal-Oxide Based Nanoparticles. *Adv. Therap.* **2018**, *1*, 1700033. [CrossRef]

21. Seil, J.T.; Webster, T.J. Antimicrobial applications of nanotechnology: Methods and literature. *Int. J. Nanomed.* **2012**, *2012*, 2767–2781. [CrossRef]

22. Brunner, T.J.; Wick, P.; Manser, P.; Spohn, P.; Grass, R.N.; Limbach, L.K.; Bruinink, A.; Stark, W.J. In Vitro Cytotoxicity of Oxide Nanoparticles: Comparison to Asbestos, Silica, and the Effect of Particle Solubility. *Environ. Sci. Technol.* **2006**, *40*, 4374–4381. [CrossRef]

23. Díez-Pascual, A.M.; Xu, C.; Luque, R. Development and characterization of novel poly(ether ether ketone)/ZnObionanocomposites. *J. Mater. Chem. B* **2014**, *2*, 3065–3078. [CrossRef]

24. Coleman, A.; Jagadish, C. Basic properties and applications of ZnO—Chapter 1. In *Zinc Oxide Bulk, Thin Films and Nanostructures*; Jagadish, C., Pearton, S., Eds.; Elsevier: Oxford, UK, 2006; pp. 1–20. ISBN 978-0-08-044722-3.

25. Díez-Pascual, A.M.; Díez-Vicente, A.L. Development of Nanocomposites Reinforced with CarboxylatedPoly(ether ether ketone) Grafted to Zinc Oxide with Superior Antibacterial Properties. *ACS Appl. Mater. Interfaces* **2014**, *6*, 3729–3741. [CrossRef]

26. Díez-Pascual, A.M.; Díez-Vicente, A.L. Poly(3-hydroxybutyrate)/ZnOBionanocomposites with Improved Mechanical, Barrier and Antibacterial Properties. *Int. J. Mol. Sci.* **2014**, *15*, 10950–10973. [CrossRef] [PubMed]

27. Hajipour, M.J.; Fromm, K.M.; Akbar Ashkarran, A.; Jimenez de Aberasturi, D.; Larramendi, I.R.D.; Rojo, T.; Serpooshan, V.; Parak, W.J.; Mahmoudi, M. Antibacterial properties of nanoparticles. *Trends Biotechnol.* **2012**, *30*, 499–511. [CrossRef] [PubMed]

28. Díez-Pascual, A.M.; Díez-Vicente, A.L. ZnO-Reinforced Poly(3-hydroxybutyrate-co-3-hydroxyvalerate) Bionanocomposites with Antimicrobial Function for Food Packaging. *ACS Appl. Mater. Interfaces* **2014**, *6*, 9822–9834. [CrossRef] [PubMed]

29. Zheng, K.; Setyawati, M.I.; Leong, D.T.; Xie, J. Antimicrobial Gold Nanoclusters. *ACS Nano* **2017**, *11*, 6904–6910. [CrossRef]

30. Padmavathy, N.; Vijayaraghavan, R. Enhanced bioactivity of ZnO nanoparticles-an antimicrobial study. *Sci. Technol. Adv. Mater.* **2008**, *9*, 035004. [CrossRef]

31. Díez-Pascual, A.M.; Díez-Vicente, A.L. High-Performance Aminated Poly(phenylene sulfide)/ZnO Nanocomposites for Medical Applications. *ACS Appl. Mater. Interfaces* **2014**, *6*, 10132–10145. [CrossRef]

32. Díez-Pascual, A.M.; Díez-Vicente, A.L. Effect of TiO2 nanoparticles on the performance of polyphenylsulfone biomaterial for orthopaedic implants. *J. Mater. Chem. B* **2014**, *2*, 7502–7514. [CrossRef]

33. Pelaez, M.; Nolan, N.T.; Pillai, S.C.; Seery, M.K.; Falaras, P.; Kontos, A.G.; Dunlop, P.S.M.; Hamilton, J.W.J.; Byrne, J.A.; O'Shea, K.; et al. A review on the visible light active titanium dioxide photocatalysts for environmental applications. *Appl. Catal. B* **2012**, *125*, 331–349. [CrossRef]

34. Fadeel, B.; Garcia-Bennett, A.E. Better safe than sorry: Understanding the toxicological properties of inorganic nanoparticles manufactured for biomedical applications. *Adv. Drug Deliv. Rev.* **2010**, *62*, 362–374. [CrossRef]

35. Díez-Pascual, A.M.; Díez-Vicente, A.L. Nano-TiO2 Reinforced PEEK/PEI Blends as Biomaterials for Load-Bearing Implant Applications. *ACS Appl. Mater. Interfaces* **2015**, *7*, 5561–5573. [CrossRef]

36. Galib, M.B.; Mashru, M.; Jagtap, C.; Patgiri, B.J.; Prajapati, P.K. Therapeutic potentials of metals in ancient India: A review through Charaka Samhita. *J. Ayurveda Inter. Med.* **2011**, *2*, 55–63. [CrossRef] [PubMed]

37. Allahverdiyev, A.M.; Kon, K.V.; Abamor, E.S.; Bagirova, M.; Rafailovich, M. Coping with antibiotic resistance: Combining nanoparticles with antibiotics and other antimicrobial agents. *Expert Rev. Anti InfecyTher.* **2011**, *9*, 1035–1052. [CrossRef] [PubMed]

38. Magdolenova, Z.; Collins, A.; Kumar, A.; Dhawan, A.; Stone, V.; Dusinska, M. Mechanisms of genotoxicity. A review of in vitro and in vivo studies with engineered nanoparticles. *Nanotoxicology* **2014**, *8*, 233–278. [CrossRef] [PubMed]

39. Richardson, H.W. Copper Compounds. In *Ullmann's Encyclopedia of Industrial Chemistry*; Wiley-VCH: Weinheim, Germany, 2000; ISBN 9783527303854.

40. Moreno, J.L.V.; Padama, A.A.B.; Kasai, H. A density functional theory-based study on the dissociation of NO on a CuO(110) surface. *Cryst. Eng. Comm.* **2014**, *16*, 2260–2265. [CrossRef]

41. Angelé-Martínez, C.; Nguyen, K.V.T.; Ameer, F.S.; Anker, J.N.; Brumaghim, J.L. Reactive oxygen species generation by copper(II) oxide nanoparticles determined by DNA damage assays and EPR spectroscopy. *Nanotoxicology* **2017**, *11*, 278–288. [CrossRef]

42. Yamamoto, K.; Kawanishi, S. Hydroxyl free radical is not the main active species in site-specific DNA damage induced by copper (II) ion and hydrogen peroxide. *J. Biol. Chem.* **1989**, *264*, 15435–15440.

43. Blaney, L. Magnetite (Fe3O4): Properties, Synthesis, and Applications. *Lehigh Rev.* **2007**, *15*, 5.

44. Xiaodi, L.; Zhiguo, Z.; Yufeng, T.; Bingyu, L. Review on the Synthesis and Applications of Fe3O4 Nanomaterials. *J. Nanomater.* **2013**, *2013*, 902538. [CrossRef]

45. Arakha, M.; Pal, S.; Samantarrai, D.; Panigrahi, T.K.; Mallick, B.C.; Pramanik, K.; Mallick, B.; Jha, S. Antimicrobial activity of iron oxide nanoparticle upon modulation of nanoparticle-bacteria interface. *Sci. Rep.* **2015**, *5*, 14813. [CrossRef]

46. Díez-Pascual, A.M.; Díez-Vicente, A.L. Development of linseed oil–TiO2 green nanocomposites as antimicrobial coatings. *J. Mater. Chem. B* **2015**, *3*, 4458–4471. [CrossRef]

47. Al-Jumaili, A.; Mulvey, P.; Kumar, A.; Prasad, K.; Bazaka, K.; Warner, J.; Jacob, M.V. Eco-friendly nanocomposites derived from geranium oil and zinc oxide in one step approach. *Sci. Rep.* **2019**, *9*, 5973. [CrossRef] [PubMed]

48. Das, B.; Mandal, M.; Upadhyay, A.; Chattopadhyay, P.; Karak, N. Bio-based hyperbranched polyurethane/ Fe3O4nanocomposites: Smart antibacterial biomaterials for biomedical devices and implants. *Biomed. Mater.* **2013**, *8*, 035003. [CrossRef] [PubMed]

49. Das, B.; Chattopadhyay, P.; Upadhyay, A.; Gupta, K.; Mandal, M.; Karak, N. Biophysico-chemical interfacial attributes of Fe3O4 decorated MWCNT nanohybrid/bio-based hyperbranched polyurethane nanocomposite: An antibacterial wound healing material with controlled drug release potential. *N. J. Chem.* **2014**, *38*, 4300–4311. [CrossRef]

50. Sharmin, E.; Zafar, F.; Akram, D.; Ahmad, S. Plant oil polyol nanocomposite for antibacterial polyurethane coating. *Prog. Org. Coat.* **2013**, *76*, 541–547. [CrossRef]

51. Bazaka, K.; Jacob, M.; Truong, V.K.; Crawford, R.J.; Ivanova, E.P. The Effect of Polyterpenol Thin Film Surfaces on Bacterial Viability and Adhesion. *Polymers* **2011**, *3*, 388–404. [CrossRef]

52. Bazaka, K.; Bazaka, O.; Levchenko, I.; Xu, S.; Ivanova, E.P.; Keidar, M.; Ostrikov, K. Plasma-potentiated small molecules—Possible alternative to antibiotics? *Nano Futures* **2017**, *1*, 025002. [CrossRef]

53. Prasad, K.; Lekshmi, G.S.; Ostrikov, K.; Lussini, V.; Blinco, J.; Mohandas, M.; Vasilev, K.; Bottle, S.; Bazaka, K.; Ostrikov, K. Synergic bactericidal effects of reduced graphene oxide and silver nanoparticles against Gram-positive and Gram-negative bacteria. *Sci. Rep.* **2017**, *7*, 1591. [CrossRef]

54. Sinha, R.; Karan, R.; Sinha, A.; Khare, S.K. Interaction and nanotoxic effect of ZnO and Ag nanoparticles on mesophilic and halophilic bacterial cells. *Bioresour. Technol.* **2011**, *102*, 1516–1520. [CrossRef]

55. Arakha, M.; Saleem, M.; Mallick, B.C.; Jha, S. The effects of interfacial potential on antimicrobial propensity of ZnO nanoparticle. *Sci. Rep.* **2015**, *5*, 9578. [CrossRef]

56. Desbois, A.; Smith, V. Antibacterial free fatty acids: Activities, mechanisms of action and biotechnological potential. *Appl. Microbiol. Biotechnol.* **2010**, *85*, 1629–1642. [CrossRef]

57. Betancourt-Galindo, R.; Berlanga Duarte, M.L.; Puente Urbina, B.A.; Rodríguez-Fernández, O.S.; Sánchez-Valdés, S. Surface Modification of ZnO Nanoparticles. *Mater. Sci Forum* **2010**, *644*, 61–64. [CrossRef]

58. Rabea, E.I.; Badawy, M.E.; Stevens, C.V.; Smagghe, G.; Steurbaut, W. Chitosan as Antimicrobial Agent: Applications and Mode of Action. *Biomacromolecules* **2003**, *4*, 1457–1465. [CrossRef] [PubMed]

59. Chung, Y.; Chen, C. Antibacterial characteristics and activity of acid-soluble chitosan. *Bioresour. Technol.* **2008**, *99*, 2806–2814. [CrossRef] [PubMed]

60. Archana, D.; Dutta, J.; Dutta, P.K. Evaluation of chitosan nano dressing for wound healing: Characterization, in vitro and in vivo studies. *Int. J. Biol. Macromol.* **2013**, *57*, 193–203. [CrossRef] [PubMed]

61. Archana, D.; Singh, B.K.; Dutta, J.; Dutta, P.K. In vivo evaluation of chitosan–PVP–titanium dioxide nanocomposite as wound dressing material. *Carbohyd. Polym.* **2013**, *95*, 530–539. [CrossRef] [PubMed]

62. Archana, D.; Singh, B.K.; Dutta, J.; Dutta, P.K. Chitosan-PVP-nano silver oxide wound dressing: In vitro and in vivo evaluation. *Int. J. Biol. Macromol.* **2015**, *73*, 49–57. [CrossRef]

63. Amin, K.A.M.; Panhuis, M.I.H. Reinforced Materials Based on Chitosan, TiO2 and Ag Composites. *Polymers* **2012**, *4*, 590–599. [CrossRef]

64. Dhillon, G.; Kaur, S.; Brar, S. Facile fabrication and characterization of chitosan-based zinc oxide nanoparticles and evaluation of their antimicrobial and antibiofilm activity. *Int. Nano Lett.* **2014**, *4*, 1–11. [CrossRef]

65. Bakhshi, H.; Yeganeh, H.; Mehdipour-Ataei, S.; Shokrgozar, M.; Yari, A.; Seyyed, N. Synthesis and characterization of antibacterial polyurethane coatings from quaternary ammonium functionalized soybean oil based polyols. *Mater. Sci. Eng. C* **2013**, *33*, 153–164. [CrossRef]

66. Díez-Pascual, A.M.; Díez-Vicente, A.L. Poly(propylene fumarate)/Polyethylene Glycol-Modified Graphene Oxide Nanocomposites for Tissue Engineering. *ACS Appl. Mater. Interfaces* **2016**, *8*, 17902–17914. [CrossRef]

67. Díez-Pascual, A.M.; Díez-Vicente, A.L. PEGylated boron nitride nanotube-reinforced poly(propylene fumarate) nanocomposite biomaterials. *RSC Adv.* **2016**, *6*, 79507–79519. [CrossRef]

68. Muñoz-Bonilla, A.; Cerrada, M.L.; Fernandez-Garcia, M. Introduction to Polymeric Antimicrobial Materials. Chapter 1. In *Polymeric Materials with Antimicrobial Activity: From Synthesis to Applications*; RSC Publishing: Oxford, UK, 2014; p. 311. ISBN 978-1-84973-807-1.

69. Wu, J.; Li, C.; Tsai, C.; Chou, C.; Chen, D.; Wang, G. Synthesis of antibacterial TiO2/PLGA composite biofilms. *Nanomed. Nanotechnol.* **2014**, *10*, e1097–e1107. [CrossRef] [PubMed]

70. Cho, M.; Chung, H.; Choi, W.; Yoon, J. Linear correlation between inactivation of E. coli and OH radical concentration in TiO2 photocatalytic disinfection. *Water Res.* **2004**, *38*, 1069–1077. [CrossRef] [PubMed]

71. Nadtochenko, V.; Denisov, N.; Sarkisov, O.; Gumy, D.; Pulgarin, C.; Kiwi, J. Laser kinetic spectroscopy of the interfacial charge transfer between membrane cell walls of E. coli and TiO2. *J. Photoch. Photobio. A* **2006**, *181*, 401–407. [CrossRef]

72. Zielecka, M.; Bujnowska, E.; Kępska, B.; Wenda, M.; Piotrowska, M. Antimicrobial additives for architectural paints and impregnates. *Prog. Org. Coat.* **2011**, *72*, 193–201. [CrossRef]

73. Turkoglu, A.; Duru, M.E.; Mercan, N.; Kivrak, I.; Gezer, K. Antioxidant and antimicrobial activities of Laetiporussulphureus (Bull.) Murrill. *Food Chem.* **2007**, *101*, 267–273. [CrossRef]

74. Martí, M.; Frígols, B.; Serrano-Aroca, A. Antimicrobial Characterization of Advanced Materials for Bioengineering Applications. *J. Vis. Exp.* **2018**, *138*, e57710. [CrossRef]

75. Pramanik, S.; Konwarh, R.; Deka, R.C.; Aidew, L.; Barua, N.; Buragohain, A.K.; Mohanta, D.; Karak, N. Microwave-assisted poly (glycidyl methacrylate)-functionalized multiwall carbon nanotubes with a 'tendrillar' nanofibrous polyaniline wrapping and their interaction at bio-interface. *Carbon* **2013**, *55*, 34–43. [CrossRef]

76. Jackson, P.; Jacobsen, N.R.; Baun, A.; Birkedal, R.; Kühnel, D.; Jensen, K.A.; Vogel, U.; Wallin, H. Bioaccumulation and ecotoxicity of carbon nanotubes. *Chem. Cent. J.* **2013**, *7*, 154. [CrossRef]

7

Preparation of the Hybrids of Hydrotalcites and Chitosan by Urea Method and their Antimicrobial Activities

Bi Foua Claude Alain Gohi [1,2], Hong-Yan Zeng [1,*], Xiao-Ju Cao [1], Kai-Min Zou [1], Wenlin Shuai [3] and Yi Diao [2]

[1] Biotechnology Institute, College of Chemical Engineering, Xiangtan University, Xiangtan 411105, China; gohibifouaca@smail.xtu.edu.cn (B.F.C.A.G.); 15773228967@163.com (X.-J.C.); Kaiminzou114634987@aliyun.com (K.-M.Z.)

[2] School of Biological and Chemical Engineering, Panzhihua University, Panzhihua 617000, China; diaoy163@163.com

[3] College of Chemistry and Chemical Engineering, Xinjiang University, Urumqi 830046, China; swlswlswl123456789@163.com

* Correspondence: hongyanzeng99@hotmail.com.

Abstract: Hybrid nano-supra molecular structured materials can boost the functionality of nano- or supra-molecular materials by providing increased reactivity and conductivity, or by simply improving their mechanical stability. Herein, the studies in materials science exploring hybrid systems are investigated from the perspective of two important related applications: healthcare and food safety. Interfacing phase strategy was applied, and ZnAl layered double hydroxide-chitosan hybrids, prepared by the urea method (U-LDH/CS), were successfully synthesized under the conditions of different chitosan(CS) concentrations with a Zn/Al molar ratio of 5.0. The structure and surface properties of the U-LDH/CS hybrids were characterized by X-ray diffraction (XRD), Fourier-transform infrared spectrometer(FTIR), scanning electronmicroscopy (SEM), ultravioletvisible (UV-Vis), and zero point charge (ZPC) techniques, where the effect of CS concentration on the structure and surface properties was investigated. The use of the U-LDH/CS hybrids as antimicrobial agents against *Escherichia coli*, *Staphylococcus aureus*, and *Penicillium cyclopium* was investigated in order to clarify the relationship between microstructure and antimicrobial ability. The hybrid prepared in a CS concentration of $1.0\,g{\cdot}L^{-1}$ (U-LDH/CS$_1$) exhibited the best antimicrobial activity and exhibited average inhibition zones of 24.2, 30.4, and 22.3mm against *Escherichia coli*, *Staphylococcus aureus*, and *Penicillium cyclopium*, respectively. The results showed that the appropriate addition of CS molecules could increase antimicrobial ability against microorganisms.

Keywords: chitosan; ZnAl hydroxide; hybrid; urea method; chitosan amount; antimicrobial activity

1. Introduction

The use of nanostructures is known to achieve levels of functionality not possible to reach when using bulk materials [1]. It is in this context that the notion of hybrids has been introduced, which is a strategy that aims to combine different structures in order to obtain a more efficient one. An infinite number of possibilities can emerge from the combination of phases in hybrid systems [2]. Different types of hybrid materials have been studied, including organic hybrids, inorganic hybrids, and organic-inorganic hydrids. Organic hybrid systems still dominate several traditional areas of chemical science and well-known applications, such as the synthesis of pharmaceutical compounds and drugs [3]. However, the mechanical properties (resistance to temperature and environmental stability) of the latter limits their application fields [4]. The applications of hybrid inorganic materials range from ion exchangers, semiconductors, adsorbent, electrochemical sensor catalysts, and catalyst support [5,6]. The toxicity and low biocompatibility of organic-inorganic hybrids also limit their application in biomedical science and drugs production. Among the techniques used for the preparation of hybrids, mention may be made, inter alia, of: Directintercalation, in-situ polymerization, intact melt-blending, and surface-modified blending. Organic-inorganic hybrids overcome the limitations of the previous two by their non-toxicity, their biocompatibility, and the strengthening of their mechanical properties [7,8]. Bacterial pathogens are one of the primary causes of human morbidity worldwide [9]. Historically, antibiotics have been highly effective against most bacterial pathogens; however, the increasing resistance of bacteria to a broad spectrum of commonly used antibiotics has become a global health-care problem. In recent years, the use of hybrid materials with a wide range of properties and applications has increased considerably. Construction of organic-inorganic hybrid materials is a rapidly expanding field in materials chemistry for the design of advanced materials with specific structure and functionality [10]. Bio-inorganic hybrids may exhibit, not only a combination of properties from the disparate components, but also further enhanced property tunability and new synergistic properties that arise from the interactions between the biological molecules and inorganic materials [11]. Among various hybrids of inorganic and organic materials, layered double hydrotalcites (LDH) and chitosan (CS) are of particular interest due to their wide applications range. LDH and CS have received considerable attention recently due to their wide applications in a variety of areas, including chemistry, physics, materials science, and the biomedical science [12–14]. Chitosan (CS) is a naturally occurring, cationic polysaccharide composed of (1,4)-linked 2-amino-2-deoxy-β-D-glucose and 2-acetamido-2-deoxy-β-Dglucose units. Chitin is the transformed base material resulting in chitosan, which is obtained from shrimp, crab, and lower plants and animals. By changing different parameters, one can achieve the desirable targeted chitosan molecule without changing the chemical compositions [15].CS has three types of functional nucleophile groups consisting of a C-2 NH_2 group, a secondary C-3 OH group, and a C-6 primary OH group. CS is a biopolymer that presents reactive functional groups that are susceptible to chemical modification, and has been shown to be a functional polymer that can covalently graft antioxidant/antimicrobial activity onto its backbone [16].CS has shown interesting antibacterial and antifungal activities against a wide range of microorganisms when compared to other polymers and biopolymers [17].Layered double hydroxides (LDH) are composed of positively charged brucite-like layers of divalent and trivalent metal hydroxides whose excess positive charge is compensated by anions and water molecules present in the interstitial position. Among the heterostructured nanomaterials, layered nanohybrids have received intense attentions in many areas due to their unique physico-chemical and mechanical properties that cannot be obtained from other analogous nanohybrids [18]. Moreover, LDH has essential properties, such as biocompatibility, null toxicity, and allergenicity [19]. They can be represented by the general formula:

$$[M(II)_{1-x}M(III)_x(OH)2]^{x+}[A^{n-}_{x/n}YH_2O]^{x-},$$

where M(II) = Mg, Ni, Co, Cu, Zn, Mn; M(III) = Al, Fe, Cr, V; A^{n-} = CO_3^{2-}, Cl-, SO_4^{2-}, etc., and x = 0.1–0.35.

ZnAl hydroxides are effective antimicrobial agents for bacteria such as *Escherichia coli* and *Staphylococcusaureus* due to the hydroxides (–OH) and the nature of the metallic cations, where Zn^{2+} is one of the most active ones, due to its strong oligodynamic features [20]. Zinc and CS have excellent antibacterial activity [21,22]. The goal of layered double hydroxide-chitosan hybrids, prepared by the urea method (U-LDH/CS) preparation was to assess the possible antimicrobial activity of pure CS macromolecules under microbial culture medium pH conditions and to explore the impact of U-LDH on CS through the inorganic-organic hybrid. Since the antimicrobial properties of CS are limited to pH values below six [23–26], hybrid materials that have the ability to kill pathogenic bacteria and prevent bacterial colonization are desired for utilization in several application areas such as food-contact materials, food packaging, textiles, water purification systems, prosthetic devices, and hospital equipment surfaces. The aim here is to activate and reinforce the antimicrobial activity of pure chitosan without taking into account both the conditions of the bacterial culture medium and its molecular weight by integrating it into a hybrid structure. In this study, using interfacing phases, a surface-modified blending process that is a strategy used to obtain a set of properties in one system that are beyond the abilities of single phases [2], the LDH/CS hybrids as antimicrobial agents were prepared in different CS concentrations by urea method, where CS was used as a soft template. In order to clarify the relationship between the microstructure and antimicrobial ability, the physico-chemical properties of the LDH/CS hybrids were characterized by XRD, FTIR, SEM, UV-Vis and pH of zero point charge (pHzpc) measurements. In addition, the antimicrobial activity was also evaluated.

2. Materials and Methods

2.1. Experimental Materials

Chitosan (CS, DA $\geq$ 91%, MW $1.5.10^5$ kD) was purchased from Sinopharm Chemical Reagent Co. Urea (CON_2H_4), $Al(NO_3).9H_2O$, and $Zn(NO_3)2\cdot6H_2O$, KOH, ethanol were purchase from Heng Xing Chemical preparation Co. Ltd. (Tianjin, China). NaOH was purchase from Xilong Chemical Co., Ltd. (Guangdong, China) and tetracycline powder (TC) was purchased from Sigma–Aldrich (Shanghai, China).

2.2. Preparation of ZnAl Hydroxide and CS Hybrids

2.2.1. Preparation of the ZnAl Hydroxide by Urea Method

For the sake of comparison, ZnAl hydroxide (U-LDH$_5$) with a Zn/Al molar ratio of 5.0 was prepared using the urea method proposed by Zeng et al. (2009) [27], with slight modifications. $Zn(NO_3)_2\cdot6H_2O$ and $Al(NO_3)_3\cdot9H_2O$ (total amount of metal ions was 0.12 mol) were mixed with a set up Zn/Al molar ratio by using 600 mL deionized water. The mixed solution was poured into a three-necked round bottomed flask, and urea (urea/NO_3^- molar ratio of 4: 1) was added. The reaction solution was magnetically stirred at 110 °C for 12 h. Then the resulting reactant was crystallized statically at 80 °C for another 12 h. The precipitate was centrifuged and washed thoroughly with deionized water and was subsequently dried at 80 °C overnight. The dried material was denoted as LDH. For convenience, the resulting sample with the Zn/Al molar ratio of 5:1 was designated as U-LDH$_5$.

2.2.2. Preparation of the U-LDH$_5$ and CS Hybrids

The hybrids (U-LDH/CS) of the LDH and chitosan (CS) were prepared under the conditions of different CS concentrations and a Zn/Al molar ratio of 5.0 by the urea method. The U-LDH$_5$ and CS hybrids were prepared using CS molecules as a soft template. CS solutions at the CS concentrations of 0.5, 1.0, 1.5, 2.0, and 3.0 g·L^{-1} were separately obtained by dissolving powered CS in the solution containing 1% (w/v) acetic acid. During the preparation of the hybrid, 400 mL of mixed salt solution containing $Zn(NO_3)_2\cdot6H_2O$ (0.15 mol·L^{-1}), $Al(NO_3)_3\cdot9H_2O$ (0.03 mol·L^{-1}), and urea (urea/NO_3^- molar ratio of 4.0) was placed into a three-necked flask. 100 mL of CS solution was dropped into the mixed salt solution under stirring (300 rpm) in pH 8.5 at room temperature. After dripping, the reaction temperature was raised to 103 °C for 12 h under stirring. After the reaction, it was crystallized at 80 °C for 18 h, filtrated, washed, and then dried at 90 °C for 6 h, and was denoted as U-LDH/CS. For convenience, the resulting products prepared in the CS solutions of 0.5, 1.0, 1.5, 2.0, and 3.0 g·L^{-1} were designated as U-LDH/CS$_{0.5}$, U-LDH/CS$_1$, U-LDH/CS$_{1.5}$, U-LDH/CS$_2$, and U-LDH/CS$_3$, respectively.

2.3. Characterization Techniques

It should be noted that 21 samples were analyzed and tested, including hybrid preparation methods and analytical measurements, all in triplicate.

2.3.1. XRD Analyses

X-ray powder diffraction (XRD) analysis was performed on a Rigaku (Tokyo, Japan) D/MAX-2500/PC with Cu Ka radiation (λ = 1.5405 Å), with an operating voltage of 40 kV, a current of 30 mA, a scan rate of $2°min^{-1}$, and a scanning range from 5 to 90 °2θ, in steps of 0.0334° with a counting time per step of 650 s.

2.3.2. FTIR Analyses

Fourier-transform infrared (FTIR) spectra of the samples were obtained with a Perkin–Elmer Spectrum One B instrument (Shanghai, China). Powder samples were molded in KBr pellets. 1 mg of the powdered samples was carefully mixed with 250 mg of KBr (infrared grade) and pelletized under a pressure of 10 t for 1 min. The pellets were analyzed to collect 32 scans in the range of 4000–400 cm^{-1} at a resolution of 2 cm^{-1}.

2.3.3. SEM Analyses

Scanning electron microscopy (SEM) was carried out using aJSM-6700F microscope (JEOL, Tokyo, Japan) operating at15 keV. Prior to analysis, the samples were covered with gold to avoid charge effects. Samples were sputtered using ion sputtering technology at a magnification of 15,000.

2.3.4. UV-Vis Analysis

UV-visible (UV-Vis) spectra were recorded by spectrophotometer (Shimadzu UV-2550, Kyoto, Japan). The range of wavelengths was 200~800 nm using high-purity $BaSO_4$ as a reference.

2.3.5. Point Zero Charge Analysis

The determination of the pH point of zero charge (pHzpc) of the materials was carried out using the potentiometric titration (PT) method described by Li et al. [28]. The pH at pHzpc was determined in NaCl solutions (inert electrolytes) with different concentrations. The experiments were carried out in a shaker at 150 rpm and 25 °C for 200 min. After the experiments, the pH in the solution was measured while a 0.1 $mol·L^{-1}$ NaOH solution was added. The adsorption amount of H^+ (Γ_{H+}) and OH^- (Γ_{OH-}) was calculated. Finally, PT curves were obtained by plotting ($\Gamma_{OH-} - \Gamma_{H+}$) versus pH in NaCl solutions with different concentrations, and the crossover point of ($\Gamma_{OH-} - \Gamma_{H+}$) ~ pHzpc curves was pHzpc, which was electrically neutral. The permanent charge density (σp) at pHzpc was as follows: [29].

$$\sigma p = F(\Gamma_{OH-} - \Gamma_{H+})zpc/S_{BET} \qquad (1)$$

where, S_{BET} and F are the specific surface area of the samples and the Faraday constant (96485 $C·m^{-2}$), respectively.

2.3.6. Rheological Analysis

Rheological compression measurements were carried out according to the norm established by Yi et al. [30]. Circular disk-like samples with a 25 mm diameter and a 2 mm thickness were prepared on a compression mold. Rheological studies were carried out on a controlled strain rheometer, (MCR 301, Anton Paar, Austria). The dynamic (frequency sweep) tests were performed at a strain of 0.5%. Storage modulus (G′) and loss modulus (G″) were measured in the frequency sweep experiments performed over a frequency range of 0.1–100 rad/s, with data collected at five points per decade.

2.3.7. Thermogravimetric Analysis

Thermogravimetric and differential thermal gravimetric analyses (TGA-DTG) for pure U-LDH and the U-LDH composites with various CS contents was conducted on a Perkin–Elmer TGA 7-thermal analyzer (Arizona, Waltham, MA, USA) under nitrogen at a purge rate of 20 mL/min, with the scanning temperature in the range from 50 to 800 °C. A platinum crucible with a heating rate of 10 and 12 °C /min and a sample weight of 63.3 mg each were used. Meanwhile, the U-LDH content in the CS composite sample was determined on the basis of the residual ash percentage and the water amount of the U-LDH/CS itself, which measured the weight loss between 200 and 400 °C.

2.4. In Vitro Antimicrobial Assay

2.4.1. Microorganism and Media

The bacteria *Escherichia coli* (*E. coli*, Gram-negative bacteria, ATCC 35218), *Staphylococcus aureus* (*S. aureus*, Gram-positive bacteria, ST398) and *Penicilinum cyclopium* (*P. cyclopium*, fungi, AS 3.4513) as target organisms were from the Xiangtan University General Microbiological Culture Collection Center. The *P. cyclopium* strain was maintained in potato dextrose (PD) medium (Qingdao Hope Bio-Technology Co., Ltd., Qingdao, China) at 28 °C for 72 h. The *E. coli* and *S. aureus* strains were maintained in mineral salt (MS) media (Boyao Biotechnology Co., Ltd., Shanghai, China) in pH 7.0 at 37 °C for 24 h. PD and MS are the most widely used media for growing fungi and bacteria [31,32]. All the samples were measured to obtain their OD_{600nm} values for calculating the bacteriostatic concentration. The controlled test contained the nutrient medium with bacterial suspension but without antimicrobial agents.

Mineral salt (MS) agar plate medium (g·L^{-1}): (NH4)$_2$SO$_4$ 2.0, NaCl 5.0, K$_2$HPO$_4$ 1.0, KH$_2$PO$_4$ 1.0, MgSO$_4$ 0.1, CaCl$_2$ 0.1, and agar 20 at initial pH 7.0. and PD agar plate medium (g·L^{-1}): potato 20.4, dextrose 20.4, and agar 20. All media were sterilized by autoclaving at 121 °C for 25 min.

2.4.2. Antimicrobial Test

The cultures of the bacteria and fungi strains were activated twice in liquid potato dextrose and MS media, respectively, and then the culture at exponential growth phase (OD_{600} around 1.2 absorbance units). All antimicrobial experiments using the three cultures were performed in 250 mL sterile shaking flasks (or 90 mm agar plates), where 100 mL cultures of *P. cyclopium*, *E. coli*, and *S. aureus* strains were transferred into the PD and MS media in sequence for the antimicrobial tests.

Antimicrobial tests were performed using an agar diffusion test method [33]. In this assay, 100 mL microbial suspensions were prepared, and 50 µL of a bacterial suspension of *E. coli*, *S. aureus*, and *P. cyclopium* was inoculated evenly onto agar plates. The thin tableting of the sample with 1.0cm diameter was placed into an agar plate with a bacterial suspension. After incubation of *E. coli* and *S. aureus* strains at 37 °C for 24h and *P. cyclopium* strain at 28 °C for 72h, the zone of inhibition around the mesh samples was measured with digital vernier calipers in millimeters (mm). The zone of inhibition of the samples was measured in four directions and reported as a mean value and calculated by Equation (1),

$$A = (D - d)/2 \tag{2}$$

where A is the zone of inhibition, D is the total diameter of the thin tableting with inhibition area after incubation, and d is the diameter of the thin tableting of the sample before incubation. In this assay, the agar plates with different bacterial suspensions were used as reference samples.

3. Results and Discussion

3.1. Characterization of the U-LDH/CS Hybrids

3.1.1. XRD Analyses

The XRD pattern of pure CS molecules is showed in Figure 1a, where there is a strong reflection in the diffractogram of chitosan at $2\theta=21.1$, corresponding to the high crystallinity of chitosan. The powder XRD patterns of the U-LDH5, U-LDH/CS$_{0.5}$, U-LDH/CS$_1$, U-LDH/CS$_{1.5}$, U-LDH/CS$_2$, and U-LDH/CS$_3$ hybrids are shown in Figure 1b. There is a typical layered double hydroxide structure with sharp and intense (003), (006), (009), (110), and (113) reflections and broadened (015) and (018) reflections with impure phase ZnO in all six samples, where only the U-LDH/CS$_{1.5}$ and U-LDH/CS$_3$ include impure phase ZnAl$_2$O$_4$. As seen in Table 1, the interlayer distances (d003 about 0.76 nm) for the samples were typical of CO$_3^{2-}$ pillar hydroxide, indicating that CS molecules were not intercalated into the interlayer spaces between the brucite sheets. All the lattice parameters of the U-LDH/CS samples were the same as those of the U-LDH$_5$, revealing that the incorporation of CS did not change the structures of the U-LDH/CS hybrids, which still maintained the native structures. The results demonstrate that the adding of CS as a soft template is conducive to the formation of a heterojunction structure, ZnO-Zn(OH)$_2$, which leads to high antimicrobial activity against bacteria combined with high adhesiveness from the CS molecules.

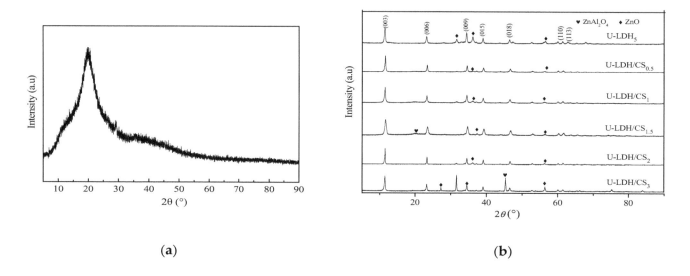

(a)   (b)

Figure 1. (a) XRD pattern of pure CS; (b) XRD patterns of the U-LDH$_5$ and U-LDH/CS hybrids.

Table 1. Calculation of lattice parameters and basal plane for the U-LDH$_5$, U-LDH/CS$_{0.5}$, U-LDH/CS$_1$, U-LDH/CS$_{1.5}$, U-LDH/CS$_2$, U-LDH/CS$_3$.

Samples Reflections	U-LDH$_5$	U-LDH/CS$_{0.5}$	U-LDH/CS$_1$	U-LDH/CS$_{1.5}$	U-LDH/CS$_2$	U-LDH/CS$_3$
d_{003} (nm)	0.7691	0.7627	0.7680	0.7596	0.7659	0.7568
d_{006} (nm)	0.3828	0.3803	0.3818	0.3795	0.3813	0.3785
d_{009} (nm)	0.2487	0.2595	0.2597	0.2594	0.2598	0.2586
d_{110} (nm)	0.1631	0.1635	0.1630	0.1635	0.1636	0.1630
FW$_{003}$ (rad)	0.4080	0.2240	0.2560	0.2990	0.260	0.5100
FW$_{110}$ (rad)	0.7770	0.4760	0.1810	0.6500	0.79	0.6700
a (nm)	0.3262	0.3204	0.3895	0.2493	0.1640	0.9613
c (nm)	2.3073	2.2471	2.2743	2.2802	2.2097	2.2174

3.1.2. FTIR Analyses

The FTIR spectra of the pure CS, U-LDH$_5$, and U-LDH/CS samples in the range of 4000–400 cm^{-1} are displayed in Figure 2. As can be seen from Figure 2, all samples show significant absorption peaks at about 3460, 1650, and 1366 cm^{-1} [34,35]. For the pure CS, these three peaks belong to the stretching vibration of –C–OH or the stretching vibration absorption peak of –N–H, the absorption peaks of amide I and amide II [36], and carboxylic acid (–C=O) or C–OH [37], respectively. Concerning U-LDH$_5$, the peeks are attributed to –OH stretching vibration and interlayer H$_2$O absorption for 3460 cm^{-1}, bending vibration of adsorbed water and inter-laminar structure water for 1650 cm^{-1}, and ν3 stretching vibration characteristic of CO$_3{}^{2-}$ at 1366 cm^{-1}. In addition to those three common peaks, pure CS at 2921 cm^{-1} has a stretching vibration peak of symmetrical or asymmetric –CH$_2$ [38] on a pyranose ring of the chitosan molecule and an oscillation absorption peak of –OH and –CH on the pyranose ring at 1435 cm^{-1} [39], and an absorption peak of –COC– on the glycosidic bond at 1082 cm^{-1}. Regarding U-LDH$_5$, its FTIR presented a vibrational absorption peak of the lattice layer M–O (Zn–O, Al–O) of the main body of the hydrotalcite in the range of 1000–400cm^{-1}. The characteristic absorption peaks of chitosan molecules appear at about 2921, 1435, and 1082 cm^{-1} in U-LDH/CS$_{0.5}$, U-LDH/CS$_1$, U-LDH/CS$_{1.5}$, U-LDH/CS$_2$, and U-LDH/CS$_3$ compared to U-LDH$_5$, indicating that chitosan and ZnAl-LDH are organically bound to CS and form hydrotalcite-chitosan composite material.

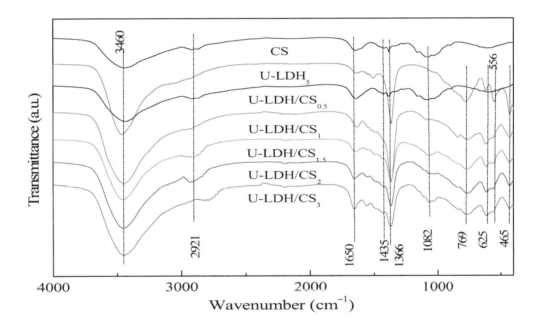

Figure 2. FTIR spectra of pure CS, U-LDH$_5$, and U-LDH/CS hybrids.

3.1.3. SEM Analysis

In order to investigate the morphology of the hybrids, the U-LDH$_5$ and U-LDH/CS hybrids were observed by SEM, and the results are shown in Figure 3. As shown in Figure 3, all samples (U-LDH$_5$ and U-LDH/CS) exhibit a typical hydrotalcite laminate structure, which is typical of LDH. There is a certain degree of aggregation, which is due to the addition of CS leading to the U-LDH$_5$ bonding accumulation. Obviously, CS has an effect on the microstructure of the U-LDH/CS hybrids.

Figure 3. SEM images of the U-LDH$_5$ and U-LDH/CS samples, ×15,000.

3.1.4. UV-Vis Analyses

UV-Vis measurement is a very simple method that is used to probe the possible changes in the molecule structure of materials. As can be seen from Figure 4, CS exhibits a strong absorption band in a wide range from 230 to 500 nm, with the electronic transition of n → σ* and n → π* belonging to –NH$_2$ at ~230 nm and –C=O or –COOH between 220 to 500 nm. U-LDH/CS samples show a weaker absorption band at 250–400 nm, which is due to the complexation of CS with U-LDH, which makes U-LDH/CS exhibit the absorption characteristic of chitosan UV-Vis. On the other hand, with the increase of the amount of chitosan, the absorption intensity of U-LDH/CS in this range also increases. At the same time, the intensity in the absorption band increases, with CS concentration arriving at the highest for U-LDH/CS$_{1.5}$. With a further increase of the CS concentration, the intensity gradually decreases. This suggests that CS has been incorporated into the U-LDH/CS hybrids and impacts the structure of the U-LDH/CS hybrids. This further confirms that CS and ZnAl-LDH have been combined to form the U-LDH/CS complex.

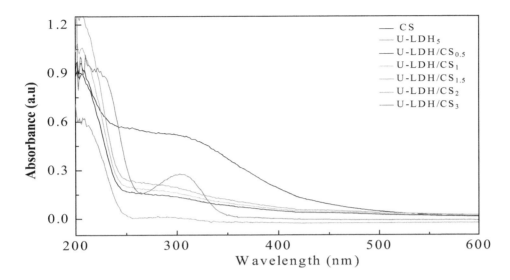

Figure 4. UV-Vis DRS spectra of the pure CS, U-LDH$_5$, and U-LDH/CS samples.

Based on the XRD, FTIR, and UV-Vis characterization analyzes, supported by studies on "Antimicrobial Chitosan and Chitosan Derivatives" by Sahariah and Masson [23] and "Polymer-inorganic supramolecular nanohybrids for red, white, green, and blue applications" by Park et al. [40], the probable structure of the hybrid could be presented as the scheme in Figure 5.

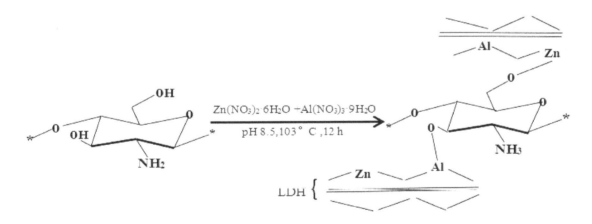

Figure 5. Synthetic scheme of U-LDH/CS hybrids.

3.1.5. Point Zero Charge Analysis

Figure 6 shows the relationship between ($\Gamma_{OH}-\Gamma_{H+}$) and the pH of U-LDH/CS. The difference between the adsorbed amount of OH$^-$ and H$^+$ on the surface of solid particles ($\Gamma_{OH}-\Gamma_{H+}$) is positive, indicating that the sample surface is alkaline with a permanent positive charge. The point zero-charge-pHpzcs of U-LDH/CS$_{0.5}$, U-LDH/CS$_1$, U-LDH/CS$_{1.5}$, U-LDH/CS$_2$, and U-LDH/CS$_3$ were 11.38, 11.31, 11.39, 11.30, and 11.00, respectively. Among them, pHpzcs of U-LDH/CS$_{1.5}$ was the highest. According to Equation (2), the surface permanent positive charge density (σp) of U-LDH/CS$_{0.5}$' U-LDH/CS$_1$, U-LDH/CS$_{1.5}$, U-LDH/CS$_2$, and U-LDH/CS$_3$ were 3.59, 3.47, 3.55, 3.71, and 3.55 C/m^{-2}' respectively. The shifts in pHpzc and σp of the U-LDH/CS hybrids to higher values were believed to be linked with the incorporation of CS molecules.

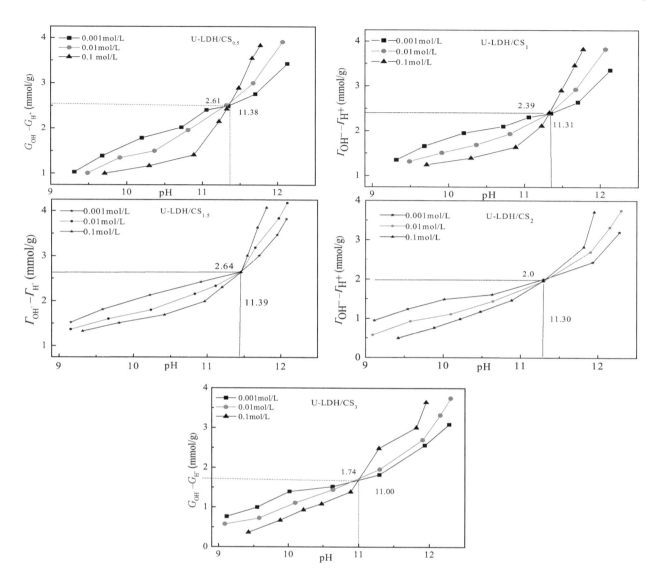

Figure 6. Potentiometric titration curves of the U-LDH/CS samples.

3.1.6. Rheological Analysis

Rheological properties are indicative of melt-processing behavior in unit operations such as injection molding. The linear viscoelastic responses of pure CS and the U-LDH hybrid in melt state were studied with the help of dynamic oscillatory shear measurements, and were reported in terms of storage modulus (G′) (Figure 7a) and complex viscosity (η*) (Figure 7b). At a fixed frequency, the (G′) values of the hybrids are higher than those of pure CS, however, the (η*) values of hybrid are lower than those of pure CS. Consequently, the decrease of the viscosity (η*) has gone up with the increase of U-LDH content. The viscous response of the material is stronger than the elastic response. These results demonstrate that the incorporation of U-LDH could reduce the viscosity properties of the CS matrix. The viscoelastic response in the low and medium frequencies region indicate that the hybrids have an obvious difference in their behavior from that of pure CS. One behavioral difference is that with the increase of CS content, the dependence of (η*) on frequency becomes weaker, while the dependence of (G′) on frequency becomes stronger in the low frequency region. This non-terminal behavior may be attributed to the fact that the intercalated U-LDH layers weaken the mobility of the CS chains and then restrict the long-range relaxation of them [41,42]. U-LDH/CS$_3$ shows a fragility to remain compact.

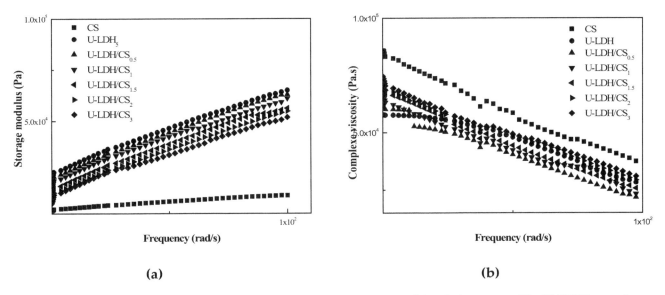

Figure 7. Storage modulus G′ (**a**) andloss modulus G″ (**b**) of the pure CS, U-LDH$_5$, and U-LDH/CS samples.

3.1.7. Thermogravimetric Analysis

Thermogravimetric Analysis (TG/DTG) is a thermal analysis technique that analyzes the composition and structure of materials and their thermal stability by measuring the relationship between mass loss and temperature during temperature control. To understand the thermal stability of CS, U-LDH$_5$, and U-LDH/CS with different CS concentrations, we performed TG/DTG analysis of each sample. Figure 8a–g shows the TG/DTG curves of the CS, U-LDH$_5$, and U-LDH/CS samples. The hydration properties of CS polysacharides depend on primary and supra macromolecular structure [43]. The decomposition of chitosan is presented in two stages, the first one, which occurred at 50 °C and extended to about 75 °C, was due to loss of water molecules, with a weight loss of about 9.3% [44], followed by stability from 75 °C to about 250 °C. Then a second stage, corresponding to the primary degradation of the pure CS, happened at 250 °C, with a percentage weight loss of about 49.3% from 250 to 550 °C, which was similar to the literature [43]. A slow degradation trend occurred after 550 °C that stabilized as it approaches 800 °C. Generally, in CS, the decomposition process of the N-acetylated compound is overlapped by the N-deacetylated unit, thereby increasing the widening process seen at temperatures up to 400°C [45], CS had a total mass loss of approximately 65.6%.

As can be seen from Figure 8b, the U-LDH$_5$ thermal decomposition process is divided into three stages [46,47], between 247~300 °C for the first phase of weight loss, which is due to the removal of water from the interlayer and the physical adsorption surface [48]. At this step, U-LDH$_5$ remained a layered structure. The second stage of weight loss, between 300 and 450 °C, was due to the decomposition of the interlayer CO_3^{2-} and the removal of the hydrotalcite-plate-OH, marking the delamination of the layered structure [49]. The last weight lost, that can be attributed to the decomposition of carbonate ions in the interlayer, started after 700 °C [50]. From Figure 8b, the mass loss of U-LDH$_5$ was 6% in the first stage, corresponding to a maximum endothermic peak of 300 °C for DTG. The mass loss of the second phase of 1.5% corresponded to an observed edge of DTG at 448 °C. The last stage of U-LDH$_5$ weight loss presented an edge at 570 °C, corresponding to a 8.2% weight loss with a total weight loss ofU-LDH$_5$ of about 16.5%. The TGA-DTA curves of U-LDH/CS at different CS concentration, shown in Figure 8c–ghave a quite similar trend with a difference in weight loss and endothermic peaks range. U-LDH/CS$_{0.5}$ showed a first mass loss of 2% at 75 °C between 50 and 102 °C, which can be accredited to the loss of adsorbed water, followed by a second stage of weight loss of U-LDH/CS$_{0.5}$ from 150 °C to 450 °C of 30%, with the maximum endothermic peak at 180.6 °C. U-LDH/CS$_{0.5}$ weight loss of the third stage was 4.07%, with the maximum endothermic peak at 650.1 °C, and the total weight loss was 36.7%. U-LDH/CS$_{0.5}$ TGA showed a gradual stabilization after 700 °C, which proceeded a new slight weight loss of mass then a definitive stabilization up to 800 °C. The first and second phases of U-LDH/CS$_1$ mass loss were 8.36% (40–190 °C) and 26.72% (200–450 °C), respectively, corresponding to the maximum endothermic peak at 163.4 and 275.9 °C. The third and last phase was 5.01% at 750 °C, corresponding to an extension of the primary degradation of the pure CS associated with decomposition of the hydrotalcite interlayer. The total weight loss of U-LDH/CS$_1$ was 40.09%. It was also noted that the total mass losses of U-LDH/CS$_{1.5}$, U-LDH/CS$_2$, and U-LDH/CS$_3$ were 46.3%, 51.3%, and 68.4%, respectively, distributed over two phases. These phases corresponded to a loss of water molecules, similar to the first phase of degradation of chitosan for the first stage, and the second stage was due to the primary degradation of the pure CS associated with decomposition of the interlayer and carbonate ions of hydrotalcite. The DTG curve of U-LDH/CS$_{1.5}$ had two maximum endothermics peaks at 75 °C over the range 50–150 °C and the second at 220 °C between 150 and 250 °C in addition to two edges at 260 and 460 °C. The DTG curves of U-LDH/CS$_2$ and U-LDH/CS$_3$ exhibited three maximum endothermic peaks in the same temperature ranges and one edge each. Their maximum endothermic peaks were at the same locations at 75 °C, 180 °C, and 280 °C in the interval ranges of 50 to 100 °C, 100 to 200 °C, and 200 to 300 °C, respectively. It was only the edges whose positions were different; one at 550 °C for U-LDH/CS$_2$ and the other at 530 °C for U-LDH/CS$_3$. TG/DTG curves showed that CS molar ratio changes had a strong effect on the thermal stability of the hybrid, indicating that even though the CS molar ratio increased, the charge density of the laminate decreased, but the hydrogen bonding between the interlayer H$_2$O and the interlayer CO$_3^{2-}$, OH- anion, the interlayer -OH, and the interlayer anion interactions were not affected much. It is worth mentioning that U-LDH/CS$_2$ and U-LDH/CS$_3$ showed a strong phase of weight loss at 200–550 °C, with a mass loss of 48.6% and 51.3%, respectively. These losses may be due to the hybrid U-LDH/CS arrangement during the formation of spinel transformation given the high concentration of CS. Obviously, increasing the concentration of CS for the preparation of U-LDH/CS induced the decrease of the thermal resistance of the U-LDH/CS hybrid.

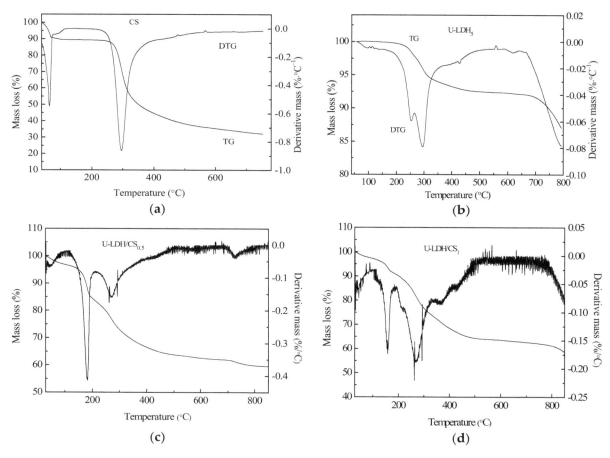

Figure 8. *Cont.*

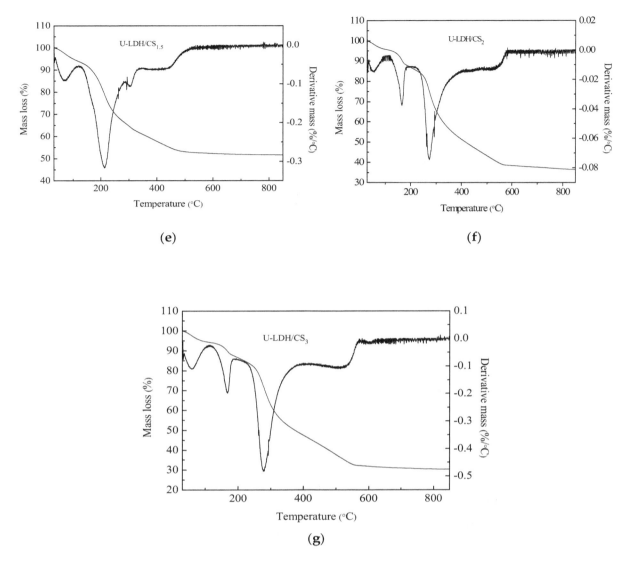

Figure 8. Thermal gravimetric and differential thermal gravimetric TGA-DTG curves of (**a**) pure CS; (**b**) U-LDH$_5$; (**c**) U-LDH/CS$_{0.5}$; (**d**) U-LDH/CS$_1$; (**e**) U-LDH/CS$_{1.5}$; (**f**) U-LDH/CS$_2$; (**g**) U-LDH/CS$_3$.

3.2. Antimicrobial Activity

The antimicrobial activities of the pure CS and U-LDH/CS hybrids prepared in different CS concentration were measured according to the inhibition zone diameter method, and the results are shown in Figure 9a–c. All samples were tested against *E. coli*, *S. aureus*, and *P. cyclopium*. As seen in Figure 9a–c and Table 2, the control, pure CS (Figure S1), and U-LDH/CS$_3$ meshes did not display antimicrobial activity, while the U-LDH/CS$_{0.5}$ and U-LDH/CS$_1$ meshes exhibited strong antimicrobial activity against the three microorganisms. In particular, U-LDH/CS$_1$ showed the highest antimicrobial ability and produced average inhibition zones of 24.2, 30.4, and 22.3mm against *E. coli*, *S. aureus*, and *P. cyclopium*, respectively. The antimicrobial activity of U-LDH/CS$_1$ was much larger than that of U-LDH$_5$ due to the incorporation of CS molecules. The results suggest that the hybrids of chitosan and ZnAl-LDH made for the improvement of antimicrobial ability.

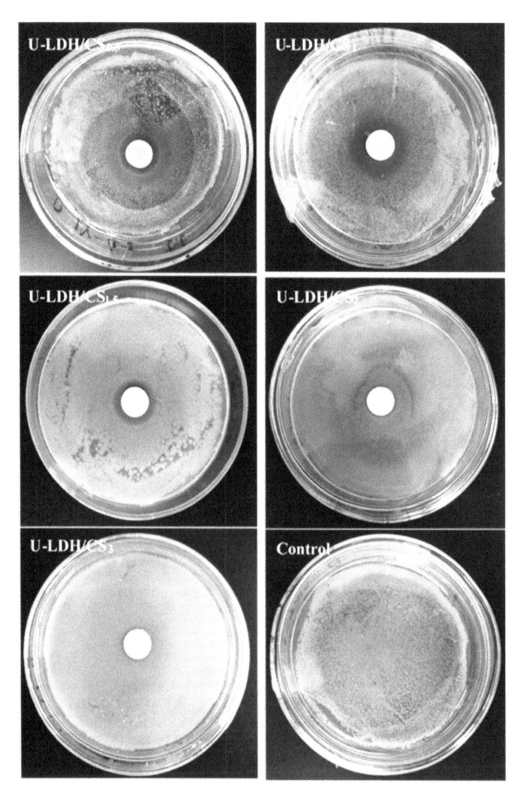

(a)

Figure 9. *Cont.*

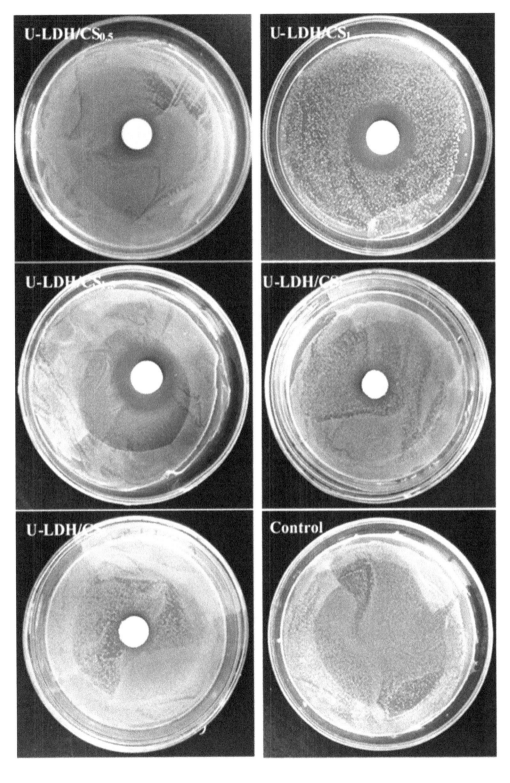

(b)

Figure 9. *Cont.*

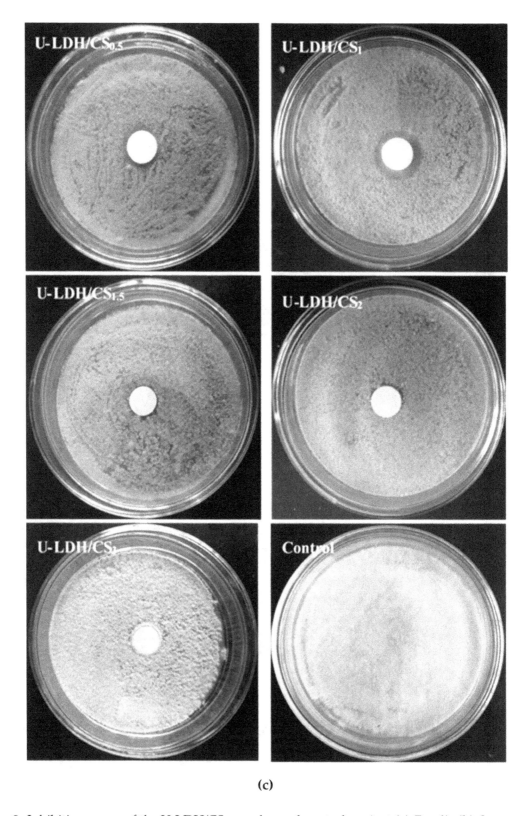

(c)

Figure 9. Inhibition zones of the U-LDH/CS samples and control against (**a**) *E. coli.*; (**b**) *S. aureus*; (**c**) *P. cyclopium.*

Table 2. Antimicrobial activities of the U-LDH/CS hybrids.

Samples g	E. coli	S. aureus	P. cyclopium
Control	0 mm	0 mm	0 mm
Pure CS	0 mm	0 mm	0 mm
U-LDH$_5$	16.5 ± 0.3 mm	20.5 ± 0.6 mm	9.8 ± 0.7 mm
U-LDH/CS$_{0.5}$	18.3 ± 0.4 mm	23.2 ± 0.2 mm	13.3 ± 0.7 mm
U-LDH/CS$_1$	24.2 ± 0.8 mm	30.4 ± 0.5 mm	22.3 ± 0.5 mm
U-LDH/CS$_{1.5}$	17.1 ± 0.3 mm	27.0 ± 0.6 mm	0mm
U-LDH/CS$_2$	0 mm	14mm	0 mm
U-LDH/CS$_3$	0 mm	0 mm	0 mm
Tetracycline*	21 ± 0.9mm	26 ± 0.9 mm	21 ± 0.8 mm

Notes: Values are mean of three replicates; * standard antibiotic.

The antimicrobial inactivity of pure chitosan was reported by Kong et al. in this term: "The antimicrobial activity for chitosan is pH dependent". The ability of chitosan to inhibit microbial growth is observed only in an acidic medium, where the polymer is soluble and carries a net positive charge [51,52]. The failure of chitosan to remain bactericidal at pH values of around six could be explained by the presence of a large majority of positively uncharged amino groups as well as by the poor solubility of chitosan [53,54]. The pH of the bacterial culture medium (pH 5.7~7.5) inhibits the chelating effect that could cause the chitosan on the bacteria to cause their lysis. Therefore, to overcome this limit, derivatization of chitosan is particularly aimed at enhancing the solubility of chitosan in aqueous medium while improving chitosan antimicrobial properties. As an example, Kong et al. [55] proposed the use of chitosan microsphere (CM) in a solid dispersing system, and even in this case, chitosan, which showed inhibitory effects, had a deacetylation degree DD of 62.6%. Above this value, and especially as reported by Kong et al. [55], in the range 83.5 to 97.5 of DD, CM showed no antimicrobial activity. Furthermore, chitosan with a lower degree of acetylation (DA) resulted in increased antimicrobial activity against various strains of fungi, Gram-positive, and Gram-negative bacteria [56]. The chitosan used in this study had a DA of 91%, which is very high. The lack of antimicrobial activity shown by chitosan could be made worse by the large size of pure chitosan particles, which prevents their entry through the pores of the bacterial membrane. Hence, the inability to cause membrane lysis as might be expected. This explains the importance of hydrolyzing chitosan in order to reduce its molecular and voluminous mass and also associate it with other compounds to protect it, thus enabling it to conserve its electrical charge. The significant antibacterial activity of the U-LDH/CS >> U-LDH >>> CS hybrid is due to the combination of the antibacterial activity of chitosan now protected by its association with the U-LDH and the antibacterial activity of the latter. Using protecting groups presents advantages, such as allowing selective modifications at the reactive centers, allowing reactions in the homogeneous medium, and giving a high degree of substitution in the products [57]. Indeed, the introduction of functional groups, such as trimethyl or quaternary alkyl groups, gives the polymer a permanent positive charge, improving its solubility in an aqueous medium and making it possible to measure the bioactivity at pH 7 [23].

In Figure 9a, against Gram-negative *E. coli*, the antibacterial activity is marked by the translucent outline around the material hybrid. This can be supported by the mechanism proposed by Dutta et al. [58]. The inactivation of *E. coli* by chitosan occurred via a two-step sequential mechanism; an initial separation of the cell wall from its cell membrane, followed by destruction of the cell membrane. This is confirmed by the mechanisms proposed by Zeng et al. and Papineau et al. [26,59]. Chitosan, through the hybrid, was reactivated and acted mainly on the outer surface of the bacteria. At a lower concentration (0.2 mg/mL), the polycationic chitosan does probably bind to the negatively charged bacterial surface to cause agglutination, while at higher concentrations the larger number of positive charges may have imparted a net positive charge to the bacterial surfaces to keep them in suspension. In Figure 9b, against Gram-positive *S. aureus*, the visible translucent areas are the inhibitions zones, which do not results in the behavior of the hybrid U-LDH/CS towards the bacteria strain. The hybrid

U-LDH/CS on the surface of the cell (*S. aureus*) can form a polymer membrane, which inhibits nutrients from entering the cell [26]. The hybrid mechanism on fungi (*P. cyclopium*), shown in Figure 8c, can be explained by the mechanism described by Bai et al. [60]. Chitosan, after being protected by U-LDH, activated a dual function: (a) to direct the interference of fungal growth and (b) to activate several defense processes. The strong antimicrobial activity shown by the U-LDH/CS hybrid results from a synergistic harmony of the proportions of ZnAl-LDH and CS compounds, and the adequation of their combination has activated the activity of the antimicrobial compounds CS, Zn^{2+}, ZnO, and Al_2O_3 contained in the structure of the hybrid. This could lead to the probable antimicrobial activity mechanisms of the hybrid U-LDH/CS described below.

The antimicrobial mechanism of the hybrid could be compared to that of the chitosan derivative as described by Hosseinnejad and Jafari [61]. It is a multiple action posed by all components of the hybrid taken together or individually. The chitosan molecules of the hybrid, positively charged (cationic NH^{3+} groups), interfere with negatively charged bacterial cell membranes. This interaction with the bacteria membrane leads to the alteration of cell permeability and membrane lysis [62–64]. The hybrid, via chitosan, could activate a chelation of nutrients, causing the inhibition of microbial growth by the essential metals [63,65]. The hybrid could also, through the generation of reactive oxygen species (ROS) through its metallic compounds, cause lysis of the bacterium [66]. The production of ROS is presented as a major contributor to the antibacterial activities of various metal oxides [67]. Such reactive species are superoxide anion (O^2), hydrogen peroxide (H_2O_2), and hydroxide (OH-). The toxicity of these species involves the destruction of cellular components such as lipids, DNA, and proteins as a result of their internalization into the bacteria cell membrane [61]. Beside these mechanisms, it also causes the release of zinc ions (Zn^{2+}) in medium containing ZnO nanoparticles and bacteria [68,69]. Released Zn^{2+} has a significant effect, causing active transport inhibition as well as the disruption of the amino acid metabolism and enzyme systems [70,71]. The probable antimicrobial activity mechanisms of hybrid U-LDH/CS is proposed in Figure 10.

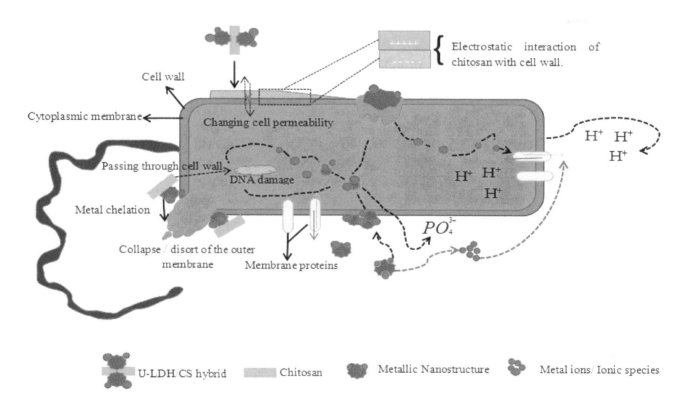

Figure 10. Schematic representation of probable mechanisms involved in U-LDH/CS antimicrobial activity.

U-LDH/CS$_2$ and U-LDH/CS$_3$ showed either no antimicrobial effect, or a decreased antimicrobial effect. The characterization analysis (UV-Vis, thermogravimetric, and rheological) showed that U-LDH/CS$_2$ and U-LDH/CS$_3$ behaved almost like pure chitosan. It seems that at about 2 g L^{-1} or more of chitosan concentration, U-LDH$_5$ is either insufficient or unable to firmly attach and protect the chitosan against the pH of the medium. LDH, which has not been attached to chitosan, has therefore not been able to release in the medium metal oxides and ionic particles. So, the chitosan molecules that are subjected to the pH of the medium diffuse this pH effect by their interconnection bond, which considerably reduces the chitosan molecules that have retained their antimicrobial activity. In addition, the large amount of chitosan also disrupts the antibacterial activity of metallic nanostructures released from LDH. This explains why, from 2 g L^{-1} of chitosan concentration, there is a remarkable reduction or loss of the antimicrobial activity of chitosan. This was confirmed by rheological, FTIR, UV-Vis, and thermogravimetric analyses, which indicated that U-LDH/CS$_3$ showed fragility to remain compact, with U-LDH/CS$_2$ and U-LDH/CS$_3$ absorption bands being more similar to those of chitosan. The amount of chitosan is a significant factor of effectiveness of the antibacterial activity of the hybrid. In the range of 0.5 to 1.5 g L^{-1} of chitosan concentration, its plays a stimulating role in the antimicrobial activity of the hybrid by producing an increase of the antimicrobial activity. This increase of the hybrid antimicrobial activity is shown by the enlargement of the inhibition zones. However, in the range of 1.5 to 3 g L^{-1}, chitosan concentration has an inhibitory effect on the antimicrobial activity of the hybrid. This results in a reduction or complete loss of the antimicrobial activity of the hybrid. The concentration of chitosan therefore has a pivotal role in the synthesis and efficacy of the antimicrobial activity of the U-LDH/CS hybrid.

4. Conclusions

The development and spread of antibiotic-resistant pathogens involves research to develop new antimicrobial agents. Hybrids derived from chitosan are newly emerging areas of research into the synthesis of antimicrobial agents. Hybridization of chitosan with ZnAl-LDH has the potential to restore the effectiveness of the antibacterial activity of chitosan at a pH of about six and above, provide novel candidates with a synergistic effect in terms of efficacy and lowered resistance selection propensity, and could confer antibacterial activity against a broad spectrum of resistant microbes. Antibacterial hybrid materials based on the ZnAl/chitosan matrix have been successfully synthesized. It was shown that the obtained hybrid tested against *E. coli*, *S. aureus*, and *P. cyclopium* demonstrated the highest antimicrobial activity toward *S. aureus*. It has been found that an increase of CS concentration around and above 1.5 g L^{-1} led to a decrease in the antibacterial activity of U-LDH/CS. The effectiveness of the antibacterial activity of U-LDH/CS hybrids suggest that the proposed preparation of such hybrid materials could be extended to the development of other organic-inorganic hybrid antimicrobial materials that may have applications in several areas of biotechnology and medical engineering.

Author Contributions: B.F.C.A.G. and H.-Y.Z. conceived and designed the experiments; B.F.C.A.G., X.-J.C. and K.-M.Z. performed the experiments; B.F.C.A.G., H.-Y.Z., W.S. and Y.D. analyzed the data; H.-Y.Z. contributed reagents/materials/analysis tools; B.F.C.A.G. wrote the paper.

Acknowledgments: This work was supported by the National Natural Science Foundation of China (21105085, 31270988) and the Key Project of Hunan Provincial Natural Science Foundation of China (12JJ2008).

References

1.   Nunes, D.; Pimentel, A.; Santos, L.; Barquinha, P.; Pereira, L.; Fortunato, E.; Martins, R. *Textbook of Metal Oxide Nanostructures: Synthesis, Properties and Applications*, 1st ed.; Elsevier: North-Holland, The Netherlands, 2018; p. 328.

2.   Follmann, H.D.; Messias, I.; Queiroz, M.N.; Araujo, R.A.; Rubira, A.F.; Silva, R. Designing hybrid materials with multifunctional interfaces for wound dressing, electrocatalysis, and chemical separation. *J. Colloid Interface Sci.* **2019**, *533*, 106–125. [CrossRef] [PubMed]

3.   Ananikov, V.P.; Eremin, D.B.; Yakukhnov, S.A.; Dilman, A.D.; Levin, V.V.; Egorov, M.P.; Karlov, S.S.; Kustov, L.M.; Tarasov, A.L.; Greish, A.A.; et al. Organic and hybrid systems: From science to practice. *Mendeleev Commun.* **2017**, *27*, 425–438. [CrossRef]

4.   Toshima, N. Recent progress of organic and hybrid thermoelectric materials. *Synth. Met.* **2017**, *225*, 3–21. [CrossRef]

5.   Sirinakorn, T.; Imwiset, K.; Bureekaew, S.; Ogawa, M. Inorganic modification of layered silicates toward functional inorganic-inorganic hybrids. *Appl. Clay Sci.* **2018**, *153*, 187–197. [CrossRef]

6.   Jo, Y.K.; Lee, J.M.; Son, S.; Hwang, S.J. 2D inorganic nanosheet-based hybrid photocatalysts: Design, applications, and perspectives. *J. Photochem. Photobiol. C Photochem. Rev.* **2018**, *40*, 150–190. [CrossRef]

7.   Zhu, J.; Wen, M.; Wen, W.; Du, D.; Zhang, X.; Wang, S.; Lin, Y. Recent progress in biosensors based on organic-inorganic hybrid nanoflowers. *Biosens. Bioelectron.* **2018**, *120*, 175–187. [CrossRef] [PubMed]

8.   Ma, Y.; Chen, L.; Ye, Y.; Wan, H.; Zhou, H.; Chen, J. Preparation and tribological behaviors of a novel organic-inorganic hybrid resin bonded solid lubricating coating cured by ultraviolet radiation. *Prog. Org. Coat.* **2019**, *127*, 348–358. [CrossRef]

9.   Sharaha, U.; Rodriguez-Diaz, E.; Riesenberg, K.; Bigio, I.J.; Huleihel, M.; Salman, A. Using Infrared Spectroscopy and Multivariate Analysis to Detect Antibiotics' Resistant *Escherichia coli* Bacteria. *Anal. Chem.* **2017**, *89*, 8782–8790. [CrossRef]

10. Gagnon, K.J.; Perry, H.P.; Clearfield, A. Chem Inform Abstract: Conventional and Unconventional Metal-Organic Frameworks Based on Phosphonate Ligands: MOFs and UMOFs. *Chem. Rev.* **2012**, *43*, 1034–1054. [CrossRef]

11. Zhang, T.; Zhou, Y.; Wang, Y.; Zhang, L.; Wang, H.; Wu, X. Fabrication of hierarchical nanostructured BSA/ZnO hybrid nanoflowers by a self-assembly process. *Mater. Lett.* **2014**, *128*, 227–230. [CrossRef]

12. Wang, X.; Lou, T.; Zhao, W.; Song, G. Preparation of pure chitosan film using ternary solvents and its super absorbency. *Carbohydr. Polym.* **2016**, *153*, 253–257. [CrossRef] [PubMed]

13. Ge, H.C.; Hua, T.T. Synthesis and characterization of poly (maleic acid)-grafted crosslinked chitosan nanomaterial with high uptake and selectivity for Hg (II) sorption. *Carbohyd. Polym.* **2016**, *153*, 246–252. [CrossRef] [PubMed]

14. Zubair, M.; Daud, M.; McKay, G.; Shehzad, F.; Al-Harthi, M.A. Recent progress in layered double hydroxides (LDH)-containing hybrids as adsorbents for water remediation. *Appl. Clay Sci.* **2017**, *143*, 279–292. [CrossRef]

15. Khan, S.; Kamal, T.; Asiri, A.; Akhtar, K.; Khan, S. Recent Development of Chitosan Nanocomposites for Environmental Applications. *Recent Pat. Nanotechnol.* **2016**, *10*, 181–188. [CrossRef] [PubMed]

16. Choi, C.; Nam, J.-P.; Nah, J.-W. Application of chitosan and chitosan derivatives as biomaterials. *J. Ind. Eng. Chem.* **2016**, *33*, 1–10. [CrossRef]

17. Thomas, V.; Yallapu, M.M.; Sreedhar, B.; Bajpai, S.K. Fabrication, Characterization of Chitosan/Nanosilver Film and Its Potential Antibacterial Application. *J. Biomater. Sci. Polym. Ed.* **2009**, *20*, 2129–2144. [CrossRef]

18. Ruiz-Hitzky, E.; Aranda, P.; Darder, M.; Rytwo, G. Hybrid materials based on clays for environmental and biomedical applications. *J. Mater. Chem.* **2010**, *20*, 9306. [CrossRef]

19. Cunha, V.R.R.; Guilherme, V.A.; De Paula, E.; De Araujo, D.R.; Silva, R.O.; Medeiros, J.V.R.; Leite, J.R.S.A.; Peterson, P.A.D.; Foldvari, M.; Petrilli, H.M.; et al. Delivery system for mefenamic acid based on the nanocarrier layered double hydroxide: Physicochemical characterization and evaluation of anti-inflammatory and antinociceptive potential. *Mater. Sci. Eng. C* **2016**, *58*, 629–638. [CrossRef]

20. Ferraris, S.; Spriano, S. Review Antibacterial titanium surfaces for medical implants. *Adv. Mater. Sci. Eng.C* **2016**, *61*, 965–978. [CrossRef]

21. Mishra, G.; Dash, B.; Pandey, S.; Mohanty, P.P. Antibacterial actions of silver nanoparticles incorporated Zn-Al layered double hydroxide and its spinel. *J. Environ. Chem. Eng.* **2013**, *1*, 1124–1130. [CrossRef]

22. Li, B.; Zhang, Y.; Yang, Y.; Qiu, W.; Wang, X.; Liu, B.; Wang, Y.; Sun, G. Synthesis, characterization, and antibacterial activity of chitosan/TiO$_2$ nanocomposite against Xanthomonas oryzae pv. oryzae. *Carbohydr. Polym.* **2016**, *152*, 825–831. [CrossRef] [PubMed]

23. Sahariah, P.; Másson, M. Antimicrobial Chitosan and Chitosan Derivatives: A Review of the Structure–Activity Relationship. *Biomacromolecules* **2017**, *18*, 3846–3868. [CrossRef] [PubMed]

24. Liu, H.; Du, Y.; Wang, X.; Sun, L. Chitosan kills bacteria through cell membrane damage. *Int. J. Food Microbiol.* **2004**, *95*, 147–155. [CrossRef] [PubMed]

25. Helander, I.; Nurmiaho-Lassila, E.-L.; Ahvenainen, R.; Rhoades, J.; Roller, S. Chitosan disrupts the barrier properties of the outer membrane of Gram-negative bacteria. *Int. J. Food Microbiol.* **2001**, *71*, 235–244. [CrossRef]

26. Zheng, L.Y.; Zhu, J.F. Study of antimicrobial activity of chitosan with different molecular weight. *Carbohyd. Polym.* **2003**, *54*, 527–530. [CrossRef]

27. Zeng, H.-Y.; Deng, X.; Wang, Y.-J.; Liao, K.-B. Preparation of Mg-Al hydrotalcite by urea method and its catalytic activity for transesterification. *AIChE J.* **2009**, *55*, 1229–1235. [CrossRef]

28. Li, S.-P.; Hou, W.-G.; Han, S.-H.; Li, L.-F.; Zhao, W.-A. Studies on intrinsic ionization constants of Fe–Al–Mg hydrotalcite-like compounds. *J. Colloid Interface Sci.* **2003**, *257*, 244–249. [CrossRef]

29. Barka, E.A.; Gognies, S.; Nowak, J.; Audran, J.-C.; Belarbi, A. Inhibitory effect of endophyte bacteria on Botrytis cinerea and its influence to promote the grapevine growth. *Biol. Control* **2002**, *24*, 135–142. [CrossRef]

30. Yi, J.-Z.; Zhang, L.-M. Studies of sodium humate/polyacrylamide/clay hybrid hydrogels. I. Swelling and rheological properties of hydrogels. *Eur. Polym. J.* **2007**, *43*, 3215–3221. [CrossRef]

31. Zheng, Z.; Levin, R.E.; Pinkham, J.L.; Shetty, K. Decolorization of polymeric dyes by a novel Penicillium isolate. *Process Biochem.* **1999**, *34*, 31–37. [CrossRef]

32. Hou, W.-G.; Su, Y.-L.; Sun, D.-J.; Zhang, C.-G. Studies on Zero Point of Charge and Permanent Charge Density of Mg-Fe Hydrotalcite-like Compounds. *Langmuir* **2001**, *17*, 1885–1888. [CrossRef]

33. Jahangirian, H.; Haron, M.J.; Shah, M.H.; Abdollahi, Y.A.; Rezayi, M.A.; Vafaei, N.A. Well diffusion method for evaluation of antibacterial activity of copper phenyl fatty hydroxamate synthesized from canola and palm kernel oils. *Dig. J. Nanomater. Biostruct.* **2013**, *8*, 1263–1270.

34. Luo, L.; Li, Q.; Xu, Y.; Ding, Y.; Wang, X.; Deng, D.; Xu, Y. Amperometric glucose biosensor based on NiFe2O4 nanoparticles and chitosan. *Sens. Actuators B Chem.* **2010**, *145*, 293–298. [CrossRef]

35. Viswanathan, N.; Meenakshi, S. Enriched fluoride sorption using alumina/chitosan composite. *J. Hazard. Mater.* **2010**, *178*, 226–232. [CrossRef] [PubMed]

36. Si, Y.; Samulski, E.T. Synthesis of water soluble grapheme. *Nano Lett.* **2008**, *8*, 1679–1682. [CrossRef] [PubMed]

37. Pawlak, A.; Mucha, M. Thermogravimetric and FTIR studies of chitosan blends. *Thermochim. Acta* **2003**, *396*, 153–166. [CrossRef]

38. Darder, M.; Colilla, M.; Ruiz-Hitzky, E. Chitosan-clay nanocomposites: Application as electrochemical sensors. *Appl. Clay Sci.* **2005**, *28*, 199–208. [CrossRef]

39. Darder, M.; Colilla, M.; Ruiz-Hitzky, E. Biopolymer-Clay Nanocomposites Based on Chitosan Intercalated in Montmorillonite. *Chem. Mater.* **2003**, *15*, 3774–3780. [CrossRef]

40. Krishnamoorti, R.; Giannelis, E.P. Rheology of End-Tethered Polymer Layered Silicate Nanocomposites. *Macromology* **1997**, *30*, 4097–4102. [CrossRef]

41. Costa, F.R.; Wagenknecht, U.; Jehnichen, D.; Goad, M.A.; Heinrich, G. Nanocomposites based on polyethylene and Mg-Al layered double hydroxide. Part II. Rheological characterization. *Polymer* **2006**, *47*, 1649–1660. [CrossRef]

42. Cárdenas, G.; Miranda, S.P. FTIR and TGA studies of chitosan composite films. *J. Chil. Chem. Soc.* **2004**, *49*, 291–295. [CrossRef]

43. Kumar, S.; Koh, J. Physiochemical, Optical and Biological Activity of Chitosan-Chromone Derivative for Biomedical Applications. *Int. J. Mol. Sci.* **2012**, *13*, 6102–6116. [CrossRef] [PubMed]

44. Taboada, E. Retención de Metales Pesados Utilizando Quitosano y Derivados. Ph.D. Thesis, Universidad de Concepción, Concepción, Chile, January 2003.

45. Vágvölgyi, V.; Palmer, S.J.; Kristóf, J.; Frost, R.L.; Horváth, E. Mechanism for hydrotalcite decomposition: A controlled rate thermal analysis study. *J. Colloid Interface Sci.* **2008**, *318*, 302–308. [CrossRef] [PubMed]

46. Tichit, D.; Rolland, A.; Prinetto, F.; Fetter, G.; Martinez-Ortiz, M.D.J.; Valenzuela, M.A.; Bosch, P. Comparison of the structural and acid???base properties of Ga- and Al-containing layered double hydroxides obtained by microwave irradiation and conventional ageing of synthesis gels. *J. Mater. Chem.* **2002**, *12*, 3832–3838. [CrossRef]

47. Ahmed, A.A.A.; Talib, Z.A.; Bin Hussein, M.Z. Thermal, optical and dielectric properties of Zn–Al layered double hydroxide. *Appl. Clay Sci.* **2012**, *56*, 68–76. [CrossRef]

48. Auwalu, A.; Linlin, T.; Ahmad, S.; Hongying, Y.; Zhenan, J.; Song, Y. Preparation and application of metal ion-doped CoMgAl-hydrotalcite visible-light-driven photocatalyst. *Int. J. Ind. Chem.* **2019**, *10*, 121–131. [CrossRef]

49. Elhalil, A.; Elmoubarki, R.; Machrouhi, A.; Sadiq, M.; Abdennouri, M.; Qourzal, S.; Barka, N. Photocatalytic degradation of caffeine by ZnO-ZnAl2O4 nanoparticles derived from LDH structure. *J. Environ. Chem. Eng.* **2017**, *5*, 3719–3726. [CrossRef]

50. Kong, M.; Chen, X.G.; Xing, K.; Park, H.J. Antimicrobial properties of chitosan and mode of action: A state of the art review. *Int. J. Food Microbiol.* **2010**, *144*, 51–63. [CrossRef]

51. Simunek, J.; Brandysova, V.; Koppova, I.; Simunek, J., Jr. The antimicrobial action of chitosan, low molar mass chitosan, and chitooligosaccharides on human colonic bacteria. *Folia Microbiol.* **2012**, *57*, 341–345. [CrossRef]

52. Jung, E.J.; Youn, D.K.; Lee, S.H.; No, H.K.; Ha, J.G.; Prinyawiwatkul, W. Antibacterial activity of chitosans with different degrees of deacetylation and viscosities. *Int. J. Food Sci. Technol.* **2010**, *45*, 676–682. [CrossRef]

53. Aiedeh, K.; Taha, M.O. Synthesis of iron-crosslinked chitosan succinate and iron-crosslinked hydroxamated chitosan succinate and their in vitro evaluation as potential matrix materials for oral theophylline sustained-release beads. *Eur. J. Pharm. Sci.* **2001**, *13*, 159–168. [CrossRef]

54. Sudarshan, N.R.; Hoover, D.G.; Knorr, D. Antibacterial action of chitosan. *Food Biotechnol.* **1992**, *6*, 257–272. [CrossRef]

55. Kong, M.; Chen, X.G.; Liu, C.S.; Yu, L.J.; Ji, Q.X.; Xue, Y.P.; Cha, D.S.; Park, H.J. Preparation and antibacterial activity of chitosan microspheres in a solid dispersing system. *Front. Mater. Sci.* **2008**, *2*, 214–220. [CrossRef]

56. Andres, Y.; Giraud, L.; Gerente, C.; Le Cloirec, P. Antibacterial Effects of Chitosan Powder: Mechanisms of Action. *Environ. Technol.* **2007**, *28*, 1357–1363. [CrossRef] [PubMed]

57. Sahariah, P.; Óskarsson, B.M.; Hjálmarsdóttir, M.A.; Másson, M. Synthesis of guanidinylated chitosan with the aid of multiple protecting groups and investigation of antibacterial activity. *Carbohydr. Polym.* **2015**, *127*, 407–417. [CrossRef]

58. Dutta, P.; Tripathi, S.; Mehrotra, G.; Dutta, J. Perspectives for chitosan based antimicrobial films in food applications. *Food Chem.* **2009**, *114*, 1173–1182. [CrossRef]

59. Papineau, A.M.; Hoover, D.G.; Knorr, D.; Farkas, D.F. Antimicrobial effect of water-soluble chitosans with high hydrostatic pressure. *Food Biotechnol.* **1991**, *5*, 45–57. [CrossRef]

60. Bai, R.-K.; Huang, M.-Y.; Jiang, Y.-Y. Selective permeabilities of chitosan-acetic acid complex membrane and chitosan-polymer complex membranes for oxygen and carbon dioxide. *Polym. Bull.* **1988**, *20*, 83–88. [CrossRef]

61. Hosseinnejad, M.; Jafaria, S.M. Evaluation of different factors affecting antimicrobial properties of chitosan. *Int. J. Biol. Macromol.* **2016**, *85*, 467–475. [CrossRef] [PubMed]

62. Severino, R.; Vu, K.D.; Donsi', F.; Salmieri, S.; Ferrari, G.; Lacroix, M. Antimicrobial effects of different combined non-thermal treatments against Listeria monocytogenes in broccoli florets. *J. Food Eng.* **2014**, *124*, 1–10. [CrossRef]

63. Chien, R.-C.; Yen, M.-T.; Mau, J.-L. Antimicrobial and antitumor activities of chitosan from shiitake stipes, compared to commercial chitosan from crab shells. *Carbohydr. Polym.* **2016**, *138*, 259–264. [CrossRef] [PubMed]

64. Li, Z.; Yang, F.; Yang, R. Synthesis and characterization of chitosan derivatives with dual-antibacterial functional groups. *Int. J. Biol. Macromol.* **2015**, *75*, 378–387. [CrossRef] [PubMed]

65. Yuan, G.; Lv, H.; Tang, W.; Zhang, X.; Sun, H. Effect of chitosan coating combined with pomegranate peel extract on the quality of Pacific white shrimp during iced storage. *Food Control* **2016**, *59*, 818–823. [CrossRef]

66. Zhang, H.; Lv, X.-J.; Li, Y.; Wang, Y.; Li, J. P25-Graphene Composite as a High Performance Photocatalyst. *ACS Nano* **2009**, *4*, 380–386. [CrossRef] [PubMed]

67. Prasad, R.; Basavaraju, D.; Rao, K.; Naveen, C.; Endrino, J.; Phani, A. Nanostructured TiO$_2$ and TiO$_2$-Ag antimicrobial thin fifilms. In Proceedings of the 2011 International Conference on Nanoscience, Technology and Societal Implications (NSTSI), Bhubaneswar, India, 8–10 December 2011.

68. Premanathan, M.; Karthikeyan, K.; Jeyasubramanian, K.; Manivannan, G. Selective toxicity of ZnO nanoparticles toward Gram-positive bacteria and cancer cells by apoptosis through lipid peroxidation. *Nanomed. Nanotechnol. Biol. Med.* **2011**, *7*, 184–192.

69.   Li, M.; Zhu, L.; Lin, D. Toxicity of ZnO nanoparticles to *Escherichia coli*: Mechanism and the inflfluence of medium components. *Environ. Sci. Technol.* **2011**, *45*, 1977–1983. [CrossRef] [PubMed]

70.   Wong, S.W.; Leung, P.T.; Djurišić, A.B.; Leung, K.M. Toxicities of nano zinc oxide to five marine organisms: Influences of aggregate size and ion solubility. *Anal. Bioanal. Chem.* **2010**, *396*, 609–618. [CrossRef] [PubMed]

71.   Wu, B.; Wang, Y.; Lee, Y.-H.; Horst, A.; Wang, Z.; Chen, D.-R.; Sureshkumar, R.; Tang, Y.J. Comparative Eco-Toxicities of Nano-ZnO Particles under Aquatic and Aerosol Exposure Modes. *Environ. Sci. Technol.* **2010**, *44*, 1484–1489. [CrossRef] [PubMed]

Antimicrobial Activity of Lignin and Lignin-Derived Cellulose and Chitosan Composites against Selected Pathogenic and Spoilage Microorganisms

Abla Alzagameem [1,2], Stephanie Elisabeth Klein [1], Michel Bergs [1], Xuan Tung Do [1], Imke Korte [3], Sophia Dohlen [3], Carina Hüwe [3], Judith Kreyenschmidt [3], Birgit Kamm [2,4], Michael Larkins [1,5] and Margit Schulze [1,*]

[1] Department of Natural Sciences, Bonn-Rhein-Sieg University of Applied Sciences, von-Liebig-Str. 20, D-53359 Rheinbach, Germany; abla.alzagameem@h-brs.de (A.A.); stephanie.klein@h-brs.de (S.E.K.); michel.bergs@h-brs.de (M.B.); xuan-tung.do@h-brs.de (X.T.D.); mclarki2@ncsu.edu (M.L.)
[2] Faculty of Environment and Natural Sciences, Brandenburg University of Technology BTU Cottbus-Senftenberg, Platz der Deutschen Einheit 1, D-03046 Cottbus, Germany; b.kamm@kplus-wood.at
[3] Rheinische Friedrich Wilhelms-University Bonn, Katzenburgweg 7-9, D-53115 Bonn, Germany; i.korte@uni-bonn.de (I.K.); sophia.dohlen@uni-bonn.de (S.D.); chuewe@uni-bonn.de (C.H.); j.kreyenschmidt@uni-bonn.de (J.K.)
[4] Kompetenzzentrum Holz GmbH, Altenberger Strasse 69, A- 4040 Linz, Austria
[5] Department of Forest Biomaterials, North Carolina State University, 2820 Faucette Drive Biltmore Hall, Raleigh, NC 27695, USA
*   Correspondence: margit.schulze@h-brs.de.

Abstract: The antiradical and antimicrobial activity of lignin and lignin-based films are both of great interest for applications such as food packaging additives. The polyphenolic structure of lignin in addition to the presence of O-containing functional groups is potentially responsible for these activities. This study used DPPH assays to discuss the antiradical activity of HPMC/lignin and HPMC/lignin/chitosan films. The scavenging activity (SA) of both binary (HPMC/lignin) and ternary (HPMC/lignin/chitosan) systems was affected by the percentage of the added lignin: the 5% addition showed the highest activity and the 30% addition had the lowest. Both scavenging activity and antimicrobial activity are dependent on the biomass source showing the following trend: organosolv of softwood > kraft of softwood > organosolv of grass. Testing the antimicrobial activities of lignins and lignin-containing films showed high antimicrobial activities against Gram-positive and Gram-negative bacteria at 35 °C and at low temperatures (0–7 °C). Purification of kraft lignin has a negative effect on the antimicrobial activity while storage has positive effect. The lignin release in the produced films affected the activity positively and the chitosan addition enhances the activity even more for both Gram-positive and Gram-negative bacteria. Testing the films against spoilage bacteria that grow at low temperatures revealed the activity of the 30% addition on HPMC/L1 film against both *B. thermosphacta* and *P. fluorescens* while L5 was active only against *B. thermosphacta*. In HPMC/lignin/chitosan films, the 5% addition exhibited activity against both *B. thermosphacta* and *P. fluorescens*.

Keywords: antimicrobial activity; antiradical activity; chitosan; hydroxypropylmethylcellulose; lignin; pathogenic microorganisms; organosolv

1. Introduction

Due to the environmental problems caused by nonbiodegradable synthetic plastic packaging materials, research has been focused to develop biodegradable packaging materials using renewable resources and biomass-derived waste. Thus, high strength oxygen-barrier films prepared from renewable forestry product waste (hot-water wood extract) were reported by Cheng et al. as an industrial, scalable, simple, and green processing approach [1]. Natural polysaccharides such as starch, cellulose, and hemicellulose have been intensively investigated as appropriate raw materials for the development of novel biodegradable food packaging materials [2–5]. In this context, the search for appropriate starting compounds is supported by the establishment of biorefineries to exploit lignocellulose feedstock (LCF) and corresponding LCF-rich biomass and/or waste [6–8].

Besides cellulose and hemicellulose, lignin attracts scientific interest as source of aromatic compounds, representing 30% of all non-fossil organic carbon on earth. Lignin is produced in large quantities as a byproduct of the pulp and paper industry wherein it is primarily burnt as a low-efficiency fuel to power paper mills [9]. Aside from its abundance and inexpensive supply, lignin is favorable for its numerous attractive properties, such as biodegradability, antioxidant activity, high carbon content, high thermal stability, and stiffness, comprehensively reviewed by Rinaldi et al. [10]. These important features of lignin can be synergistically combined with the advanced functionalities of well-defined polymers via covalent bond linkages [11]. Concerning the antibacterial properties of lignin, it seems to be also a promising green replacement for fossil-based agents useful against several dangerous microorganisms. Its biocidal activity makes it a more attractive compound than silver nanoparticles due to its reduced environmental impact. Some studies have shown that lignin has an antimicrobial effect [12]. The phenolic hydroxyl and methoxy groups contained in lignin have been reported to be biologically active. Depending on biomass source and pulping process, lignins vary in their 3D structure and possess different antimicrobial, antioxidant, and UV absorption properties. Various investigations have suggested that lignins can be applied to stabilize food and feedstuffs because of their antioxidant, antifungal, and antiparasitic properties [12–16]. Thus, Guo and others found that lignin extracts show considerable antimicrobial activity against *Listeria innocua* (a Gram-positive bacteria) [17].

As a polyphenol, lignin has the potential as an antioxidant to prevent oxidation reactions in biofuels, animal feeds, and polymeric composite materials. Structurally, lignin is a randomly crosslinked polymer consisting of three different phenylpropane derivatives mainly linked by ether bonds (Figures 1 and 2). Numerous studies were reported to elucidate the detailed 3D structure of lignin, including the formation of more complex bonds (as shown in Figure 2, last row) during biosynthesis, comprehensively reviewed by Lupoi et al. [18].

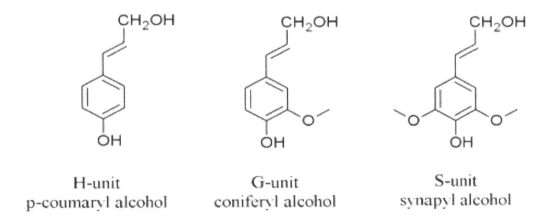

Figure 1. Lignin units H, G, and S derived from corresponding cinnamoyl alcohols.

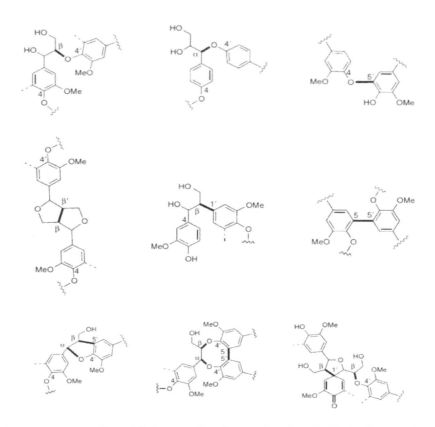

Figure 2. Most common monolignol linkages. First line: ether bonds (ß-O-4′, α-O-4′, 4-O-5′); second line: C–C bonds (ß-ß′, ß-1′, 5-5′), and third line: more complex linkages (ß-5′/ α-O-4′, 5-5′/ ß-O-4′/ α-O-4′, ß-1′/ ß-O-4′). Reprinted from [15] under open access license.

Due to this rather complex mixture of differently connected monolignols, most of the properties of lignins are dependent on the final 3D structure. Antioxidant capacity mainly is caused by hydrogen transfer and single electron transfer reactions and the capacity varies as soon as the number of available functional groups, such as OH and OCH$_3$ changes. So far, lignin is not transferred to chemical industry as a commercial antioxidant due to this inhomogeneity [19,20]. Very recently, it was shown that plant portions such as stem- and leaf-derived lignins significantly differ in their monolignol composition and connectivity [21,22]. In addition to their UV absorption properties, the free radical scavenging ability of phenolic groups gives lignin an excellent antioxidant property and can increase thermal and oxidation stability of polymers in blends [23]. Sugarcane bagasse lignin and modified lignins can serve as natural, safer and less expensive potential antioxidant substitutes for synthetic antioxidants such as BHT. Also, unmodified lignin from sugarcane bagasse and epoxy lignin could be used as natural antibacterials [24].

For a long time, natural cellulose and related composites have been prepared and applied in many different fields including packaging, optics, and sensor technologies. Within the last five years, first examples of lignin/cellulose-based coatings have been reported [25]. In the context of food preservation, few studies have been reported on the use of organic active principles (vegetal) against bacteria/fungi plant pathogens, while a large amount of literature is available on how these active agents can be incorporated in biobased polymeric matrices. It is important to remember how crop damage and the economic losses of agriculture production caused by plant pathogens both represent a serious worldwide problem that needs innovative and green strategies able to preserve important fresh productions during their transport and storage. Recently, the possibility of obtaining various functionalities by using the combination of different nanofillers, including antimicrobial activity, was also considered; for example, cellulose nanocrystals and lignin nanoparticles were added to PLA films to both act as promising bioactive packaging for the preservation of fresh food products against food borne pathogens and to reinforce the PLA nanocomposites [26].

Among the many (polysaccharidic) biopolymers used for the preparation of edible films and coatings, hydroxypropylmethylcellulose (HPMC) has been the focus of many studies due to its availability, edibility, good film-forming properties (resulting in transparent films with suitable mechanical performance) and excellent gas and grease barrier properties, as well as its ability to retain a large amount of active compounds [27].

Chitosan, composed of 2-glucosamine and *N*-acetyl-2-glucosamine monomer units, is mainly produced from crustacean shells (i.e., crab, shrimp). Chitosans are biodegradable and biocompatible, furthermore they are attractive due to their antioxidant and antimicrobial activity. However, some disadvantages, such as dissolution in highly acidic solutions, low surface area, high cost, and poor thermal and mechanical properties, requires the combination with other components such as polyvinyl alcohol (PVA) to overcome the weaknesses [28].

In this work, lignin and chitosan were introduced in both binary (HPMC/lignin) and ternary (HPMC/lignin/chitosan) composite systems with different lignin types and weight loading using a solvent casting approach. The antiradical and antimicrobial activities of the prepared films were studied as well as the antimicrobial activities of added lignins.

2. Materials and Methods

2.1. Lignin Isolation and Purification

Industrial black liquor was delivered by from the Zellstoffwerk Blankenstein GmbH (Blankenstein, Germany). Gradual acidic precipitation of black liquor was performed for lignin extraction using HCl and H_2SO_4 with varying pH, temperature, and time of stirring. Corresponding yields were determined. L1 was extracted under the following conditions: H_2SO_4 with stirring at room temperature, pH = 2 for 90–180 min L2 was prepared via soaking of L1 with diethyl ether. Selective extraction with acetone and ethanol, respectively, produced L3 and L4. Diethyl ether was used to precipitate all samples. Monitoring was performed via thin layer chromatography (TLC). The organosolv lignins were isolated from spruce/pine (L5), beech wood (L6), and Miscanthus (L7) according to an earlier published procedure [29].

2.2. Lignin Purity, Ash and Sugar Content via NREL Measurements

The chemical composition (%, *w/w*) was determined according to the standard analytical procedures published by National Renewable Energy Laboratory (NREL) [30]. NREL measurements were performed by the BIOPOS Research Institute (Teltow-Seehof, Germany). HPLC analysis was conducted using water at a flow rate of 0.4 mL/min in a column (300 × 7.8 mm, Machery-Nagel, Düren, Germany) at a constant temperature of 90 °C. Structural carbohydrates in biomass and lignin samples were determined following NREL procedures [31–34].

2.3. 2D Heteronuclear Single Quantum Coherence (HSQC) NMR Analysis

According to the procedure described in [29], HSQC spectra were recorded with a spectral width of 7211 Hz, a receiver gain of 2050, and a total acquisition time of 0.28 s. O1 was set to 5 ppm (^{1}H) and 80 ppm (^{13}C).

2.4. ^{31}P NMR Analysis

Analogue to a procedure given in [29], ^{31}P NMR spectra were obtained using ^{1}H-^{31}P decoupling experiment (Avance III 600, Bruker, Karlsruhe, Germany), 131,000 points were recorded at 12,175.324 Hz.

2.5. Antioxidant Activity (DPPH Assay)

The DPPH inhibition was assessed by the procedure described by Alzagameem et al. using a Jasco V-630 spectrophotometer [13]. Shortly, dioxane/water (90:10, *v/v*) at a concentration of 1 g/L was used to dissolve the samples; a mixture was prepared of 0.1 mL of the sample and 3.9 mL of a 6×10^{-5} M

DPPH solution. Finally, the absorbance was determined after 15 and 30 min, resp., at a wavelength of 518 nm.

2.6. Antimicrobial Activity of Lignins (Zone of Inhibition Test)

Inoculums of *Staphylococcus aureus* ssp. *Aureus* (DSM No 799), *L. monocytogenes* (DSM No 19094), and *E. coli* (DSM No 1576) were prepared by transferring a frozen culture to 10 mL of nutrient broth (Roth, Karlsruhe, Germany). Afterwards the broth was incubated at 37 °C for 24 h. In the beginning of each trial, the inoculum was diluted in physiological saline solution with tryptone (Oxoid, Hampshire, UK) to a final concentration of 10^7 cfu/mL. 1 mL of the suspension was spread with a sterile spatula over the surface of a plate count agar plate (Roth, Karlsruhe, Germany). Three filter papers were impregnated with the antimicrobial agent, 0.1 g of lignin was dissolved in 1 mL DMSO and dried over a circular filter paper (0.5 cm of diameter) and three blank filter papers as references were applied on the inoculated agar plate. The agar plates were incubated for 24 h at 37 °C. A clear zone without microorganism growth (zone of inhibition) is related to the level of antimicrobial activity of the agent.

2.7. Film Formation

HPMC film-forming solutions were prepared using the procedure described by Sebti et al. by dissolving 3 parts of HPMC in 200 parts of 0.01 mol L^{-1} HCl solution, 100 parts of absolute ethanol, and 10% (w/w HPMC) of PEG 400 [35]. Twenty-five mL of the solution was plated onto a glass and dried at room temperature (RT) for 36 h. HPMC-lignin film-forming solutions were prepared by dissolving 5, 10, 15, 20, 25, and 30% (w/w HPMC) of the first fraction of the purification of kraft lignin (L1) in the smallest amount of DMSO then added to the HPMC-film forming solution mentioned before, and stirred for 15 min at RT. 25 mL of the polymer solution was plated onto a glass and dried at RT for 36 h. HPMC-Organosolv lignin (L5, L6 and L7) samples were prepared the same way. HPMC-lignin-chitosan film-forming solutions were prepared by dissolving 5% of 85% deacetylated chitosan in absolute ethanol and added to the HPMC film forming solution, while stirring, 5, 10, 15, 20, 25, and 30% of (L1, L5, L6, L7) dissolved in the smallest amount of DMSO was added to the HPMC-chitosan film forming solution, and stirred for 15 min at RT. Twenty-five mL of the polymer solution were plated onto a glass and dried at RT for 36 h.

2.8. Antioxidant Activity (DPPH Test)

The DPPH radical scavenging activity for the studied films was determined according to the method proposed by Yang et al., with a slight modification [12]. Films (0.1 g) were cut into small pieces and immersed in 2 mL of methanol for 24 h at room temperature then centrifuged for 2 min. The obtained supernatant was analyzed for evaluation of DPPH radical scavenging activity: an aliquot of methanol extract (0.5 mL) was mixed with 0.5 mL of DPPH in methanol (50 mg L^{-1}). The mixture was maintained at room temperature in the dark for 60 min. The absorbance was measured at 517 nm using a UV–Vis spectrometer (Jasco V-630 spectrophotometer, Silver Spring, MD, USA). The mixture solution of methanol extracted from neat HPMC and DPPH methanol was used as control. DPPH radical scavenging activity was calculated by using Equation (1):

$$(RSA, \%) = [(Acontrol - Asample)/ Acontrol] \times 100 \tag{1}$$

where Asample was the absorbance of sample and Acontrol was the absorbance of the control.

2.9. Antimicrobial Activity of the Films

The antimicrobial activity of the samples was analyzed by modifying the Japanese Industrial Standard (JIS) Z 2801:2000 in order to test conditions typical for perishable products. The JIS is based on a comparison of bacteria counts (*S. aureus*, *E. coli*) in saline solution on reference and sample materials after a defined storage temperature and time (35 °C, 24 h) [36,37]. The material that shows a calculated log10-reduction $\geq$ 2.0 log10 units after 24 h is considered as an effective antimicrobial agent (JIS Z 2801:2000). If log10-reduction $\emptyset \geq 2$ log10 units is reached, the antimicrobial activity against *B. thermosphacta* and *P. fluorescens* is then tested. The test will be conducted at a constant temperature of 7 °C for 24 h. In case the $\emptyset$ log10-reduction is ≥ 2 log10, highly concentrated meat extract solution (18 µg mL^{-1}) will be chosen as the reference media for perishable foods [36–40].

3. Results and Discussion

3.1. Antimicrobial Activity of Lignin

Lignin purification fractions were tested against both Gram-positive (*S. aureus* and *L. monocytogenes*) and Gram-negative bacteria (*E. coli*) to test the effect of purification on the antibacterial activity. Unmodified lignins were found to be the most effective against *Bacillus* sp. (Gram-positive bacteria) than against *Klebsiella* sp. (Gram-negative bacteria) [24]. Rocca and others discussed the antibacterial activity of different lignins and nanocomposites containing lignin that have higher sugar content against *S. aureus* (known to present a thicker peptidoglycan layer) as a representative of Gram-positive bacteria, and *E. coli* (contains more fatty acids) for Gram-negative. They concluded that the interaction of the nanocomposites with the bacterial cell wall can be governed by the lignin structure helping not only the stability of the particles but also their selectivity towards different types of bacteria [41]. The disk diffusion method was applied on lignin purification fractions L1 to L4. The samples were dissolved first in ethanol and tested, then in acetone and tested, and finally in DMSO and tested. The results showed that the inhibition zone of L1 increased from ethanol to acetone to DMSO. L2, L3 and L4 show no activity at all using ethanol and acetone while L2 showed slight activity against *L. monocytogenes* when dissolved in DMSO. This is due to the solubility of lignin in each solvent, where it is 100% soluble in DMSO, 60% soluble in acetone, and 40% soluble in ethanol. This allows a small fraction to dissolve, causing the antibacterial activity.

The effect of storage on the antibacterial activity of L1 was also studied, where a fresh sample of L1 was tested quantitatively against *S. aureus*. A fresh sample of L1 with a concentration of 0.1 g mL^{-1} gave a log10 reduction of 4.4, and 2.1 at the concentration of 0.001 g mL^{-1}. After storing L1 for 6 months, L1 showed better activity with a log10 reduction of 4.5 at 0.1 g mL^{-1} and remained the same at 0.001 g mL^{-1}. This is demonstrated in Alzagameem et al., 2018 through the analytical analysis of fresh and stored fractions and the DPPH inhibitions as well as the total phenol contents of the studied fractions [13]. Lignin degradation by time produces low molecular weight species that have both antioxidant and antibacterial capacities.

Based on the effect of solvent on the antibacterial activity, further studies on lignin samples were done using DMSO to guarantee that all active species were dissolved and introduced to the tested platelets. Table 1 shows the result of the disk diffusion method of the studied purification fractions in addition to two softwood-based organosolv lignins: from spruce/pine (L5) and from beech (L6) as well as a grass-based organosolv lignin: from Miscanthus L7 for comparison purposes.

The reference, which does not contain lignin, does not have any observed inhibiting effects on the bacteria, but is overgrown with colonies. *L. monocytogenes* has a higher antimicrobial efficacy than *S. aureus*. This can be seen by L1, L2, L5, and L6, in which inhibition zones around the platelets have formed on the *L. monocytogenes* plates. The tests show, as well, that the first fraction of the purification (L1 to L4) has the highest activity. Organosolv lignins obtained from softwood have certain activities while the lignin obtained from grass L7 has none. The beech based (L6) sample specifically has higher activity that the spruce/pine-based samples (L5); it actually has the highest activity among all the lignins studied. L1, on the other hand, as a kraft lignin, has higher activity than L5. The same with Organosolv lignins: Gram-positive bacteria were inhibited more than Gram-negative. In regard to *E. coli*, all studied lignins show no activity at all.

Table 1. Inhibition zones of lignin samples against *S. aureus, L. monocytogenes*, and *E. coli* obtained by disk diffusion method.

Lignin Platelets	Bacteria	
	S. aureus	*L. monocytogenes*
Reference DMSO	− −1x contaminated with yeast	− −
L1	+ (max 1–2 mm)	++ (max 5 mm)
L2	−	+ (max 1 mm)
L3	−	+ (max 1 mm), slight growth in inhibition zone
L4	−	−
L5	+ (max 1 mm)	+ (max 1–2 mm)
L6	++ (max 2–3 mm)	+ (max 7 mm)
L7	−	−

− − = no inhibition, overgrowth of platelets; − = no inhibition; + = Inhibition; ++ = strong inhibition; max = maximum: measured at the greatest distance from the plate.

The antimicrobial mechanism may involve the generation of localized heat and reactive oxygen species (ROS) upon light irradiation. Rocca et al. postulated the sugar content of the lignin might cause and/or support the adhesion to the bacterial membrane. For example, the peptidoglycan layer of bacterial cell walls is able interact with sugar molecules, thereby increasing the activity against *S. aureus* [41]. The NREL results of the lignins L1 to L4 in Table 2 show that L1 has the lowest acid insoluble lignin content and the highest acid soluble content. Glucan content (polysaccharide) is the highest for L1. Also, only L1 and L2 have arabinan content. Xylan content for L1 and L2 as well was much higher than that for L3 and L4, but it seems that the effect of the acid soluble lignin content has the main role in the antibacterial activity since L1 was more active than L2 and it correlates with the antibacterial activity results. This supports the finding of Rocca et al. [41]. PH2 SK was extracted according to Klein et al., 2018 [15] at pH = 2 which has almost the same NREL values as L2.

Table 2. Compositional analysis according to National Renewable Energy Laboratory (NREL) procedure of kraft lignin fractions (L1 to L4, purified via solvent extraction and PH2SK, isolated at pH2 without further purification).

Fraction	AIL [%]	ASL [%]	Total Lignin [%]	Ash [%]	Glucan [%]	Xylan [%]	Galactan [%]	Arabinan [%]	Rhamnan [%]	Mannan [%]	Sum [%]
L1	86.86	15.14	102.00	1.01	0.45	0.59	0.00	0.04	0.00	0.00	104.09
L2	93.37	9.76	103.13	0.88	0.31	0.76	0.00	0.06	0.00	0.00	105.15
L3	91.43	9.46	100.89	0.24	0.26	0.04	0.00	0.00	0.00	0.00	101.42
L4	95.17	2.42	97.59	0.40	0.36	0.07	0.00	0.00	0.00	0.00	98.42
PH2SK	86.24	13.28	99.52	0.85	0.26	0.85	0.00	0.07	0.00	0.00	101.55

AIL = acid insoluble lignin; ASL = acid soluble lignin.

The HSQC analysis of the purification fractions in Figure 3 shows low molecular weight species in L1 at chemical shift $\delta C/\delta H$ 10.0–37.0/2.5–0.7. The number of those species decreased going through the purification from L1 to L4. It indicates the carbohydrate content which is also supported by the NREL testing.

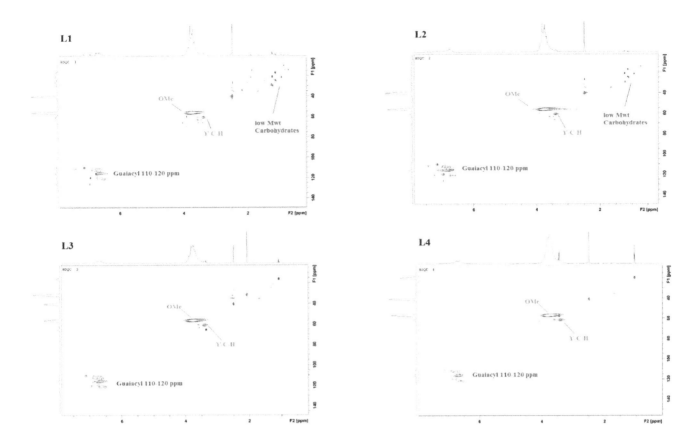

Figure 3. HSQC of lignin purification fractions: L1, L2, L3, and L4.

In addition, it was recently shown that 2D HSQC NMR in combination with SEC measurements can be used to determine the molecular weights (Mw) of biopolymers using multivariate data analysis (i.e., principal component analysis) to simulate Mw data [42]. The ^{31}P NMR analysis, on the other hand, shows the presence of aliphatic OH groups in all the purification fractions confirming data reported for lignins obtained from other sources and isolated using different methods [43–45].

The number of OH groups in the lignin fractions (Table 3) was obtained from the ^{31}P NMR analysis, showing that the aliphatic OH number of the L1 samples was the highest, followed by L2, L3, and L4 which have half this number. Also, the number of the carboxylic acid OH groups for L1 was also the highest, demonstrating the antibacterial activity as it is usually related to its origin, and specifically due to the presence of phenolic compounds and different functional groups containing oxygen (methoxyl and epoxy groups) in its structure [18,19,24]. PH2SK was extracted according to Klein et al., 2018 [15]. Previous studies confirmed the degradation of lignin with temperature around 60 °C [13,19]. During the extraction of PH2SK, the black liquor was heated to 50–60 °C. The heating could have produced phenolic carboxylic acids, containing fragments which are demonstrated by the high number of phenolic carboxylic acid OH groups in Table 3. Primary antimicrobial activity tests showed certain activity of PH2SK against Gram-positive bacteria which also matched with the NREL results.

In a previous study, X-ray diffraction (XRD) measurements confirmed the increase of the purification level from L1 to L4 [13]. For kraft lignin samples purified by selective extraction, the storage effects were observed to result in depolymerization being initiated by temperature and UV irradiation, respectively [46,47].

Table 3. Number of OH in the lignin fractions as determined by ^{31}P NMR analysis.

	Aliphatic OH	Condensed-OH	G and Dimethylated-OH	Carboxylic Acids-OH
L1	8.70	5.04	6.30	0.59
L2	8.52	0.28	0.71	0.00
L3	3.81	2.99	2.91	0.16
L4	3.17	0.48	1.79	0.15
L5 *	3.02	0.21	0.18	0.02
PH2SK	7.32	7.11	9.73	3.10

* L5 has S–OH: 0.01.

3.2. Preparation of Lignin-Derived Composites

Film preparation was based on the antibacterial activity results as well as the antioxidant activities of the lignins; the polymerization of HPMC-based films was limited to the selected active lignin samples: L1, L5, and L6 samples to be tested against Gram-positive and Gram-negative bacteria. The antioxidant activity of the lignins was studied using the DPPH assay [11]. Table 4 shows the DPPH inhibitions of the lignin samples. All of the lignin samples show a specific activity that varies from high (68.2% of L4) to low (31% of L7). Accordingly, lignins from L1 to L7 were added to HPMC and the antioxidant activity of the resulted films was then investigated.

Table 4. The DPPH inhibitions of kraft lignin fractions (L1 to L4) and organosolv lignins obtained from spruce/pine (L5), beech (L6), and *Miscanthus x giganteus* (L7).

	L1	L2	L3	L4	L5	L6	L7
DPPH Inhibition (%)	65.1 ± 3.7	66.8 ± 6.6	62.2 ± 9.5	68.2 ± 3.6	42 ± 1.9	64 ± 2.6	31 ± 1.0

The lignin was added to the HPMC during polymerization in addition to polyethylene glycol 400 (PEG) that improved the flexibility and impact strength of the blend and act as a compatibilizer with the polymer. As mentioned, the polymers were produced based on the antioxidant activities as well as the antibacterial activities of the lignins (L1 to L7). HPMC-film is transparent and water soluble. Water solubility decreases with the addition of lignin. The color of the films was light honey and became denser with the addition of lignin as shown in Figure 4. The brittleness appeared after the addition of 25% of the total weight of the sample in lignin to both L1 and L7, at 30% of L5, and at 20% of L6. Lignin release started at 30% of L1 and 15% of L5, while L7 films did not exhibit any release. L6 had lignin release at 20% and 25% but had none at 30%. Primary antimicrobial tests on the films against Gram-positive bacteria confirmed the activity of the films against *S. aureus*. The films were not active against *E. coli*, and hence, the addition of chitosan was the key. Eighty-five percent chitosan (deacetylated) was added to the HPMC-lignin film solutions in 5% increments, the lignin additions to the HPMC and chitosan were chosen based on the best lignin-HPMC combination with no lignin release and no brittleness. Lignin release appeared at lower percentages starting at 5% in L1 films, at 5% and 10% in L5 films, and at 15% in L6 films, while chitosan release happened at 20% and 30% for L1 and at 15% and 25% for L6 while HPMC-L5-chitosan films exhibited no chitosan release. The lignin and/or chitosan release depended on the compatibility of the three components in the film in addition to the lignin type. In HPMC-lignin films, the increment of the lignin added to the HPMC solution improved the mechanical properties of the HPMC film and decreased the water solubility. The lignin release for the softwood based organosolv lignins started at lower lignin percentages compared to kraft lignin and grass-based organosolv lignin. The HPMC-lignin-chitosan films are more complex: the lignin started the release at lower concentrations while the chitosan started the release at higher lignin concentration. The released lignin/chitosan were non-reacted species that failed in the competition to bond in the film cluster. At low lignin concentrations, the chitosan won the competition and bonded fully in the film cluster, leaving the rest of the lignin to release. At high lignin concentrations, the lignin won the competition and bonded fully to the film cluster, leaving the rest of the chitosan out of the polymer structure. This may be due to the increased number of the phenolic OH groups introduced by the lignin at higher concentrations.

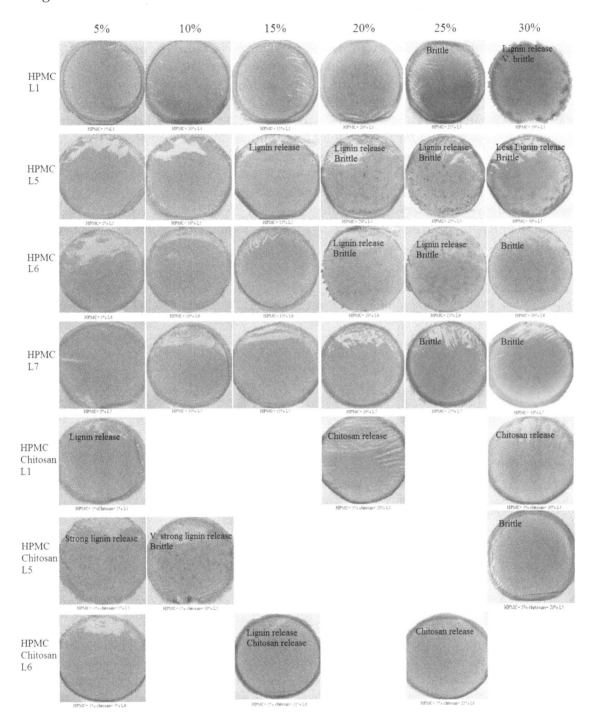

Figure 4. HPMC/lignin and HPMC/lignin/chitosan composite films.

3.3. Antiradical Activity of Lignin-Derived Films

Radicals originating from oxygen exist naturally in the atmosphere or can be created by the thermal processing or irradiation of packaging and food. These radicals act as initiators of the chain oxidation of lipids. It is therefore advantageous to eliminate theses radicals from the headspace and the bulk of the food, as an alternative route to eliminate oxygen from the package. The radical scavenging efficiency of an antiradical substance depends on the rate of hydrogen atom abstraction from the phenyl group and also on the stability of the resulting radical. Strategies employed in active packaging generally include: (1) the design of active compound releasing systems and (2) undesired compound scavenging systems. In the first strategy, the generation of low molecular weight substances inside the packaging film and promotion of their migration into the food stuff is of interest. In the second strategy, one of the advantages of radical scavengers is their efficiency upon contact without the need for the release of active compounds. This has been shown for hydroxyl radicals in the gas phase scavenged by essential oils supported on active packaging films containing essential oils [12]. The phenolic mobile hydrogen atom (ArO–H) plays the main role in the ability of monomeric phenolic compounds to scavenge free radicals (R •). The process occurs according to the following scheme: ArO–H + R • → ArO • + R–H [13]. In the plant cell wall, lignin is often associated with carbohydrate polymers. The nonphenolic carbohydrate impurities, which remain strongly associated with lignin during its isolation and purification, can decrease the concentration of active phenolic OH groups and negatively influence their reactivity (by increasing the O–H bond dissociation enthalpy). The phenoxy radicals in some lignins can take part in the reaction with oxidant radicals as secondary antioxidants [45]. Figure 5 shows the scavenging activity (SA) of free radical DPPH of the HPMC-based films. The films studied depended on the percentage of the lignin added; the lowest addition (5%), the highest addition (30%), and the lignin addition that produces the best film for each lignin sample.

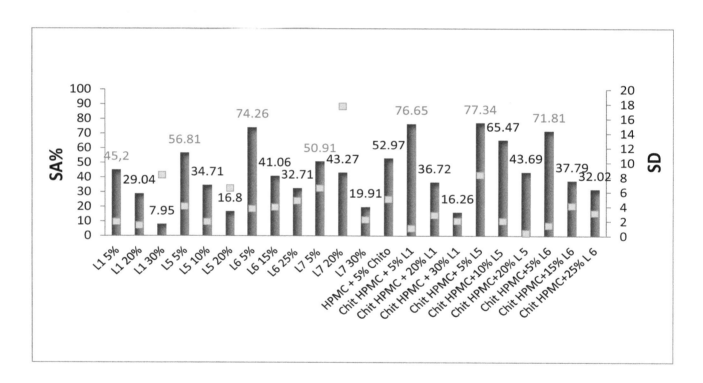

Figure 5. Antioxidant activity of HMPC/lignin and HPMC/lignin/chitosan films. The measurements were held in triplets. The red data labels indicate the highest activity for each lignin type. The yellow points relate to the standard deviation.

The results showed that the 5% addition of lignin had the highest SA% and the 30% had the lowest in which the UV absorption increased from the 5% to the 30% and the SA decreased. During the polymerization, lignin reacted with HPMC to form ether or ester bonds when 5% of lignin was added. Lignin is reactive to an extent where some phenolic OH groups are left free, causing the SA; however, not every phenolic OH group takes part in the reaction with the free radical [48]. For the 30% addition, the high concentration of lignin increased the connectivity of HPMC-lignin, where it converted the phenolic OH groups to an ether or ester bond or even self-reacted; this decreased the number of active phenolic OH groups against oxidants and hence decreased the SA. Chitosan-lignin-HPMC films, in general, have higher SA than lignin-HPMC films because simply, the activity of neat chitosan, which is basically due to the hydroxyl and the amine groups [49], was added to the lignin-HPMC activity; also, the reaction between the components could happen on the α-position of the phenolic group in lignin, where an electron donating group (like methyl or methylene) increases the SA and an electron withdrawing group (like carbonyl or carboxyl) decreases the SA by increasing the polarization of the –OH bond and complicating its hemolytic dissociation [50,51]. The type of lignin added also affects the SA value. The trend of the SA of all the samples, based on the lignin type, was: organosolv of softwood > kraft of softwood > Organosolv of grass. The Organosolv lignins, in general, had a higher methoxyl content than kraft lignins which may act as a secondary antiradical and the carbohydrates content is higher in kraft lignin than organosolv [52], resulting in a decreased SA value. Table 4 shows the DPPH inhibition of the added lignins, where L1 has the highest and grass-based organosolv lignin has lowest. This basically could be the reason why films with lignins from grass have lower antioxidant activity.

Small angle X-ray scattering (SAXS) studies are ongoing to obtain more detailed information regarding the film morphology including micro- and/or nanostructures as observed for lignin-derived polyurethanes [16,53].

3.4. Antimicrobial Activity of HPMC/Lignin and HPMC/Lignin/Chitosan Composites

The prepared films were tested against *S. aureus* and *E. coli* at 35 °C for 24 h in pure culture. In regard to Dohlen and coworkers, those films that reached a log10-reduction Ø ≥ 2 log10 were tested against *B. thermosphacta* (a Gram-positive bacteria that is one of the most abundant spoilage organisms of fresh and cured meats, fish, and fish products) due to its tolerance to high-salt and low-pH conditions, its ability to grow at refrigeration temperatures (4 °C), and its production of organoleptically unpleasant compounds and *P. fluorescens* (Gram-negative bacteria that poses a significant spoilage problem in refrigerated (0–7 °C) meat and dairy products) [38,54]. A log10-reduction Ø ≥ 2 log10 units allowed the adaption of temperature and inoculated solution. The test was conducted at a constant temperature of 7 °C for 24 h. In case the Ø log10-reduction was ≥ 2 log10, highly concentrated meat extract solution (18 μg mL^{-1}) will be chosen as the reference media for perishable foods. Figure 6 shows the antimicrobial activities of the studied films against *S. aureus* and *E. coli*; Figure 7 shows them against spoilage bacteria at low temperatures (*B. thermosphacta* and *P. fluorescens*).

The results in Figure 6 correlate with the inhibition zone test results (Table 1), in which lignin films are active against Gram-positive bacteria more than Gram-negative bacteria. For *S. aureus*, the activity of films containing organosolv lignins in general was higher than kraft lignins. This could be because the carbohydrates in the kraft lignin could have reacted with the HPMC and entered the film structure, leaving the film without the active part of the sugar content that participated in the antimicrobial activity of the lignin alone. The activity of the lignin films increased with increasing lignin concentration in the film which introduced more active functional groups (aliphatic OH, carbonyl CO, COOH) into the film structure. For *E. coli*, the kraft lignin films exhibited no activity at low concentration. Only the 30% addition had an activity, which is justified by the introduction of more active functional groups into the film and the lignin release could be the main reason of the activity. HPMC-L5 films exhibited high activity at higher percentages, which also could be enhanced by the released lignins; the 5% addition had the lowest activity due to the concentration factor. The 25% L6-HPMC film was the only active one among HPMC-L6 films. The chitosan addition to the HPMC-lignin combination improved the activity since chitosan is known as an antibacterial compound against both Gram-positive and Gram-negative bacteria. It gains its activity due to the presence of the amino group and the OH functionality in its structure [55,56].

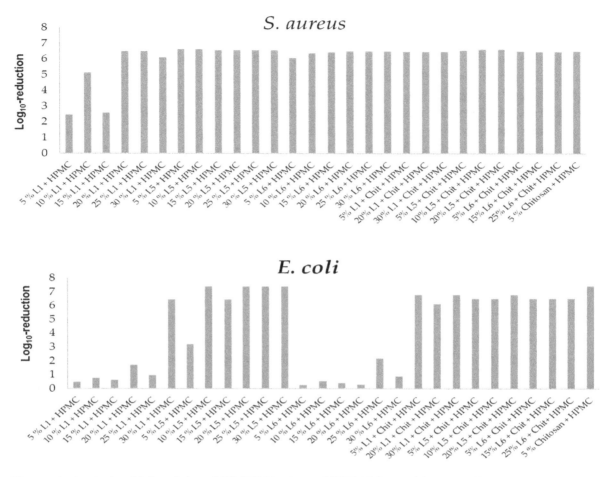

Figure 6. Antimicrobial activity of HMPC/lignin and HPMC/lignin/chitosan films against *S. aureus* (above) and *E. coli* (below). All films are active against *S. aureus*. Columns in orange relate to nonactive films against *E. coli*. Chitosan incorporation increased the activity. L5 films showed highest activities against both bacteria.

The chitosan release did not have a significant effect on the samples and the activities of the chitosan films were close to each other. Also, the chitosan activity of the corresponding films dominates over the lignin activity. Thus, can be clearly seen from the chitosan-HPMC activity compared to the chitosan-HPMC-lignin activities, with the exception of the slight negative effect of lignin on the activity of the films in which more amino or OH groups of chitosan and/or lignin are involved in bonding with the HPMC blocking the antimicrobial activity for some of those functionalities.

The films were further tested against spoilage bacteria that grow at low temperatures (0–7 °C): *B. thermosphacta* (Gram-positive) and *P. fluorescens* (Gram-negative). Figure 7 shows that the 30% addition of all lignins in HPMC-lignin films was active against *B. thermosphacta*. On the other hand, the 5% addition of the lignins in the HPMC-lignin-chitosan films was the active one. The activities were close to each other and showed no effect of the lignin type.

In HPMC-lignin films, the 30% addition of L1 is the only active film against *P. fluorescens*; L5 and L6 showed no activity. In HPMC-lignin-chitosan films, the 5% addition of L5 had the highest activity. Strangely, activity of L1 against *P. fluorescens* was higher than that against *B. thermosphacta*; this may be due to the lignin release (which has to be clarified in ongoing studies).

Differences in the level of antimicrobial activity could be explained by a variety of factors, such as the Gram characteristics of the bacteria, the ability of building exopolysaccharides, differences in the zeta potential, differences in the permeability of the outer membrane which itself is influenced by temperature, different fatty acids, etc. These factors combined with different charges of the surfaces could have caused the variation in the antimicrobial activity against different bacteria. As an example, in different studies, a higher activity of high molecular weight chitosan against Gram-positive bacteria (*S. aureus*) compared with Gram-negative bacteria, *E. coli* and *Ps. fluorescens* was observed [36,37,39]. In our results, *P. fluorescens* is even more resistant than *E. coli*, which is due to higher production of exopolysaccharides by *P. fluorescens*.

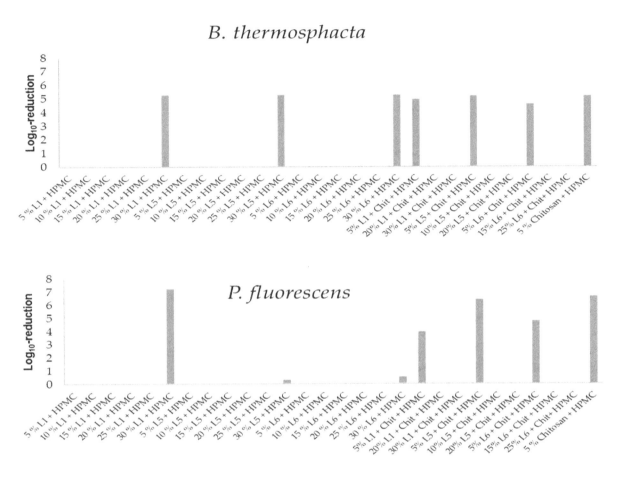

Figure 7. Antimicrobial activity of HMPC/lignin and HPMC/lignin/chitosan films against *B. thermosphacta* (above) and *P. fluorescens* (below). Columns in orange relate to nonactive films against *P. fluorescens*.

4. Conclusions

In addition to antioxidant activity, lignins of different origins show distinct antimicrobial effects. Due to the obtained results, purification of kraft lignin via solvent extraction has a negative effect on the antimicrobial activity against *S. aureus* and *L. monocytogenes* most obviously due to the presence of carbohydrate content and aliphatic OH groups which was supported by both NREL and NMR analysis. The biomass source (hard and soft wood, grasses) and pulping process influence both antioxidant and antimicrobial activity. Organosolv lignins, in general, have a higher methoxyl content than kraft lignins which may act as a secondary antiradical and the carbohydrate content is higher in kraft lignin than organosolv. In addition, the antimicrobial activity of the lignins also depends on the solvent polarity (as shown for ethanol, acetone, and DMSO). Storage led to an increase in antimicrobial activity against *S. aureus* due to the degradation of lignin over time. The scavenging activity of both binary (HPMC/lignin) and ternary (HPMC/lignin/chitosan) systems was affected by the percentage of the added lignin. The lignin release in the produced films positively affected the activity and the chitosan addition enhances the activity even more for both Gram-positive and Gram-negative bacteria. Spoilage bacteria that grow at low temperatures, such as *B. thermosphacta* and *P. fluorescens*, were affected by lignin/chitosan composite films. The detailed mechanism regarding the bioactivity is the focus of ongoing studies and might be correlated to bacterial surfaces charges and zeta potential.

Author Contributions: A.A. mainly contributed to the manuscript, performed the isolation and main parts of the analytical experiments, processed all data and contributed main parts of the manuscript; S.E.K.; M.B. and X.T.D. contributed analytical data (i.e., the organosolv sample L7 and lignin-polyurethane sample PH2 SK); I.K., S.D., C.H. and J.K. contributed antimicrobial analyses; M.L. performed parts of the antioxidant experiments and as a native speaker contributed in manuscript writing; B.K. contributed NREL data; M.S. conceived and designed the experimental studies and contributed in writing the manuscript.

Acknowledgments: Bonn-Rhein-Sieg University/Graduate Institute for scholar ship (A.A., M.B.) and Erasmus-Mundus Avempace-II scholar ship (A.A.); Bonn-Rhein-Sieg University/TREE institute (S.E.K.); North Carolina State University in conjunction with the DAAD RISE Scholarship Program (M.L.).

References

1. Chen, G.-G.; Fu, G.-Q.; Wang, X.-J.; Gong, X.-D.; Niu, Y.-S.; Peng, F.; Yao, C.-L.; Sun, R.-C. Facile synthesis of high strength hot-water wood extract films with oxygen-barrier performance. *Sci. Rep.* **2017**, *7*, 41075. [CrossRef]

2. Dilkes-Hoffman, L.S.; Pratt, S.; Lant, P.A.; Laycock, B. The Role of Biodegradable Plastic in Solving Plastic Solid Waste Accumulation. In *Plastics to Energy*; Fuel, Chemicals, and Sustainability Implications Plastics Design Library 2019; William Andrew Publishing: Norwich, NY, USA, 2019; pp. 469–505.

3. Muthuraj, R.; Misra, M.; Kumar, A. Biodegradable compatibilized polymer blends for packaging applications: A literature review. *J. Appl. Polym. Sci.* **2018**, *135*, 45726. [CrossRef]

4. Abdul Khalil, H.P.S.; Tye, Y.Y.; Leh, C.P.; Saurabh, C.K.; Ariffin, F.; Fizree, H.M.; Mohamed, A.; Suriani, A.B. Cellulose Reinforced Biodegradable Polymer Composite Film for Packaging Applications. In *Bionanocomposites for Packaging Applications*; Jawaid, M., Swain, S., Eds.; Springer: Cham, The Netherlands, 2018.

5. Shankar, S.; Reddy, J.P.; Rhim, J.-W. Effect of lignin on water vapor barrier, mechanical, and structural properties of agar/lignin composite films. *Int. J. Biolog. Macromol.* **2015**, *81*, 267–273. [CrossRef] [PubMed]

6. Kamm, B.; Kamm, M.; Hirth, T.; Schulze, M. Lignocelluloses Based Chemical Products and Product Family Trees. In *Biorefineries-Industrial Processes and Products*; Kamm, M., Kamm, B., Gruber, P.C., Eds.; Wiley-VCH: Weinheim, Germany, 2006; pp. 97–150, ISBN 3-527-31027-4.

7. Kamm, B.; Gruber, P.R.; Kamm, M. Biorefineries-Industrial Processes and Products. In *Ullmann's Encyclopedia of Industrial Chemistry*; Wiley-VCH: Weinheim, Germany, 2016; ISBN 9783527306732.

8. Su, Y.; Yang, B.; Liu, J.; Sun, B.; Cao, C.; Zou, X.; Lutes, R.; He, Z. Prospects for Replacement of Some Plastics in Packaging with Lignocellulose Materials: A Brief Review. *BioResources* **2018**, *13*, 4550–4576. [CrossRef]

9. Ko, F.K.; Goudarzi, A.; Lin, L.-T.; Li, Y.; Kadla, J.F. Lignin-Based Composite Carbon Nanofibers. *Lignin Polym. Compos.* **2016**, 167–194. [CrossRef]

10. Rinaldi, R.; Jastrzebski, R.; Clough, M.T.; Ralph, J.; Kennema, M.; Bruijnincx, P.C.A.; Weckhuysen, B.M. Paving the Way for Lignin Valorisation: Recent Advances in Bioengineering, Biorefining and Catalysis. *Angew. Chem. Int. Ed.* **2016**, *55*, 2–54. [CrossRef]

11. Liu, H.L.; Chung, H.Y. Lignin-Based Polymers via Graft Copolymerization. *J. Polym. Sci.* **2017**, *55*, 3515–3528. [CrossRef]

12. Yang, W.; Fortunati, E.; Dominici, F.; Kenny, J.M.; Giovanale, G.; Mazzaglia, A.; Balestra, G.M.; Puglia, D. Effect of cellulose and lignin on disintegration, antimicrobial and antioxidant properties of PLA active films. *Int. J. Biol. Macromol.* **2016**, *89*, 360–368. [CrossRef] [PubMed]

13. Alzagameem, A.; El Khaldi-Hansen, B.; Büchner, D.; Larkins, M.; Kamm, B.; Witzleben, S.; Schulze, M. Lignocellulosic Biomass as Source for Lignin-Based Environmentally Benign Antioxidants. *Molecules* **2018**, *23*, 2664. [CrossRef]

14. Witzler, M.; Alzagameem, A.; Bergs, M.; El Khaldi-Hansen, B.; Klein, S.E.; Hielscher, D.; Kamm, B.; Kreyenschmidt, J.; Tobiasch, E.; Schulze, M. Lignin-Derived Biomaterials for Drug Release and Tissue Engineering. *Molecules* **2018**, *23*, 1885. [CrossRef]

15. Klein, S.E.; Rumpf, J.; Kusch, P.; Albach, R.; Rehahn, M.; Witzleben, S.; Schulze, M. Utilization of Unmodified Kraft Lignin for the Preparation of Highly Flexible and Transparent Polyurethane Coatings. *RSC Adv.* **2018**, *8*, 40765. [CrossRef]

16. Klein, S.E.; Rumpf, J.; Rehahn, M.; Witzleben, S.; Schulze, M. Biobased Flexible Polyurethane Coatings Prepared from Kraft Lignin: One-Pot Synthesis and Antioxidant Activity. *J. Coat. Technol. Res.* **2019**, in press.

17. Guo, M.; Jin, T.; Nghiem, N.P.; Fan, X.; Qi, P.X.; Jang, C.H.; Shao, L.; Wu, C. Assessment of Antioxidant and Antimicrobial Properties of Lignin from Corn Stover Residue Pretreated with Low-Moisture Anhydrous Ammonia and Enzymatic Hydrolysis Process. *Appl. Biochem. Biotechnol.* **2018**, *184*, 350–365. [CrossRef]

18. Lupoi, J.S.; Singh, S.; Parthasarathi, R.; Simmons, B.A.; Henry, R.J. Recent innovations in analytical methods for the qualitative and quantitative assessment of lignin. *Renew. Sustain. Energy Rev.* **2015**, *49*, 871–906. [CrossRef]

19. Alzagameem, A.; El Khaldi-Hansen, B.; Kamm, B.; Schulze, M. Lignocellulosic biomass for energy, biofuels, biomaterials, and chemicals. In *Biomass and Green Chemistry*, 1st ed.; Vaz, S., Jr., Ed.; Springer International Publishing: Basel, Switzerland, 2018; pp. 95–132, ISBN 978-3-319-66736-2.

20. Hansen, B.; Kamm, B.; Schulze, M. Qualitative and quantitative analysis of lignins from different sources and isolation methods for an application as a biobased chemical resource and polymeric material. In *Analytical Techniques and Methods for Biomass Products*; Vaz, S., Jr., Seidl, P., Eds.; Springer: Berlin, Germany, 2017; pp. 15–44, ISBN 978-3-319-41414-0.

21. Bergs, M. Einfluss von Miscanthus-Genotyp und Erntezeit auf Gehalt und Struktur von Lignin aus Organosolv-Verfahren. Ph.D. Thesis, Rheinische Friedrich-Wilhelms-University, Bonn, Germany, 2018.

22. Bergs, M.; Völkering, G.; Kraska, T.; Do, X.T.; Kusch, P.; Monakhova, Y.; Konow, C.; Pude, R.; Schulze, M. Miscanthus *X giganteus* Stem versus Leave-derived Lignins Differing in Monolignol Ratio and Linkage. *Int. J. Mol. Sci.* **2019**, *20*, 1200. [CrossRef]

23. Sadeghifar, H.; Venditti, R.; Jur, J.; Gorga, R.E.; Pawlak, J.J. Cellulose-Lignin Biodegradable and Flexible UV Protection Film. *ACS Sustain. Chem. Eng.* **2017**, *5*, 625–631. [CrossRef]

24. Kaur, R.; Uppal, S.K.; Sharma, P. Antioxidant and Antibacterial Activities of Sugarcane Bagasse Lignin and Chemically Modified Lignins. *Sugar Tech.* **2017**, *19*, 675–680. [CrossRef]

25. Hambardzumyan, A.; Foulon, L.; Bercu, N.B.; Pernes, M.; Maigret, J.E.; Molinari, M.; Chabbert, B.; Aguié-Béghin, V. Organosolv lignin as natural grafting additive to improve the water resistance of films using cellulose nanocrystals. *Chem. Eng. J.* **2015**, *264*, 780–788. [CrossRef]

26. Yang, W.; Fortunati, E.; Dominici, F.; Giovanale, G.; Mazzaglia, A.; Balestra, G.M.; Kenny, J.M.; Puglia, D. Synergic effect of cellulose and lignin nanostructures in PLA based systems for food antibacterial packaging. *Eur. Polym. J.* **2016**, *79*, 1–12. [CrossRef]

27. Rhimi, W.; Boulila, A.; Gheribia, R.; Khwaldia, K. Development, characterization and application of hydroxypropylmethylcellulose films enriched with cypress seed extract. *RCS Adv.* **2018**, *8*, 23615–23622. [CrossRef]

28. Yang, W.; Owczarek, J.S.; Fortunati, E.; Kozanecki, M.; Mazzaglia, A.; Balestra, G.M.; Kenny, J.M.; Torre, L.; Puglia, D. Antioxidant and antibacterial lignin nanoparticles in polyvinyl alcohol/chitosan films for active packaging. *Ind. Crops Prod.* **2016**, *91*, 800–811. [CrossRef]

29. Hansen, B.; Kusch, P.; Schulze, M.; Kamm, B. Qualitative and quantitative analysis of lignin produced from beech wood by different conditions of the Organosolv process. *J. Polym. Environ.* **2016**, *24*, 85–97. [CrossRef]

30. Determination of Structural Carbohydrates and Lignin in Biomass. Available online: https://www.nrel.gov/docs/gen/fy13/42618.pdf (accessed on 10 October 2018).

31. Sluiter, A.; Hames, B.; Ruiz, R.; Scarlata, C.; Sluiter, J.; Templeton, D.; Crocker, D. Determination of structural carbohydrates and lignin in biomass. *Lab. Anal. Proced.* **2008**, *1617*, 1–16.

32. Sluiter, A.; Ruiz, R.; Scarlata, C.; Sluiter, J.; Templeton, D. Determination of Extractives in Biomass (NREL/TP-510-42619). *Natl. Renew. Energy Lab. Gold.* **2008**, *2008*, 42619.

33. Sluiter, A.; Hames, B.; Hyman, D.; Payne, C.; Ruiz, R.; Scarlata, C.; Sluiter, J.; Templeton, D.; Wolfe, J. Determination of Total Solids in Biomass and Total Dissolved Solids in Liquid Process Samples (NREL/TP-510-42621). *Natl. Renew. Energy Lab. Gold.* **2008**, *2008*, 42621.

34. Sluiter, A.; Hames, B.; Ruiz, R.; Scarlata, C.; Sluiter, J.; Templeton, D. Determination of ash in Biomass (NREL/TP-510-42622). *Natl. Renew. Energy Lab. Gold.* **2008**, *2008*, 42622.

35. Sebti, I.; Chollet, E.; Degraeve, P.; Noel, C.; Peyrol, E. Water Sensitivity, Antimicrobial, and Physicochemical Analyses of Edible Films Based on HPMC and/or Chitosan. *J. Agric. Food Chem.* **2007**, *55*, 693–699. [CrossRef]

36. Dohlen, S.; Braun, C.; Brodkorb, F.; Fischer, B.; Ilg, Y.; Kalbfleisch, K.; Kreyenschmidt, M.; Lorenz, R.; Kreyenschmidt, J. Effect of different packaging materials containing poly-[2-(tert-butylamino) methylstyrene] on the growth of spoilage and pathogenic bacteria on fresh meat. *Int. J. Food Microbiol.* **2017**, *257*, 91–100. [CrossRef]

37. Hüwe, C.; Schmeichel, J.; Brodkorb, F.; Dohlen, S.; Kalbfleisch, K.; Kreyenschmidt, M.; Lorenz, R.; Kreyenschmidt, J. Potential of antimicrobial treatment of linear low-density polyethylene with poly((tert-butyl-amino)-methyl-styrene) to reduce biofilm Formation in the Food industry. *Biofouling* **2018**, *34*, 378–387. [CrossRef]

38. Dohlen, S.; Braun, C.; Brodkorb, F.; Fischer, B.; Ilg, Y.; Kalbfleisch, K.; Kreyenschmidt, M.; Lorenz, R.; Robers, O.; Kreyenschmidt, J. Potential of the polymer poly-[2-(tert-butylamino) methylstyrene] as antimicrobial packaging material for meat products. *J. Appl. Microbiol.* **2016**, *4*, 1059–1070. [CrossRef]

39. Braun, C.; Dohlen, S.; Ilg, Y.; Brodkorb, F.; Fischer, B.; Heindirk, P.; Kalbfleisch, K.; Richter, T.; Robers, O.; Kreyenschmidt, M.; et al. Antimicrobial Activity of Intrinsic Antimicrobial Polymers Based on Poly((tertbutyl-amino)-methyl-styrene) Against Selected Pathogenic and Spoilage Microorganisms Relevant in Meat Processing Facilities. *J. Antimicrob Agents* **2017**, *3*, 1000136. [CrossRef]

40. Strotmann, C.; Göbel, C.; Friedrich, S.; Kreyenschmidt, J.; Ritter, G.; Teitscheid, P. A Participatory Approach to Minimizing Food Waste in the Food Industry—A Manual for Managers. *Sustainability* **2017**, *9*, 66. [CrossRef]

41. Rocca, D.M.; Vanegas, J.P.; Fournier, K.; Becerra, M.C.; Scaiano, J.C.; Lanterna, A.E. Biocompatibility and photo-induced antibacterial activity of lignin-stabilized noble metal nanoparticles. *RSC Adv.* **2018**, *8*, 40454–40463. [CrossRef]

42. Monakhova, Y.; Diehl, B.W.K.; Do, X.T.; Witzleben, S.; Schulze, M. Novel method for the determination of average molecular weight of natural polymers based on 2D DOSY NMR and chemometrics: Example of heparin. *J. Pharm. Biomed. Anal.* **2018**, *149*, 128–132. [CrossRef]

43. Gilca, I.A.; Ghitescu, R.E.; Puitel, A.C.; Popa, V.I. Preparation of lignin nanoparticles by chemical modification. *Iran. Polym. J.* **2014**, *23*, 355–363. [CrossRef]

44. Argyropoulos, D.S. Quantitative phosphorus 31 NMR analysis of lignins: A new tool for the lignin chemist. *J. Wood Chem. Technol.* **1994**, *14*, 45–63. [CrossRef]

45. Sun, S.-N.; Cao, X.-F.; Xu, F.; Sun, R.-C.; Jones, G.L. Structural Features and Antioxidant Activities of Lignins from Steam-Exploded Bamboo (*Phyllostachys pubescens*). *J. Agric. Food Chem.* **2014**, *62*, 5939–5947. [CrossRef]

46. Do, X.T.; Nöster, J.; Weber, M.; Nietsch, A.; Jung, C.; Witzleben, S.; Schulze, M. Comparative Studies of Lignin Depolymerisation: Photolysis *versus* Ozonolysis in Alkaline Medium. In Proceedings of the Annual Conference of the GDCh Division Sustainable Chemistry, Aachen, Germany, 17–19 September 2018.

47. Do, X.T.; Nietzsch, A.; Jung, C.; Witzleben, S.; Schulze, M. Lignin-Depolymerisation via UV-Photolysis and Titanium Dioxide Photocatalysis. *Preprints* **2017**, *2017*, 100128. [CrossRef]

48. Ponomarenko, J.; Lauberts, M.; Dizhbite, T.; Lauberte, L.; Jurkjane, V.; Telysheva, G. Antioxidant activity of various lignins and lignin-related phenylpropanoid units with high and low molecular weight. *Holzforschung* **2015**, *69*, 795–805. [CrossRef]

49. Rajalakshmi, A.; Krithiga, N.; Jayachitr, A. Antioxidant Activity of the Chitosan Extracted from Shrimp Exoskeleton. *Middle-East J. Sci. Res.* **2013**, *16*, 1446–1451. [CrossRef]

50. Shen, D.K.; Gu, S.; Luo, K.H.; Wang, S.R.; Fang, M.X. The pyrolytic degradation of wood-derived lignin from pulping process. *Bioresour. Technol.* **2010**, *101*, 6136–6146. [CrossRef]

51. Dizhbite, T.; Telysheva, G.; Jurkjane, V.; Viesturs, U. Characterization of the radical scavenging activity of lignins natural antioxidants. *Bioresour. Technol.* **2004**, *95*, 309–317. [CrossRef]

52. El Mansouri, N.-E.; Salvadó, J. Structural characterization of technical lignins for the production of adhesives: Application to lignosulfonate, kraft, soda-anthraquinone, organosolv and ethanol process lignins. *Ind. Crops Prod.* **2006**, *24*, 8–16. [CrossRef]

53. Witzleben, S.T.; Walbrück, K.; Klein, S.E.; Schulze, M. Investigation of Temperature Dependency of Morphological Properties of Thermoplastic Polyurethane Using WAXS and SAXS Monitoring. *J. Chem. Chem. Eng.* **2016**, *9*, 494–499.

54. Stanborough, T.; Fegan, N.; Powell, S.M.; Tamplin, M.; Chandry, P.S. Insight into the Genome of Brochothrix thermosphacta, a Problematic Meat Spoilage Bacterium. *Appl. Environ. Microbiol.* **2017**, *83*, 1–20. [CrossRef]

55. Garrido-Maestu, A.; Ma, Z.; Paik, S.-Y.-R.; Chen, N.; Ko, S.; Tong, Z.; Jeong, K.C.C. Engineering of chitosan-derived nanoparticles to enhance antimicrobial activity against foodborne pathogen Escherichia coli O157:H7. *Carbohydr. Polym.* **2018**, *197*, 623–630. [CrossRef]

56. Kurniasih, M.; Dewi, R.S. Toxicity tests, antioxidant activity, and antimicrobial activity of chitosan. In Proceedings of the 12th Joint Conference on Chemistry, Semarang, Indonesia, 19–20 September 2017.

9

Use of Orange Oil Loaded Pectin Films as Antibacterial Material for Food Packaging

Tanpong Chaiwarit [1], Warintorn Ruksiriwanich [1], Kittisak Jantanasakulwong [2] and Pensak Jantrawut [1,*]

[1] Department of Pharmaceutical Sciences, Faculty of Pharmacy, Chiang Mai University, Chiang Mai 50200, Thailand; Tanpong.c@gmail.com (T.C.); Yammy109@gmail.com (W.R.)

[2] School of Agro-Industry, Faculty of Agro-Industry, Chiang Mai University, Chiang Mai 50100, Thailand; jantanasakulwong.k@gmail.com

* Correspondence: pensak.j@cmu.ac.th.

Abstract: This study aims to develop orange oil loaded in thin mango peel pectin films and evaluate their antibacterial activity against *Staphylococcus aureus*. The mango peel pectin was obtained from the extraction of ripe Nam Dokmai mango peel by the microwave-assisted method. The thin films were formulated using commercial low methoxy pectin (P) and mango pectin (M) at a ratio of 1:2 with and without glycerol as a plasticizer. Orange oil was loaded into the films at 3% *w/w*. The orange oil film containing P and M at ratio of 1:2 with 40% *w/w* of glycerol (P_1M_2GO) showed the highest percent elongation (12.93 ± 0.89%) and the lowest Young's modulus values (35.24 ± 3.43 MPa). For limonene loading content, it was found that the amount of limonene after the film drying step was directly related to the final physical structure of the film. Among the various tested films, P_1M_2GO film had the lowest limonene loading content (59.25 ± 2.09%), which may be because of the presence of numerous micropores in the P_1M_2GO film's matrix. The inhibitory effect against the growth of *S. aureus* was compared in normalized value of clear zone diameter using the normalization value of limonene content in each film. The P_1M_2GO film showed the highest inhibitory effect against *S. aureus* with the normalized clear zone of 11.75 mm but no statistically significant difference. This study indicated that the orange oil loaded in mango peel pectin film can be a valuable candidate as antibacterial material for food packaging.

Keywords: antibacterial activity; food packaging; orange oil; pectin film

1. Introduction

Currently, environmental problems and food safety have been of great public concern. Using packaging materials consisting of biopolymer is considered environmentally advantageous. Active packing has been defined as 'a type of packaging in which the package, the product, and the environment interact to extend shelf life or improve safety and convenience or sensory properties, while maintaining the quality and freshness of the product [1]. One of the major concerns of the food industry is the spoilage of foods and food poisoning caused by microbial contamination, which occurs mainly on the surface as a result of the post processing and food handling process [2]. Thus, packaging possessing an antimicrobial property might be novel and gain interest for active packaging to be used in the food industry.

Pectin, one of the main structural water-soluble polysaccharides derived from plant cell walls, has been utilized in several fields, such as the pharmaceutic, cosmetic, and food industries, because of its excellent gelling property. Moreover, pectin can be extracted by simple methods from various plant materials. Mango peel, which is an agro-waste substantially arisen from fruit processing products in Thailand, was found to be a rich source of pectin [3–5].

Natural aromatic compounds and flavors such as fruit and vegetable essential oils are also extensively used in the food industry. For one clear example, orange oil exhibiting a significant bacterial inhibitory effect is one of the most beneficial and frequently used essential oils. Limonene, a major constituent found in orange oil, has been an active compound implicated in an antimicrobial property [6]. Limonene is a clear liquid at 25 °C with a citrus-like taste and odor. It is slightly soluble in glycerol, soluble in ethanol and carbon tetrachloride, and miscible with fixed oil. Limonene has demonstrated antibacterial activity as has been shown to inhibit the growth of many bacterial species in vitro, for example, *Staphylococcus aureus*, *Salmonella enteritidis*, *Escherichia coli*, *Klebsiella pneumoniae*, and *Proteus vulgaris* [7–9]. Limonene, a lipophilic compound, is able to penetrate through lipids of a bacterial cell, distribute cell structure, and render them more permeable. Then, it causes cell death by extensive leakage of intracellular fluid and ions [10]. There were some previous studies of film loading bioactive compounds that showed antimicrobial activity for preservative packaging. For example, gelatin film containing bergamot and lemongrass essential oils against *Escherichia coli*, *Listeria monocytogenes*, *Staphylococcus aureus*, and *Salmonella typhimurium*. Hydroxy methylcellulose film incorporated with kiam wood extract against *E. coli*, *S. aureus* and *L. monocytogenes* and starch film containing saponin against *E. coli*, *S. typhi*, and *E. erogenous* [11–13]. In the current study, the bio-based packaging materials have been developed by incorporating selected components, that is, essential oil derived from oranges, into a mango pectin film packaging model. The developed orange oil loaded pectin films were characterized and evaluated for their antibacterial activity against *S. aureus*.

2. Materials and Methods

2.1. Materials

Commercial low methoxy pectin (LMP; Unipectine OF300C; DE = 30% and DA = 0%) was purchased from Cargill^{TM}, Saint Germain, France. Orange oil was purchased from Thai-China flavors and Fragrance Industry Co., Ltd., Nonthaburi, Thailand. Calcium chloride ($CaCl_2$) was purchased from Merck, Damstadt, Germany. Mueller Hinton agar (MHA) was purchased from Becton Dickinson, Holdrege, NE, USA. Tryptic soy broth (TSB) was obtained from HiMedia Laboratories Pty. Ltd., Mumbai, India. *S. aureus* (ATCC 25923) was obtained from the BIOTEC, Manassas, VA, USA. Distilled water served as the solvent for preparing film solutions.

2.2. Mango Pectin

Mangoes (cv. Nam Dokmai) were collected from Chiang Mai province in Thailand, during June to August 2018. The mangoes were washed and the peels were stripped with a peeling knife, and then dried in a hot air oven at 50 ± 2 °C. The dried peel was ground to a fine powder using a high-speed food processor, and was then screened using a sieve (number 30; diameter 0.6 mm). Pectin from mango peel was extracted using the microwave-assisted method [14]. First of all, mango peel powder was mixed with acidified water pH 1.5. The mixture was extracted using microwave oven, irradiated at a frequency of 2450 MHZ at 500 W for 20 min. Then, it was centrifuged at 9000 rpm for 10 min. After that, supernatant was precipitated with ethanol and the precipitated pectin was washed by ethanol three times. The mango peel pectin was dried in hot air oven at 40 °C. The dried mango peel pectin was ground to a fine powder and kept in desiccator for further experiment.

2.3. Identification of Major Compounds of Orange Oil

The major compounds of orange oil used in this study were analyzed by gas chromatography mass spectrometer (GC-MS analyzer) from Agilent-Technologies (Santa Clara, CA, USA). A split ratio of 1:650 was used to inject the sample of 1.0 µL. The compounds of the sample were separated using an HP-5 MS capillary column (30 m × 0.25 mm, film thickness 0.25 µm). The carrier gas was helium and the flow rate was set as 1.5 mL/min. The data obtained from GC-MS analysis were compared with the National Institute of Standards and Technology mass spectral library for compound identification.

2.4. Preparation of Films Loaded with Orange Oil

The minimum inhibitory concentration (MIC) of orange oil against *S. aureus* was evaluated before loading orange oil into the film formulations. The MIC was evaluated using the broth dilution method [8]. Orange oil was prepared in various concentrations from 0.5 to 150 mg/mL by dilution with Tween® 80. The orange oil concentration above MIC was then selected and used for loading in the prepared film, using the ionotropic gelation with solvent casting techniques [15]. Briefly, film forming solution 3% *w/v* was prepared using commercial low methoxyl pectin (P) mixed with the extracted mango peel pectin (M) in various ratios with and without 40% glycerol (G), basic on dry pectin weight as the plasticizer. Orange oil (3% *w/w*) was added into pectin solution and then mixed until a homogeneous mixture was obtained. After that, the steps that were previously described were followed [15]. The film without mango pectin nor glycerol was used as a control film (P_3M_0O) in order to focus on the effect of mango pectin in thin film formulation. Furthermore, only mango pectin cannot form gel by this technique. Thus, the combinations of both pectin in the film formulation were prepared.

2.5. Characterization of Films Loaded with Orange Oil

The mechanical properties of the films loaded with orange oil were tested by a texture analyzer TX.TA plus (Stable Micro Systems, Surrey, UK). Each experiment was repeated six times. The tensile strength, percent elongation, and Young's modulus was calculated [15]. The film morphology was examined using scanning electron microscopy (SEM, JSM-5410LV, JEOL Ltd., Peabody, MA, USA). The SEM were performed at 15 kV under low vacuum mode. The morphology of film's matrix at magnifications of ×500 was evaluated.

2.6. Limonene Loading Content

The limonene loading content in the films was determined using GC-MS, which was previously described [16]. The limonene loading was calculated by the following equation: %Limonene loading = the actual quantity of limonene/the theoretical quantity of limonene × 100.

2.7. Film Sterility Test

Films containing orange oil were cut into a circle with a diameter of 4.0 mm, which is the same size as the disc that was used in the anti-bacterial activity test. The cut films were sterilized by ethylene oxide and tested for sterility by the direct inoculation in culture medium method [17,18]. Tryptic soy broth media and its containing *S. aureus* were used as negative and positive control, respectively. Film samples were placed in the media and evaluated the microbial growth sign by visual observation after incubation at 37 ± 2 °C for seven days.

2.8. Anti-Bacterial Activity of Films Loaded with Orange Oil

In vitro anti-bacterial activity of films was determined using the agar disc diffusion method [19]. *S. aureus* was activated in TSB and incubated for 6–10 h. Absorbance was then measured at 600 nm. The activated *S. aureus* turbidity had to equal 10^8 cfu/mL and was used and cultivated on MHA plates. The films loaded with orange oil were put onto the MHA plate. Ampicillin was also used as a positive control. After that, the MHA plates were incubated at 37 °C for 24 to 48 h, and the diameter of the clear zone was measured, including the diameter of the film or disc, using a Vernier caliper. Each experiment was performed in triplicate.

2.9. Statistical Analysis

An analysis of variance (ANOVA) was carried out with SPSS software version 16.0 (SPSS Inc., Chicago, IL, USA).

3. Results and Discussion

3.1. Identification of Orange Oil

GC-MS analysis of orange oil showed 19 known compounds including hydrocarbons, aldehydes, alcohol, and ketone (Table 1). The major compound of the oil was limonene (84.57%), which was exhibited at the retention time of 4.801 min. Other compounds were also found in the oil, such as *cis*-limonene oxide (1.86%), β-myrcene (1.06%) and *trans*-limonene oxide (0.88%). Thus, limonene was used as the marker for further experiments in this study. Previous studies have also identified limonene to be the major component (67.44–94.50%) in orange oil [20–22].

Table 1. Identified compounds of orange oil. GS-MC—gas chromatography mass spectrometer.

Compound	GC-MS RT (min)	Peak Area (%)
2-Thujene	3.623	0.46
3-p-Menthene	3.708	0.06
β-Myrcene	3.874	1.06
Octanal	4.097	0.21
3-Carene	4.309	0.06
Limonene	4.801	84.57
1,2-Dimethylcyclobutane	5.597	0.13
Linaolool oxide	5.705	0.08
Linalool	6.375	0.77
cis-Limonene oxide	7.313	1.86
trans-Limonene oxide	7.433	0.88
α-Terpineol	8.990	0.25
Decanol	9.396	0.33
Carveol	9.842	0.67
Carvone	10.592	0.76
Farnesol	13.350	0.49
9,11-Dodecadien-1-ol	15.719	0.10
1,5-Cyclooctadiene, 1,5-dimethyl	16.125	0.06
Valencene	18.225	0.10

3.2. Morphology and Mechanical Properties of Film Loaded with Orange Oil

All film formulations' compositions are shown in Table 2. The film without orange oil (P_1M_2 film) exhibited a smooth surface with dense matrix, whereas orange oil films showed rough surface with the presence of micropores inside the film's matrix. When orange oil was loaded, we observed the increasing of the thickness of orange oil films, which was related to the size and amount of micropores inside the film's matrix (Table 2 and Figure 1). The more orange oil evaporated, the greater the number and size of micropores, which resulted in a loose structure of the matrix and increased film thickness. Our results are consistent with Jouki et al. 2014, which found that quince seed film showed the loose film's matrix when oregano oil was incorporated [23]. Furthermore, in a previous study, the effect of grape pomace extract in chitosan film was investigated. Film structure discontinuities were induced by incorporation of wax or oil into the polysaccharide matrix [24]. Moreover, this study found that the greater the mango peel pectin ratio, the greater the number of micropores as seen in P_3M_0O film dense film matrix with less micropores than P_1M_2O film. The mechanical properties of films loaded with orange oil are shown in Table 3. Young's modulus of P_1M_2GO film significantly decreased to 35.24 MPa compared with 145.34 MPa for this film without orange oil. These results may be because of the amount of micropores in the polymer matrix. A previous study showed that quine seed mucilage films incorporated with the highest amount of oregano essential oil, which had many micropores, exhibited the best mechanical properties with the lowest tensile strength and Young's modulus, as well as the highest percent elongation. The works of [23,25] also found that percent elongation of citrus pectin film incorporated with clove bud oil increased as a result of oils existing in the film matrix in the form of oil droplets, which can be easily deformed and improve the film's extensibility.

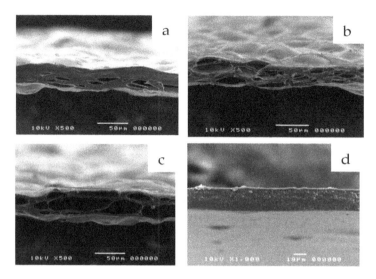

Figure 1. Scanning electron microscopy (SEM) of film's matrix of P_3M_0O (**a**), P_1M_2O (**b**), P_1M_2GO (**c**), and P_1M_2 (**d**) films.

Table 2. Film composition and thickness. LMP—low methoxy pectin.

Film Code	Composition				Thickness (μm)
	LMP (% *w/v* of Pectin Solution)	Mango Peel Pectin (% *w/v* of Pectin Solution)	Glycerol (% *w/w* Based on Pectin Weight)	Orange Oil (% *w/w* of Pectin Solution)	
P_3M_0O	3	0	-	3	52.94 ± 0.21
P_1M_2O	1	2	-	3	58.82 ± 0.23
P_1M_2GO	1	2	40	3	70.58 ± 0.18
P_1M_2	1	2	-	-	25.71 ± 0.11

Table 3. Orange oil loading content, tensile strength, elongation, and Young's modulus of films containing orange oil.

Film Code	Orange Oil Loading (%)	Tensile Strength (MPa)	Elongation (%)	Young's Modulus (MPa)
P_3M_0O	86.17 ± 3.41 [a]	6.12 ± 1.18 [a]	2.28 ± 0.15 [a]	266.04 ± 35.39 [a]
P_1M_2O	70.10 ± 1.03 [a]	3.12 ± 1.23 [b]	2.57 ± 0.10 [a]	126.91 ± 42.32 [b]
P_1M_2GO	59.25 ± 2.09 [b]	4.54 ± 0.18 [c]	12.93 ± 0.89 [b]	35.24 ± 3.43 [c]
P_1M_2	-	4.98 ± 0.54 [c]	6.15 ± 0.88 [c]	145.34 ± 31.77 [b]

Note: For each test, the different letters are statistically different ($p < 0.05$).

3.3. Limonene Oil Loading Content

The limonene loading contents of films loaded with orange oil are shown in Table 3. The highest limonene loading content (86.17%) was obtained from P_3M_0O film, whereas P_1M_2GO film exhibited the lowest limonene loading content (59.25%). This result found that the texture of the orange oil film played an important role in limonene loading contents. Micropores indicated that orange oil had evaporated from the polymer matrix. Film with a lower amount and smaller size of micropores, like P_3M_0O, can encapsulate more limonene content. While the P_1M_2GO film with larger amount and size of micropores showed the lowest limonene content.

3.4. Film Sterility and Anti-Bacterial Activity

The product sterility test was used to confirm that the film samples were sterile, before testing the anti-bacterial activity. In sterility test, there was no microbial growth in any of the film samples. Growth of *S. aureus* was observed in the positive control, while no sign of *S. aureus* growth was observed in the negative control. The results of examination of the anti-microbial activity of films against *S. aureus* using the agar disk diffusion method are shown in Table 4. All films loaded with limonene exhibited a clear zone. In the direct comparisons of the clear zone diameter of film formulations containing different limonene loading content, the P_3M_0O film, which had the highest limonene loading content (86.17%), showed the widest clear zone diameter (10.02 mm), whereas P_1M_2GO film, which exhibited the lowest limonene loading content (59.25%), showed the smallest clear zone diameter (8.08 mm). Thus, the anti-bacterial activity of the films in this study was dependent on the amount of limonene remaining in the film. Generally, in a hydrophilic polymer such as pectin, water molecules from agar penetrate into the polymer matrix resulting in swelling of film; thus gradually widening the meshes of the polymer network and leading to greater release of active compounds into the surroundings [22]. However, in order to normalize limonene content in each film, the normalized clear zone was calculated and presented in Table 4. Interestingly, the normalized clear zone dimeter of P_1M_2GO film (11.75 mm) was higher than the others, but no statistically significant difference existed between the films. It may be concluded that when we consider film composition in the same amount of limonene loading content, the P_1M_2GO film, which showed better mechanical properties, tended to show higher anti-bacterial activity. In addition, our recent study found that loading orange oil in the form of microemulsion was able to reduce micropores in the film's matrix and increase limonene loading capacity as well as anti-bacterial activity [16].

Table 4. Anti-bacterial activity of films against *S. aureus*.

Sample	Clear Zone Diameter (mm)	Normalized clear Zone (mm)
P_3M_0G film	ND	ND
P_3M_0O film	10.02 ± 0.03	10.02
P_1M_2O film	8.76 ± 0.01	10.77
P_1M_2GO film	8.08 ± 0.01	11.75
P_1M_2 film	ND	-
Orange oil (100 mg/disc)	9.64 ± 0.01	-
Tween® 80	ND	-
Ampicillin (6.25 µg/disc)	9.82 ± 0.01	-

Note: ND = not detected. Normalized clear zone (nm) = (limonene content of P_3M_0O/limonene content of each film) × clear zone diameter of each film.

4. Conclusions

Films loaded with orange oil were prepared and their morphology, tensile properties, loading content, and anti-bacterial activity were investigated. The addition of orange oil decreased the Young's modulus value, which appears to be related to the amount of micropores in the film's matrix. Interestingly, the P_1M_2GO film, which had the best mechanical properties, exhibited the lowest limonene loading content (59.25%). For the test of anti-microbial against *S. aureus*, the P_1M_2GO film showed the highest normalized clear zone (11.75 mm), but no significant difference existed. This study indicated that the orange oil loaded in mango peel pectin film can be a valuable candidate as an antibacterial material for food packaging. However, this orange oil film needs further development, especially the increasing of the limonene loading content. Other technologies such as preparation of orange oil in micro-emulsion or nano-emulsion could be used in order to enhance the limonene loading content.

Author Contributions: The study was designed by all of authors. The all of experiments were conducted by T.C., W.R., and K.J. under suggestions of P.J. The manuscript was written by T.C. and P.J.

Acknowledgments: The authors gratefully acknowledge the Franco-Thai Cooperation Programme in Higher Education and Research/Franco-Thai Mobility Programme/PHC SIAM 2018-2019, Graduate School Chiang Mai University (GSCMU) and Faculty of Pharmacy, Chiang Mai University, Chiang Mai, Thailand for the financial support. The authors would like to thank Assist. Sasithorn Sirilun for all of the anti-bacterial activity testing support.

References

1. Vermeiren, L.; Devlieghere, F.; Beest, M.V.; Kruijf, N.D.; Debevere, J. Developments in the active packaging of foods. *Trends Food Sci. Technol.* **1999**, *10*, 77–86. [CrossRef]

2. Perez-Perez, C.; Regalado-Gonzalez, C.; Rodriguez-Rodriguez, C.A.; Barbosa-Rodriguez, J.R.; Villasenor-Ortega, F. Incorporation of antimicrobial agents in food packaging films and coatings. In *Advances in Agricultural and Food Biotechnology*; Research Signpost: Kerala, India, 2006; pp. 193–216, ISBN 8177362690.

3. Pedroza-Islas, R.; Aguilar-Esperanza, E.; Vernon-Carter, E.J. Obtaining pectins from solids wastes derived from mango (*Mangifera indica*) processing. *AIChE Symp. Ser.* **1994**, *90*, 38–41.

4. Tandon, D.K.; Garg, N. Mango waste: A potential source of pectin, fiber, and starch. *Indian J. Environ. Prot.* **1999**, *19*, 1924–1927.

5. Müller-Maatsch, J.; Bencivenni, M.; Caligiani, A.; Tedeschi, T.; Bruggeman, G.; Bosch, M.; Petrusan, J.; Van Droogenbroeck, B.; Elst, K.; Sforza, S. Pectin content and composition from different food waste stream. *Food Chem.* **2016**, *15*, 37–45.

6. Magwa, M.L.; Gundidza, M.; Gweru, N.; Humphrey, G. Chemical composition and biological activities of essential oil from the leaves of *Sesuvium portulacastrum*. *J. Ethnopharmacol.* **2006**, *103*, 85–89. [CrossRef] [PubMed]

7. Matiz, G.; Osorio, M.R.; Camacho, F.; Atencia, M.; Herazo, J. Effectiveness of antimicrobial formulations for acne based on orange (*Citrus sinensis*) and sweet basil (*Ocimum basilicum* L) essential oils. *Biomedica* **2012**, *32*, 125–133. [CrossRef] [PubMed]

8. Vimal, M.; Vijaya, P.P.; Mumtaj, P.; Farhath Seema, M.S. Antibacterial activity of selected compounds of essential oils from indigenous plants. *J. Chem. Pharm. Res.* **2013**, *5*, 248–253.

9. Orchard, A.; Vuuren vann, S. Commercial essential oils as potential antimicrobials to treat skin diseases. *Evid. Based Complement. Alternat. Med.* **2017**. [CrossRef] [PubMed]

10. Ahmad, M.; Benjakul, S.; Prodpran, T.; Agustini, T.W. Physico-mechanical and antimicrobial properties of gelatin film from the skin of unicorn leatherjacket incorporated with essential oils. *Food Hydrocoll.* **2012**, *28*, 189–199. [CrossRef]

11. Chana-Thaworn, J.; Chanthachum, S.; Wittaya, T. Properties and antimicrobial activity of edible films incorporated with kiam wood (*Cotyleobium lanceotatum*) extract. *LWT Food Sci. Technol.* **2011**, *44*, 284–292. [CrossRef]

12. Assefa, Z.; Admassu, S. Development and characterization of antimicrobial packaging films. *J. Food Process. Technol.* **2013**, *4*, 235–241. [CrossRef]

13. Prabuseenivasan, S.; Jayakumar, M.; Ignacimuthu, S. In vitro antibacterial activity of some plant essential oils. *BMC Complement. Altern. Med.* **2006**, *6*, 39–46. [CrossRef] [PubMed]

14. Wang, S.; Chen, F.; Wu, J.; Wang, Z.; Liao, X.; Hu, X. Optimization of pectin extraction assisted by microwave from apple pomace using response surface methodology. *J. Food. Eng.* **2007**, *78*, 693–700. [CrossRef]

15. Jantrawut, P.; Chaiwarit, T.; Jantanasakulwong, K.; Brachais, H.C.; Chambin, O. Effect of plasticizer type on tensile property and in vitro indomethacin release of thin films based on low-methoxyl pectin. *Polymers* **2017**, *9*, 289–303. [CrossRef]

16. Jantrawut, P.; Boonsermsukcharoen, K.; Thipnan, K.; Chaiwarit, T.; Hwang, K.M.; Park, E.S. Enhancement of antibacterial activity of orange oil in pectin thin film by microemulsion. *Nanomaterials* **2018**, *8*, 545–556. [CrossRef] [PubMed]

17. Mahajan, H.S.; Deshmukh, S.R. Development and evaluation of gel-forming ocular films based on xyloglucan. *Carbohydr. Polym.* **2015**, *122*, 243–247. [CrossRef] [PubMed]

18. United States Pharmacopeial Convention. Sterlility Test/Microbiological Test. In *The United State Pharmacopeia*, 40th ed.; United Book Press: Gwynn Oak, MD, USA, 2017; Volume 1, pp. 137–142.

19. Lee, T.W.; Kim, J.C.; Hwang, S.J. Hydrogel patches containing triclosan for acne treatment. *Eur. J. Pharm. Biopharm.* **2003**, *56*, 407–412. [CrossRef]

20. Högnadóttir, Á.; Rouseff, R.L. Identification of aroma active compounds in orange essence oil using gas chromatography–olfactometry and gas chromatography–mass spectrometry. *J. Chromatogr. A* **2003**, *998*, 201–211. [CrossRef]

21. Qiao, Y.; Xie, B.J.; Zhang, Y.; Zhang, Y.; Fan, G.; Yao, X.L.; Pan, S.Y. Characterization of aroma active compounds in fruit juice and peel oil of Jinchen sweet orange fruit (*Citrus sinensis* (L.) Osbeck) by GC-MS and GC-O. *Molecules* **2008**, *13*, 1333–1344. [CrossRef] [PubMed]
22. Tao, N.G.; Liu, Y.J.; Zhang, J.H.; Zeng, H.Y.; Tang, Y.F.; Zhang, M.L. Chemical composition of essential oil from the peel of Satsuma mandarin. *Afr. J. Biotechnol.* **2008**, *7*, 1261–1264.
23. Jouki, M.; Yazdi, F.T.; Mortazavi, S.A.; Koocheki, A. Quince seed mucilage films incorporated with oregano essential oil: Physical, thermal, barrier, antioxidant and anti-bacterial properties. *Food Hydrocoll.* **2014**, *36*, 9–19. [CrossRef]
24. Ferreira, A.S.; Nunes, C.; Castro, A.; Ferreira, P.; Coimbra, M.A. Influence of grape pomace extract incorporation on chitosan films properties. *Carbohydr. Polym.* **2014**, *113*, 490–499. [CrossRef] [PubMed]
25. Nisar, T.; Wang, Z.C.; Yang, X.; Tian, Y.; Iqbal, M.; Guo, Y. Characterization of citrus pectin films integrated with clove bud essential oil: Physical, thermal, barrier, antioxidant and anti-bacterial properties. *Int. J. Biol. Macromol.* **2018**, *106*, 670–680. [CrossRef] [PubMed]

10

Preparation and Characterization of Polymer Composite Materials Based on PLA/TiO₂ for Antibacterial Packaging

Edwin A. Segura González [1,2], **Dania Olmos** [2,*], **Miguel Ángel Lorente** [2], **Itziar Vélaz** [3] **and Javier González-Benito** [2,*]

[1]   Universidad Interamericana de Panamá, Research Direction (DI-UIP 6338000), Av. Ricardo J. Alfaro, Panama City, Panama; edwin_segura@uip.edu.pa

[2]   Department of Materials Science and Engineering and Chemical Engineering, Instituto de Química y Materiales Álvaro Alonso Barba (IQMAA), Universidad Carlos III de Madrid, Leganés 28911, Madrid, Spain; malorente@ing.uc3m.es

[3]   Departamento de Química, Facultad de Ciencias, Universidad de Navarra, 31080 Pamplona, Navarra, Spain; itzvelaz@unav.es

*   Correspondence: dolmos@ing.uc3m.es (D.O.); javid@ing.uc3m.es (J.G.-B.).

Abstract: Polymer composite materials based on polylactic acid (PLA) filled with titanium dioxide (TiO₂) nanoparticles were prepared. The aim of this work was to investigate the antibacterial action of TiO₂ against a strain of *E. coli* (DH5α) to obtain information on their potential uses in food and agro-alimentary industry. PLA/TiO₂ systems were prepared by a two-step process: Solvent casting followed by a hot-pressing step. Characterization was done as a function of particle size (21 nm and <100 nm) and particle content (0%, 1%, 5%, 10%, and 20%, wt %). Structural characterization carried out by X-ray diffraction (XRD) and Fourier Transformed Infrared spectroscopy (FTIR) did not reveal significant changes in polymer structure due to the presence of TiO₂ nanoparticles. Thermal characterization indicated that thermal transitions, measured by differential scanning calorimetry (DSC), did not vary, irrespective of size or content, whereas thermogravimetric analysis (TGA) revealed a slight increase in the temperature of degradation with particle content. Bacterial growth and biofilm formation on the surface of the composites against DH5α *Escherichia coli* was studied. Results suggested that the presence of TiO₂ nanoparticles decreases the amount of extracellular polymeric substance (EPS) and limits bacterial growth. The inhibition distances estimated with the Kirby-Bauer were doubled when 1% TiO₂ nanoparticles were introduced in PLA, though no significant differences were obtained for higher contents in TiO₂ NPs.

Keywords: polylactic acid (PLA); TiO₂ nanoparticles; polymer nanocomposites; antibacterial packaging

1. Introduction

Nowadays, the impact of plastic waste is a worldwide concern in our society. Consequently, the research on biodegradable materials is a response to a global need. Polylactic acid (PLA), ($-[CH-(CH_3)-COO]_n-$), belongs to the family of aliphatic polyesters and is an environmentally friendly polymer that has been widely used for producing biodegradable, biocompatible, and compostable materials [1,2]. PLA is a thermoplastic polymer with tunable mechanical properties (depending on the crystalline to amorphous fractions) that, thanks to its biodegradability, it can replace other non-degradable polymers in several applications in such a way that it improves the environmental side-effects of non-degradable polymers. PLA is present in many different applications. For example, in biomedical industry is used for sutures, films, implants, or scaffolds for tissue engineering applications [3,4]. In agro-alimentary industry, PLA is present in food packaging containers, in the manufacture of greenhouse films, or for biodegradable yard-waste bags [5,6].

In the field of agro-alimentary industry, for example for food packaging applications, apart from the biodegradable character of the material, it would be worth producing materials that inhibit bacterial growth and biofilm development. Biofilms are resistant to many antibiotics and other antimicrobial agents. To avoid or to reduce the possible degradation of PLA, one alternative could be the incorporation of some additives such as TiO_2 nanoparticles (NPs). The interest in TiO_2 nanoparticles is twofold. First, TiO_2 NPs are able to absorb most of the UV radiation, thus preventing polymer degradation from environmental aging due to the exposure of plastics to UV light. Secondly, TiO_2 inhibits to certain extent bacterial growth. Therefore, the introduction of TiO_2 nanoparticles may be an efficient tool, not only for modifying some polymer properties, but also for hindering its degradation from bacteria or from environmental aging in outdoor materials.

Titanium dioxide, TiO_2, has attracted the interest in many fields for its photocatalytic and bacteriostatic activity [7]. In the presence of light, TiO_2 is able to produce the transition of an electron towards the conductive band favoring the oxidative capacity of other species by generating active agents like radicals. Keeping in mind that bacteria, when subjected to oxidative stress, are able to unleash a specific self-destruction mechanism; it is easy to understand that TiO_2 is a material with bactericide activity which behavior is enhanced in the presence of light. Apart from the well-known bactericidal properties of TiO_2 [8–15], as well as the fact that its biocompatibility and small size when TiO_2 is used in the form of nanoparticles, it improves the catalytic effect of such materials [16–19], having a great potential in applications related with environment purification, decomposition of carbonic acid gas into hydrogen gas, etc. This filler is usually applied as pigment, adsorbent, catalyzer support, filter, coatings, and dielectric materials. Moreover, due to its bacteriostatic behavior, the TiO_2 is also useful for the inhibition of odors and can be part of a self-cleaning system for specific surfaces. Such advantages make the TiO_2 an inorganic filler ideal for the development of nanocomposite materials resistant to UV radiation, probably resistant to thermal degradation, and may inhibit the formation of harmful biofilms. For that reason, the addition of TiO_2 nanoparticles could provide the final material some of the functional properties of the proper TiO_2 like the UV radiation protection and the bactericidal activity.

Apart from the functional properties that the filler itself confers the final material, in most polymer nanocomposite materials, particle size is an important factor affecting the final behavior of composite materials and influencing the physical properties of the material [20]. For the same amount of particles, the ones with a smaller size would provide a larger surface to be in contact with the polymeric matrix and, for that reason, a larger interfacial region is formed. This effect is explained in detail in a recent article by J. Gonzalez-Benito et al. [21] in which the influence of TiO_2 particles size in the thermal expansion coefficient of nanocomposites materials based on a poly(ethylene-co-vinylacetate)matrix, was studied.

In this work polymer nanocomposites based on PLA/TiO_2 were prepared and characterized. TiO_2 nanoparticles with different particle sizes, 21 nm and <100 nm, were selected. The nanoparticles were mixed and dispersed in the PLA matrix by solvent casting and the final materials were obtained after hot pressing the casted films. The effect of particle size and content in structural and thermal properties and materials behavior against bacterial growth and biofilm development of a strain of *E. coli* (DH5α) was investigated.

2. Experimental

2.1. Materials

Polylactic acid (PLA) was provided by Resinex Spain, SL, and manufactured by Nature Works LLC (Blair, NE, USA) (Ref. code: PLA Polymer 7032D; glass transition temperature, T_g = 55–60 °C; melting temperature, T_m = 160 °C; and processing temperature 200–220 °C). To prepare the nanocomposites, TiO_2 nanoparticles with two different particle sizes, 21 nm and <100 nm, were used (Sigma Aldrich, St. Louis, MO, USA, with reference numbers 718467, and 634662, respectively). The solvent used to prepare the polymer solutions and particles suspensions was dichloromethane (purity 99.9%, Sigma Aldrich).

2.2. Sample Preparation

PLA/TiO_2 nanocomposites films were prepared with different particle content (0%, 1%, 5%, 10%, and 20% weight percentages, wt %). Films of ca. 2.0 g were obtained according to the protocol described in Reference [22]. A solution of 10% wt/vol of PLA in dichloromethane (CH_2Cl_2) was mixed with a suspension of TiO_2 NPs in CH_2Cl_2 (previously sonicated). This mixture (PLA+TiO_2 in CH_2Cl_2) is stirred for 1 h at room temperature and then casted on a Petri dish (φ = 60 mm) to obtain pre-films. After drying at 40 °C for 24 h, the pre-films were placed between two Kapton® sheets inside a hot plate press (FONTIJE PRESSES, TP400 model, Fontijne Presses, Barendrecht, The Netherlands) and processed at 30 kN and 160 °C for 10 min, obtaining 10 × 10 cm^2 films with an average thickness of 200 μm. The prepared films were stored in a desiccator. In Figure 1, an example of a pre-film obtained after casting and the corresponding film obtained after the hot-pressed step are shown.

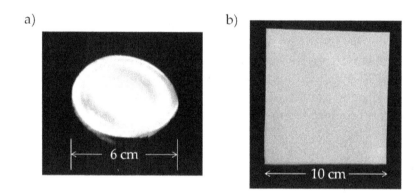

a)   b)

Figure 1. Example of a pre-film of PLA/TiO_2 composite obtained (**a**) after casting and (**b**) after the hot-pressing step.

Sample labeling follows the notation: PLA/TiO_2-Particle size-particle content, in weight percentage. For example, a sample with the following code: PLA/TiO_2-21-5 refers to a composite of PLA filled with TiO_2 nanoparticles with a particle size of 21 nm and a particle content of 5% by weight.

2.3. Characterization Techniques

Structural characterization of the samples was done by X-ray diffraction (XRD) and Fourier Transformed Infrared Spectroscopy (FTIR). X-Ray diffraction experiments were done in a Phillips X'Pert X ray diffractometer (Malvern Panalytical Ltd, Malvern, UK) in the range 2θ = 3–70° using the $K_{\alpha1}$ radiation from copper with a wavelength, λ = 0.15406 nm. The working conditions were set at 40 kV and 40 mA. Fourier transformed infrared (FTIR) spectra were recorded using an FTIR Nicolette Avatar 360 (Analytical Instruments Brokers LLC, Minneapolis, MN, USA) equipped with a Golden Gate ATR accessory (diamond window), from 600 to 4000 cm^{-1} with a resolution of 2 cm^{-1} and averaging 32 scans, at room temperature with the software OMNIC ESP v5.1 (ThermoFisher Scientific Inc., Waltham, MA USA).

Thermal characterization of the materials was done using differential scanning calorimetry (DSC) and thermogravimetric analysis (TGA). DSC experiments were carried out in a Mettler Toledo DSC822^e instrument (Mettler Toledo, Greifensee, Switzerland). The thermal cycle was: (i) heating scan from 40 to 200 °C at 100 °C ·min^{-1}; (ii) 5 min at 200 °C; (iii) cooling scan from 200 to 10 °C at 20 °C ·min^{-1}; (iv) 5 min at 10 °C; and (v) a final heating scan from 10 to 200 °C at 20 °C ·min^{-1}. Thermal transitions of the polymer, glass transition, melting and crystallization were analyzed from the heating and cooling scans. Thermal degradation of the samples was studied by TGA. The experiments were done in a TGA-SDTA 851 Mettler Toledo (Mettler Toledo, Greifensee, Switzerland) from 30 to 750 °C at a heating rate of 10 °C·min^{-1} under a nitrogen atmosphere (gas flow of 20 mL·min^{-1}).

2.4. Biofilm Development and Bacterial Growth

The behavior of the materials against bacterial growth and biofilm development was studied using a strain of *Escherichia coli* (DH5α) following two different approaches. First, bacterial cultures on the surface of the films were done and biofilm development was studied. To do these experiments, an aliquot of the *E. coli* strain was heated and 90 μL of bacteria were mixed with 2910 μL Luria Bertani media (LB). The mixture was stirred at 200 rpm and incubated for 12 h at 37 °C. After that, the suspension was diluted 1/100 to a final volume of 20 mL. Cultures on the surfaces of the PLA/TiO$_2$ materials were grown in a 24 microwell plate (ThermoFischer Scientific, Waltham, MA, USA) using the DH5α *E. coli* strain. Square samples of approximately 0.5 cm^2 were glued on stainless steel discs with a diameter of 10 mm using cyanoacrylate-based glue. The samples were sterilized with a 70% (wt %) solution of ethanol directly sprayed on the surface and dried in a laminar flow hood. All the processes were done in a sterile environment. Then, 1 mL of the 1/100 dilution previously prepared was added to each well plate and incubated in aerobic conditions for 3 h at 37 °C with a continuous agitation at 150 rpm. After incubation, the LB medium containing the bacteria was removed from the multiwell plate and rinsed using 1 mL of physiological saline solution (NaCl 0.9% wt) to eliminate the poorly adhered cells from the surface of the materials. For examining biofilm growth on the surface of the PLA/TiO$_2$ systems, a scanning electron microscope, SEM, Philips XL30 (FEI Europe Ltd., Eindhoven, The Netherlands) was used. Micrographs at different magnifications (50×, 5000× and 8000×) were collected. The voltage was set at 10 kV and the working distance at ~10 mm. To avoid charge accumulation, the samples were gold coated by sputter deposition.

The second approach to study the antibacterial behavior of the nanocomposites consists of a modification of the Kirby-Bauer diffusion test [23,24] following the protocol described in a previous work [25]. For these studies, a seed of the *E. coli*, DH5α competent cells, from ThermoFischer Scientific (Waltham, MA, USA) was used. In this case, 100 μL of bacteria were mixed with 900 μL of Luria Bertani media and stirred for 1 h at 37 °C and 200 rpm. From this suspension, 200 μL were seed in an LB agar plate. Then, square samples of ~1 cm^2 were placed in the agar plate and incubated at 37 °C overnight. The inhibition distances were measured using Olympus optical microscope image analysis software (analySIS getIT, Olympus, Tokyo, Japan).

3. Results and Discussion

3.1. Structural Characterization

The structural characterization of the PLA/TiO$_2$ films was done by X-ray diffraction. X-ray diffractograms of the samples (Figure S1) showed the diffraction maxima of the PLA at 2θ = 16.8° and 19.1° assigned to (200)/(110) and (203) reflexions. The XRD spectra of TiO$_2$ NPs and the corresponding PLA/TiO$_2$ nanocomposites showed diffraction peaks that can be assigned to both polymorphs of TiO$_2$, anatase (JCPDS 89-4921) and rutile (JCPDS 89-4920) [20]. In Table S1, the assignment of XRD peaks of both polymorphs of TiO$_2$ is given. From XRD patterns, it can be concluded that the structure of PLA was not affected by the presence of titania nanoparticles.

In Figure 2 are shown the ATR-FTIR spectra (mid-infrared) corresponding to PLA-0 and the different PLA/TiO$_2$ composites as a function of particle content for PLA/TiO$_2$ filled with the two types of nanoparticles.

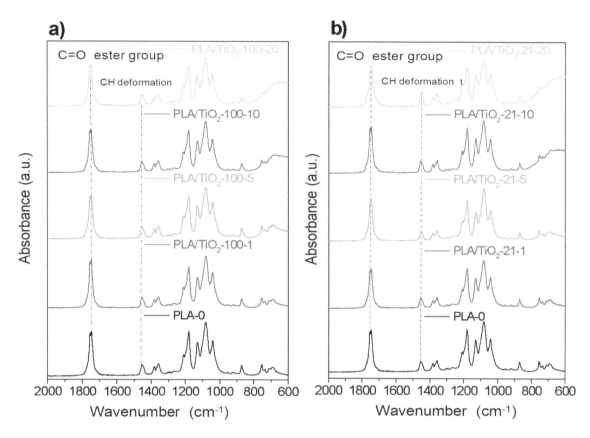

Figure 2. ATR-FTIR spectra for the samples with TiO_2-100 nm (**a**) and TiO_2-21 nm (**b**) as a function of the content of nanoparticles.

In the region between 600 and 800 cm^{-1}, it is possible to observe the peak due to the absorption bands corresponding to the stretching vibrations of Ti–O and Ti–O–Ti (TiO_2) [26,27] which increases with the content of TiO_2 nanoparticles. The typical bands of PLA appear also in this region: Namely, carbonyl groups, C=O, at 1755 cm^{-1} and bending of –CH$_3$ (antisymmetric at 1454 cm^{-1} and symmetric at 1361 cm^{-1}) [28]. Likewise, the stretching vibrations of C–O groups are at 1225 cm^{-1} (symmetric) and at 1090 cm^{-1} (antisymmetric). Focusing on the polymer bands with smaller intensities, it is possible to identify at 920 and 956 cm^{-1} those corresponding to the main chain vibrations (rocking of CH$_3$), and also at 871 and 756 cm^{-1}, which can be assigned to the amorphous and crystalline phases of PLA, respectively. Furthermore, there are no significant variations in terms of intensity or band shifting as a function of nanoparticles content and/or size for the processing conditions used. The detailed band assignment for the most representative vibrations is given in Table S2. Additionally, a full characterization of the PLA spectra in NIR region is described in Reference [22].

3.2. Thermal Characterization

To investigate the effect of TiO_2 nanoparticles in the thermal properties of the materials, DSC experiments were done. Thermal properties were obtained from the second heating scan (the first heating scan was done to erase thermal history of the materials and it is not included here for discussion). The characteristic temperatures corresponding to the different transitions measured, glass transition, crystallization, and melting, are collected in Table S3.

Results showed that the main transitions associated to the typical thermal behavior of the pure PLA were present. Glass transition temperature was observed at T_g = 64–65 °C, similar to literature values [29,30]. Cold crystallization of the sample was present in all the samples during the heating scan with a peak temperature, T_c at ~137 °C. Melting process was observed as an endothermic peak with a melting temperature, T_m, at approximately 167 °C. Considering the data collected in Table S3 (see Supplementary Material), it can be observed that the addition of TiO_2 nanoparticles to the polymer matrix did not produce significant changes in the characteristic temperatures of the PLA matrix, indicating that TiO_2 NPs caused little effect on the dynamics of the macromolecular chains in the samples under study, producing similar effects on the crystalline regions of the nanocomposites samples under the same processing conditions [7,31,32]. On the other hand, the composite samples of PLA/TiO_2 crystallize in a temperature range of 100–130 °C, which would correspond mainly to an ordered crystalline α-phase structure [32].

Thermal characterization of the samples was completed by thermogravimetric analysis (TGA), to determine the degradation temperatures of the different materials. Figure 3a illustrates the weight loss, as a percentage, as a function of heating temperature and Figure 3b the first derivative of the weight loss as a function of temperature (differential thermogravimetric analysis, DTGA curve) for the PLA/TiO_2 nanocomposites filled with titania particles of < 100 nm. Similar plots were obtained for PLA/TiO_2 systems filled with TiO_2 NPs of 21 nm (Figure S2).

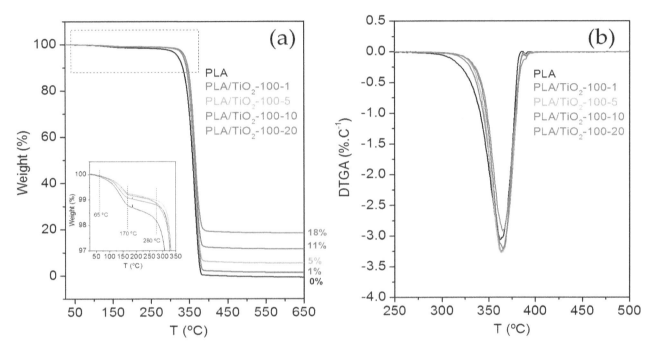

Figure 3. (a) Thermogravimetric analysis curve and (b) Differential thermogravimetric Analysis, DTGA, curve for the composites based on PLA/TiO_2-100.

In all the samples, similar TGA curves were observed (Figure 3a). First, there is an initial mass loss between 65–170 °C, attributed to the loss of water from moisture. After that, a significant mass loss between 280–390 °C is observed, which corresponds to the decomposition of the PLA. Finally, from 390 °C to 500 °C thermal analysis curves slow down to complete the decomposition of the PLA matrix [33] until a constant mass is reached. The constant mass remaining at the end of each TGA experiment (see Figure 3 for PLA/TiO_2-100 and Figure S2 for PLA/TiO_2-21) corresponds to the inorganic material, i.e. the TiO_2 NPs, which is very close to the theoretical amount of particles in the composites.

From the DTGA (Figure 3b), the degradation temperatures corresponding to a 5% and 95% mass loss of PLA, T_5 = 331.9 °C and T_{95} = 382.9 °C, were estimated. The DTGA curves showed that in the nanocomposite materials (both filled with TiO_2-100 nm and TiO_2-21 nm) the degradation begins at slightly higher temperatures, as can be confirmed by T_5 and T_{95} (Table 1). Therefore, nanocomposites have slightly higher stability than pure PLA, i.e., their thermal degradation occurs at higher temperatures than that of pure PLA. A similar trend was observed for the temperature at which the degradation rate is maximum, the peak temperature of the DTGA curve, T_p, which for PLA-0 sample is T_p = 361.8 °C, whereas for the nanocomposite samples showed higher peak values. The characteristic degradation temperatures T_5, T_{95}, and T_p are collected in Table 1 for all the samples under study. Previous studies have also shown slightly higher degradation temperatures of the polymer matrices when nanoparticles are present (LDPE/AgNPs) [25].

Table 1. Characteristic degradation temperatures for the PLA/TiO$_2$ systems determined by TGA.

Sample	T_5 (°C)	T_{95} (°C)	T_p (°C)
PLA-0	331.9	382.9	361.8
PLA/TiO$_2$-100-1	337.4	385.9	363.0
PLA/TiO$_2$-100-5	338.9	385.1	363.1
PLA/TiO$_2$-100-10	341.2	384.6	364.9
PLA/TiO$_2$-100-20	341.2	385.9	364.7
PLA/TiO$_2$-21-1	335.8	383.3	359.5
PLA/TiO$_2$-21-5	340.7	385.2	364.0
PLA/TiO$_2$-21-10	341.9	385.1	366.1
PLA/TiO$_2$-21-20	338.5	390.0	366.9

3.3. Antimicrobial Behaviour

3.3.1. Study of Biofilm Development on the Surface of the Materials

A set of micrographs obtained by SEM of the surfaces of the materials after culturing the samples in the presence of *E. coli* DH5α for the PLA/TiO$_2$ nanocomposites are shown in Figures 4 and 5, corresponding to materials filled with 21 nm particles (Figure 4) and with <100 nm particles (Figure 5).

In the case of PLA (Figures 4 and 6), only some bacteria are clearly visible which could induce to think that their proliferation was minimal. However, careful observations at higher magnifications of 5000× and 8000× it is possible to identify some hollow areas where it is clearly seen that bacteria are below a material that hides them (See dashed circled area in Figure 6c). Therefore, it is possible to conclude that above pure PLA there is a larger bacteria proliferation; indeed, such proliferation rate ends up in the generation of a biofilm that has a great amount of extracellular polymeric substance, EPS [34–37], which surrounds the bacteria. For that reason, at the end, such bacteria are hidden when the samples are studied using SEM. It is important to bear in mind that SEM technique when using the backscattered electrons signal (BSE), allows the visualization of surfaces and morphologies associated to compositional changes in which elements with different atomic numbers are involved. In this study, both the bacteria and the EPS are mainly formed by carbon, and consequently they are indistinguishable by using the backscattered electron signal from a scanning electron microscope.

Regarding the results of nanocomposite materials, bacteria can be seen more easily as individual entities (Figure 4, Figure 5, Figure S3 and S4). This result may be due to a smaller amount extracellular polymeric substance (EPS) coating the biofilm formed on the surface of the nanocomposite materials. Comparing the images of the nanocomposite materials at different magnifications as a function on the content of TiO$_2$ nanoparticles (Figures S3 and S4) no significant differences are seen. In some cases, it is possible to recognize more clearly the lack of continuity of the biofilm formed; but this effect may be basically due to the region selected for inspection.

In general, *E. coli* are approximately 500-600 nm width, with a length varying between 2–3 μm depending on the bipartition state in which they are found [25,38]. Slight differences among bacteria can be seen, though (Figure 7). Apparently, there are larger bacteria oblong shaped (Figure 7a) and smaller ones that tend to have a cylindrical geometry (Figure 7b) and some other bacteria are stretched and elongated (Figure 7c). These last two geometries are associated to bacteria that have grown on the surface of nanocomposite materials: The first one (Figure 7b) corresponds to bacteria grown above nanocomposites with 21 nm-sized nanoparticles, and the second geometry corresponds to bacteria grown on nanocomposites with <100 nm sized nanoparticles.

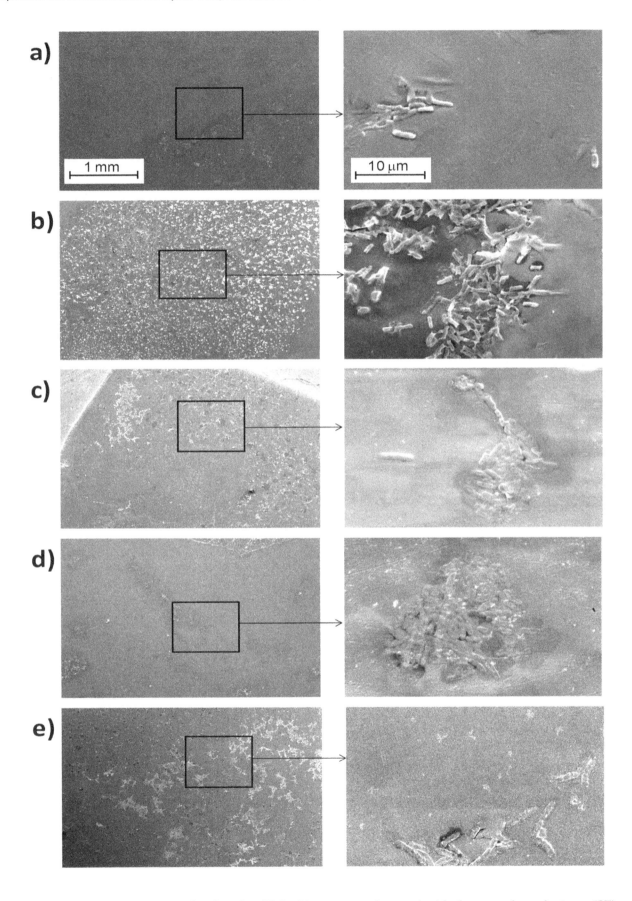

Figure 4. SEM micrographs for the PLA/TiO$_2$-21-x system observed with the secondary electrons (SE) detector: (**a**) PLA-0, (**b**) PLA/TiO$_2$-21-1, (**c**) PLA/TiO$_2$-21-5, (**d**) PLA/TiO$_2$-21-10 and (**e**) PLA/TiO$_2$-21-20 obtained at different magnifications (left side at 50× and right side at 5000×).

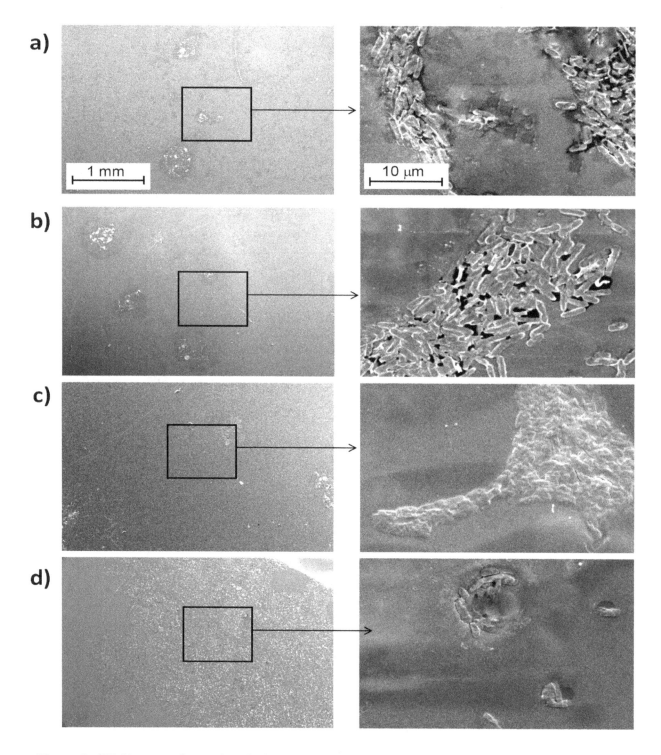

Figure 5. SEM images obtained with the secondary electrons (SE) detector for: (**a**) PLA/TiO$_2$-100-1, (**b**) PLA/TiO$_2$-100-5, (**c**) PLA/TiO$_2$-100-10 and, (**d**) PLA/TiO$_2$-100-20 obtained at different magnifications (left 50×, right 5000×) (Images for PLA-0 correspond to Figure 4a).

The initial conclusion that can be drawn from all these results is that the presence of TiO$_2$ nanoparticles decreases bacterial growth and biofilm development of *E. coli* (DH5α) (Figure 6, Figures S3 and S4). These effects may be due to a direct interference on bacterial metabolism. The bacteria observed under the presence of nanoparticles appear to be smaller in size or elongated so it seems that their growth is altered. Moreover, bacteria grown on the nanocomposites containing particles with smaller diameter, 21 nm (Figure S3b), are even smaller compare to those grown on the composites containing the <100 nm nanoparticles. This observation may be related with the surface-to-volume ratio of the particles, which might vary the oxidative catalytic behavior of titanium oxide towards the organic material associated to both EPS and bacteria themselves.

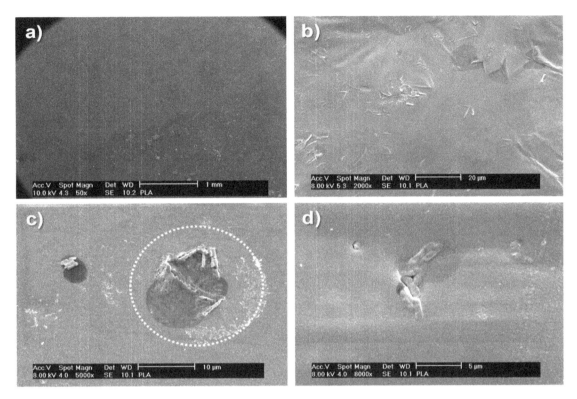

Figure 6. SEM micrographs of the biofilm generated on the surface of pure PLA (PLA-0) observed at different magnifications: (**a**) 50×; (**b**) 2000×; (**c**) 5000× and (**d**) 8000×.

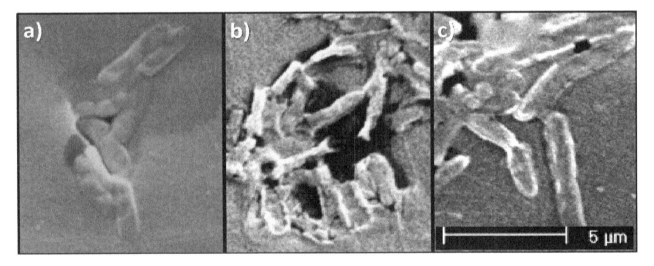

Figure 7. Morphology of bacteria as a function of the material on which they develop: (**a**) PLA-0; (**b**) PLA/TiO$_2$ systems ($\varphi \sim 21$ nm) and (**c**) PLA/TiO$_2$ systems ($\varphi < 100$ nm).

3.3.2. Kirby-Bauer Diffusion Test

Figure 8 shows the optical micrographs associated to the interfaces without bacteria identified in the materials after the experiments of the modified Kirby-Bauer diffusion test. First, it should be noted that PLA itself gives rise to a small region in which there has been no bacterial growth, so it may be concluded that PLA itself has some antibacterial behavior. It can be observed that the addition of only 1% of TiO$_2$ nanoparticles increases the size of this region. To gather more precise information, measurements of the inhibition distances for each PLA/TiO$_2$ system were done.

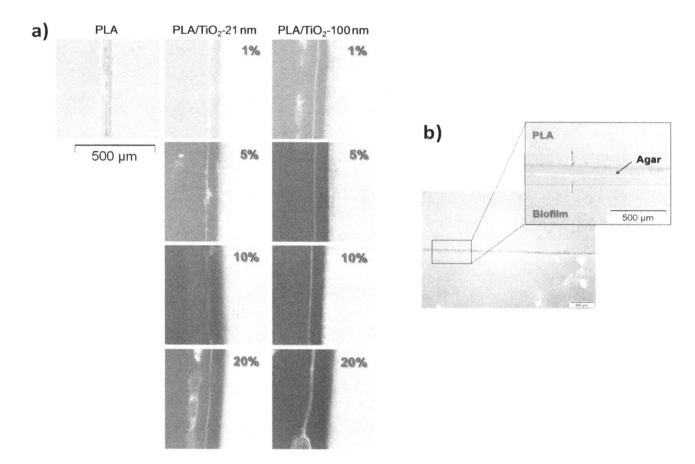

Figure 8. (**a**) Optical micrographs corresponding to the interphases without bacteria obtained with the Kirby-Bauer experiment; (**b**) zoomed area illustrating the measurement of the inhibition distances after the Kirby-Bauer experiment.

The average value of these inhibition distances calculated from the Kirby-Bauer diffusion test is represented in Figure 9 as a function of the TiO_2 particle size and content.

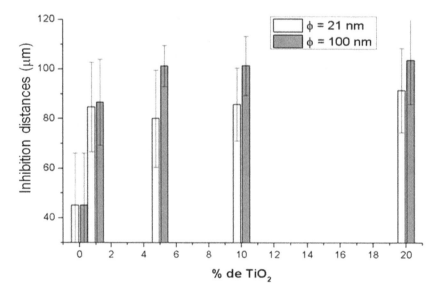

Figure 9. Inhibition distances calculated as a function of TiO_2 particle size (21 nm, in yellow and <100 nm in green) and as a function of particle content (0%, 1%, 5%, 10% and 20%, wt %).

Results from Figure 9 show that, when pure PLA is compared with the sample with 1% TiO_2 nanoparticles (PLA/TiO_2-1) the inhibition distance increases from ~45 μm to slightly less than 90 μm.

Therefore, inhibition distance is almost doubled with only 1% TiO_2 nanoparticles. As particle content in TiO_2 nanoparticles increases from 1 to 20 wt %, average inhibition distances close to 90 µm were found, irrespective of particle content or size. While for composites with 5%, 10%, and 20% in TiO_2 NPs it may seem that inhibition distances are longer for larger TiO_2 nanoparticles, the variations observed in the experimental data are not high enough to consider such variations in inhibition distances to be significant, thus concluding that particle size, in these samples, do not affect inhibition distances.

4. Conclusions

In this work PLA/TiO_2 nanocomposite materials were prepared and characterized to evaluate their potential uses as antibacterial materials. The variables considered for this research were particle size (21 nm and <100 nm) and particle content (0%, 1%, 5%, 10%, and 20%, wt %). The effect of particle size and content on the structural, thermal behavior and antibacterial behavior of a PLA matrix were considered. The presence of titania nanoparticles did not seem to exert high structural changes in the polymer matrix. Regarding thermal behavior, thermal transitions of the PLA matrix (glass transition, cold crystallization and melting) occurred at similar temperatures irrespective of the presence of the particles. The presence of the particles in the different PLA/TiO_2 nanocomposites slightly increased the thermal degradation temperature of the materials as compared with pure PLA in TGA experiments.

The behavior of the materials against bacterial growth and biofilm development revealed that the presence of TiO_2 nanoparticles considerably decreases the amount of extracellular polymeric substancd (EPS) and slightly alters the size of bacteria. Finally, in relation to the effect of particle size on the effectiveness of antibacterial action of the nanoparticles, the results obtained here from the Kirby-Bauer diffusion test were not conclusive. Therefore, future experiments should be done to clarify this issue.

Supplementary Materials: Figure S1: X-Ray Diffraction patterns for the samples with TiO_2-100 nm (left) and TiO_2-21nm (right) as a function of the content in TiO_2 nanoparticles: (a) PLA-0; (b) PLA/TiO_2-1; (c) PLA/TiO_2-5; (d) PLA/TiO_2-10;(e) PLA/TiO_2-20 and (f) pure TiO_2-100 nm (left) or pure TiO_2-21 nm (right), Figure S2: (a) Thermogravimetric analysis curve and (b) DTGA curve for the systems based on PLA/TiO_2-21, Figure S3: From top to bottom: Each row corresponds to SEM micrographs obtained at different magnifications: $50\times$; $2000\times$; $5000\times$ and $8000\times$ PLA/TiO_2 nanocomposite materials ($\varphi \sim 21$ nm) as a function of the content in TiO_2 nanoparticles in each column: (a) 1%; (b) 5%; (c) 10% and (d) 20% (wt %), Figure S4: From top to bottom: Each row corresponds to SEM micrographs obtained at different magnifications: $50\times$; $2000\times$; $5000\times$ and $8000\times$ PLA /TiO_2 nanocomposite materials ($\varphi < 100$ nm) as a function of the content in TiO_2 nanoparticles in each column: (a) 1%; (b) 5%; (c) 10% and (d) 20% (wt %), Table S1: XRD peaks of both polymorphs of TiO_2, Table S2: Band assignment for the PLA infrared spectrum in the MID-IR region, Table S3: Characteristic transition temperatures (T_g, T_c and T_m) obtained from the second heating scan in DSC experiments.

Author Contributions: Conceptualization, D.O. and J.G.-B.; Data curation, E.A.S.G., D.O. and J.G.-B.; Formal analysis, M.A.L.; Funding acquisition, E.A.S.G. and J.G.-B.; Investigation, E.A.S.G., D.O. and J.G.-B.; Methodology, E.A.S.G., D.O., M.A.L., I.V.and J.G.-B.; Project administration, J.G.-B.; Writing– original draft, D.O. and M.A.L.; Writing—review & editing, E.A.S.G., D.O., M.A.L., I.V. and J.G.-B.

Acknowledgments: Authors gratefully acknowledge G. González-Gaitano for his kind help in the ART-FTIR and TGA experiments. Also, special thanks to J.L. Jorcano and Angélica Corral for their collaboration with the cultures.

References

1.   Farah, S.; Anderson, D.G.; Langer, R. Physical and mechanical properties of PLA, and their functions in widespread applications—A comprehensive review. *Adv. Drug Deliv. Rev.* **2016**, *107*, 367–392. [CrossRef] [PubMed]

2.   Murariu, M.; Dubois, P. PLA composites: From production to properties. *Adv. Drug Deliv. Rev.* **2016**, *107*, 17–46. [CrossRef] [PubMed]

3.   Saini, P.; Arora, M.; Kumar, M.N.V.R. Poly(lactic acid) blends in biomedical applications. *Adv. Drug Deliv. Rev.* **2016**, *107*, 47–59. [CrossRef] [PubMed]

4.   Raquez, J.M.; Habibi, Y.; Murariu, M.; Dubois, P. Polylactide (PLA)-based nanocomposites. *Prog. Polym. Sci.* **2013**, *38*, 1504–1542. [CrossRef]

5.   Siracusa, V.; Blanco, I.; Romani, S.; Tylewicz, U.; Rocculi, P.; Rosa, M.D. Poly(lactic acid)-modified films for food packaging application: Physical, mechanical, and barrier behavior. *J. Appl. Polym. Sci.* **2012**, *125*. [CrossRef]

6.   Arrieta, M.P.; Samper, M.D.; Aldas, M.; López, J. On the use of PLA-PHB blends for sustainable food packaging applications. *Materials* **2017**, *10*, 1008. [CrossRef]

7.   Buzarovska, A.; Grozdanov, A. Biodegradable poly(L-lactic acid)/TiO$_2$ nanocomposites: Thermal properties and degradation. *J. Appl. Polym. Sci.* **2012**, *123*, 2187–2193. [CrossRef]

8.   Nieto Pozo, I.; Olmos, D.; Orgaz, B.; Božanić, D.K.; González-Benito, J. Titania nanoparticles prevent development of Pseudomonas fluorescens biofilms on polystyrene surfaces. *Mater. Lett.* **2014**, *127*, 1–3. [CrossRef]

9.   Bahloul, W.; Mélis, F.; Bounor-Legaré, V.; Cassagnau, P. Structural characterisation and antibacterial activity of PP/TiO$_2$ nanocomposites prepared by an in situ sol-gel method. *Mater. Chem. Phys.* **2012**, *134*, 399–406. [CrossRef]

10. Robertson, J.M.C.; Robertson, P.K.J.; Lawton, L.A. A comparison of the effectiveness of TiO$_2$ photocatalysis and UVA photolysis for the destruction of three pathogenic micro-organisms. *J. Photochem. Photobiol. A Chem.* **2005**, *175*, 51–56. [CrossRef]

11. Rincón, A.G.; Pulgarin, C. Photocatalytical inactivation of E. coli: Effect of (continuous-intermittent) light intensity and of (suspended-fixed) TiO$_2$ concentration. *Appl. Catal. B Environ.* **2003**, *44*, 263–284. [CrossRef]

12. Trapalis, C.C.; Keivanidis, P.; Kordas, G.; Zaharescu, M.; Crisan, M.; Szatvanyi, A.; Gartner, M. TiO$_2$(Fe$_3$+) nanostructured thin films with antibacterial properties. *Thin Solid Films* **2003**, *433*, 186–190. [CrossRef]

13. Arroyo, J.M.; Olmos, D.; Orgaz, B.; Puga, C.H.; San José, C.; González-Benito, J. Effect of the presence of titania nanoparticles in the development of Pseudomonas fluorescens biofilms on LDPE. *RSC Adv.* **2014**, *4*, 51451–51458. [CrossRef]

14. Joost, U.; Juganson, K.; Visnapuu, M.; Mortimer, M.; Kahru, A.; Nõmmiste, E.; Joost, U.; Kisand, V.; Ivask, A. Photocatalytic antibacterial activity of nano-TiO$_2$(anatase)-based thin films: Effects on Escherichia coli cells and fatty acids. *J. Photochem. Photobiol. B Biol.* **2015**, *142*, 178–185. [CrossRef] [PubMed]

15. De Falco, G.; Porta, A.; Petrone, A.M.; Del Gaudio, P.; El Hassanin, A.; Commodo, M.; Minutolo, P.; Squillace, A.; D'Anna, A. Antimicrobial activity of flame-synthesized nano-TiO$_2$ coatings. *Environ. Sci. Nano* **2017**, *4*, 1095–1107. [CrossRef]

16. Wang, R.M.; Wang, B.Y.; He, Y.F.; Lv, W.H.; Wang, J.F. Preparation of composited Nano-TiO$_2$ and its application on antimicrobial and self-cleaning coatings. *Polym. Adv. Technol.* **2010**, *21*, 331–336. [CrossRef]

17. Chawengkijwanich, C.; Hayata, Y. Development of TiO$_2$ powder-coated food packaging film and its ability to inactivate Escherichia coli in vitro and in actual tests. *Int. J. Food Microbiol.* **2008**, *123*, 288–292. [CrossRef]

18. Guo, C.; Zhou, L.; Lv, J. Effects of expandable graphite and modified ammonium polyphosphate on the flame-retardant and mechanical properties of wood flour-polypropylene composites. *Polym. Polym. Compos.* **2013**, *21*, 449–456. [CrossRef]

19. Man, C.; Zhang, C.; Liu, Y.; Wang, W.; Ren, W.; Jiang, L.; Reisdorffer, F.; Nguyen, T.P.; Dan, Y. Poly (lactic acid)/titanium dioxide composites: Preparation and performance under ultraviolet irradiation. *Polym. Degrad. Stab.* **2012**, *97*, 856–862. [CrossRef]

20. Thamaphat, K.; Limsuwan, P.; Ngotawornchai, B. Phase characterization of TiO$_2$ powder by XRD and TEM. *Kasetsart J. (Nat. Sci.)* **2008**, *42*, 357–361.

21. González-Benito, J.; Castillo, E.; Caldito, J.F. Coefficient of thermal expansion of TiO$_2$ filled EVA based nanocomposites. A new insight about the influence of filler particle size in composites. *Eur. Polym. J.* **2013**, *49*, 1747–1752. [CrossRef]

22. González, E.A.S.; Teno, J.; González-Benito, J.; Olmos, D. Accurate Evaluation of Dynamics and Specific Interactions in PLA/TiO$_2$ Nanocomposites. *Sci. J. Mol. Phys.* **2017**, *1*, 1–13.

23. Bauer, A.W.; Kirby, W.M.M.; Sherris, J.C.; Turck, M. Antibiotic susceptibility testing by a standardized single disk method. *Am. J. Clin. Pathol.* **1966**, *36*, 49–52. [CrossRef]

24. Bauer, A.W.; Perry, D.M.; Kirby, W.M. Single-disk antibiotic-sensitivity testing of staphylococci. *AMA Arch. Intern. Med.* **1959**, *104*, 208–216. [CrossRef] [PubMed]

25. Olmos, D.; Pontes-Quero, G.; Corral, A.; González-Gaitano, G.; González-Benito, J. Preparation and Characterization of Antimicrobial Films Based on LDPE/Ag Nanoparticles with Potential Uses in Food and Health Industries. *Nanomaterials* **2018**, *8*, 60. [CrossRef] [PubMed]

26. Buasri, A.; Chaiyut, N.; Kristsanakun, C.; Phatkun, C.; Khunsri, T. Preparation and properties of nanocomposites based ond poly(lactic acid) and modified TiO$_2$. *Adv. Mater. Res.* **2012**, *463–464*, 519–522. [CrossRef]

27. Luo, Y.B.; Wang, X.L.; Xu, D.Y.; Wang, Y.Z. Preparation and characterization of poly(lactic acid)-grafted TiO$_2$ nanoparticles with improved dispersions. *Appl. Surf. Sci.* **2009**, *255*, 6795–6801. [CrossRef]

28. Chieng, B.W.; Ibrahim, N.A.; Yunus, W.M.Z.W.; Hussein, M.Z. Poly(lactic acid)/poly(ethylene glycol) polymer nanocomposites: Effects of graphene nanoplatelets. *Polymers* **2014**, *6*, 93–104. [CrossRef]

29. Pillin, I.; Montrelay, N.; Bourmaud, A.; Grohens, Y. Effect of thermo-mechanical cycles on the physico-chemical properties of poly(lactic acid). *Polym. Degrad. Stab.* **2008**, *93*, 321–328. [CrossRef]

30. Carrasco, F.; Pagès, P.; Gámez-Pérez, J.; Santana, O.O.; Maspoch, M.L. Processing of poly(lactic acid): Characterization of chemical structure, thermal stability and mechanical properties. *Polym. Degrad. Stab.* **2010**, *95*, 116–125. [CrossRef]

31. Luo, Y.B.; Li, W.D.; Wang, X.L.; Xu, D.Y.; Wang, Y.Z. Preparation and properties of nanocomposites based on poly(lactic acid) and functionalized TiO$_2$. *Acta Mater.* **2009**, *57*, 3182–3191. [CrossRef]

32. Wang, W.W.; Man, C.Z.; Zhang, C.M.; Jiang, L.; Dan, Y.; Nguyen, T.P. Stability of poly(l-lactide)/TiO$_2$ nanocomposite thin films under UV irradiation at 254 nm. *Polym. Degrad. Stab.* **2013**. [CrossRef]

33. Liu, M.; Cheng, Z.; Yan, J.; Qiang, L.; Ru, X.; Liu, F.; Ding, D.; Li, J. Preparation and characterization of TiO$_2$ nanofibers via using polylactic acid as template. *J. Appl. Polym. Sci.* **2013**. [CrossRef]

34. Czaczyk, K.; Myszka, K. Biosynthesis of extracellular polymeric substances (EPS) and its role in microbial biofilm formation. *Pol. J. Environ. Stud.* **2007**, *16*, 799–806.

35. Sheng, G.P.; Yu, H.Q.; Li, X.Y. Extracellular polymeric substances (EPS) of microbial aggregates in biological wastewater treatment systems: A. review. *Biotechnol. Adv.* **2010**, *28*, 882–894. [CrossRef]

36. Liang, Z.; Li, W.; Yang, S.; Du, P. Extraction and structural characteristics of extracellular polymeric substances (EPS), pellets in autotrophic nitrifying biofilm and activated sludge. *Chemosphere* **2010**, *81*, 626–632. [CrossRef] [PubMed]

37. Ni, B.J.; Fang, F.; Xie, W.M.; Sun, M.; Sheng, G.P.; Li, W.H.; Yu, H.Q. Characterization of extracellular polymeric substances produced by mixed microorganisms in activated sludge with gel-permeating chromatography, excitation-emission matrix fluorescence spectroscopy measurement and kinetic modeling. *Water Res.* **2009**, *43*, 1350–1358. [CrossRef] [PubMed]

38. Zhukova, L.V.; Kiwi, J.; Nikandrov, V.V. TiO$_2$ nanoparticles suppress Escherichia coli cell division in the absence of UV irradiation in acidic conditions. *Colloids Surf. Biointerfaces* **2012**, *97*, 240–247. [CrossRef]

Poly(3-hydroxybutyrate) Modified by Nanocellulose and Plasma Treatment for Packaging Applications

Denis Mihaela Panaitescu [1,*], **Eusebiu Rosini Ionita** [2], **Cristian-Andi Nicolae** [1], **Augusta Raluca Gabor** [1], **Maria Daniela Ionita** [2], **Roxana Trusca** [3], **Brindusa-Elena Lixandru** [4], **Irina Codita** [4,5] **and Gheorghe Dinescu** [2]

[1] National Institute for Research & Development in Chemistry and Petrochemistry-ICECHIM, Polymer Department, 202 Spl. Independentei, 060021 Bucharest, Romania; cristian.nicolae@icechim-pd.ro (C.-A.N.); ralucagabor@yahoo.com (A.R.G.)

[2] National Institute for Laser, Plasma and Radiation Physics, Atomistilor 409, Magurele-Bucharest, 077125 Ilfov, Romania; ionita.rosini@infim.ro (E.R.I.); daniela.ionita@infim.ro (M.D.I.); dinescug@infim.ro (G.D.)

[3] Science and Engineering of Oxide Materials and Nanomaterials, University Politehnica of Bucharest, 1-7 Gh. Polizu Street, 011061 Bucharest, Romania; truscaroxana@yahoo.com

[4] "Cantacuzino" National Medical-Military Institute for Research and Development, 103 Spl. Independentei, 050096 Bucharest, Romania; brandusa_lixandru@yahoo.com (B.E.L.); icodita@cantacuzino.ro (I.C.)

[5] Carol Davila University of Medicine and Pharmacy, Bulevardul Eroii Sanitari 8, 050474 Bucharest, Romania

* Correspondence: panaitescu@icechim.ro.

Abstract: In this work, a new eco-friendly method for the treatment of poly(3-hydroxybutyrate) (PHB) as a candidate for food packaging applications is proposed. Poly(3-hydroxybutyrate) was modified by bacterial cellulose nanofibers (BC) using a melt compounding technique and by plasma treatment or zinc oxide (ZnO) nanoparticle plasma coating for better properties and antibacterial activity. Plasma treatment preserved the thermal stability, crystallinity and melting behavior of PHB-BC nanocomposites, regardless of the amount of BC nanofibers. However, a remarkable increase of stiffness and strength and an increase of the antibacterial activity were noted. After the plasma treatment, the storage modulus of PHB having 2 wt % BC increases by 19% at room temperature and by 43% at 100 °C. The tensile strength increases as well by 21%. In addition, plasma treatment also inhibits the growth of *Staphylococcus aureus* and *Escherichia coli* by 44% and 63%, respectively. The ZnO plasma coating led to important changes in the thermal and mechanical behavior of PHB-BC nanocomposite as well as in the surface structure and morphology. Strong chemical bonding of the metal nanoparticles on PHB surface following ZnO plasma coating was highlighted by infrared spectroscopy. Moreover, the presence of a continuous layer of self-aggregated ZnO nanoparticles was demonstrated by scanning electron microscopy, ZnO plasma treatment completely inhibiting growth of *Staphylococcus aureus*. A plasma-treated PHB-BC nanocomposite is proposed as a green solution for the food packaging industry.

Keywords: bionanocomposites; polyhydroxybutyrate; bacterial cellulose; antimicrobial activity; thermal properties; packaging; morphology

1. Introduction

Synthetic polymers have been used in packaging for many decades and at least half of their consumption is by the food packaging industry [1,2]. The increasing amount of waste from non-biodegradable plastics and the ecological problems they cause have encouraged manufacturers to find new environmentally friendly, safe and nontoxic packaging materials. Biopolymers from renewable resources are the most promising alternatives to petroleum-based polymers in food packaging [2,3]. The large application of biopolymers in this sector has huge advantages in terms of a decreased dependence on fossil fuel reserves and limitation of the environmental problems associated with synthetic polymer pollution. Polyhydroxyalkanoates (PHAs), which are obtained by bacterial synthesis, are considered as very promising candidates for food packaging and biomedical applications [4–6]. Their hydrolytic degradation in the environment and in living systems leads to oligomeric byproducts that are further processed by biochemical pathways [5,7]. Poly (3-hydroxybutyrate) (PHB) is the most studied of the PHAs and some of its properties are comparable to those of the petrochemical-based polymers [3]. However, shortcomings in the mechanical performance and melt processing behavior of PHB, i.e., high brittleness, poor thermal stability and difficult processing [3,8], along with insufficient barrier properties, limit its widespread use. Many attempts are being made to improve its properties for packaging application [9–13]. Copolyesters with higher ductility such as poly(3-hydroxybutyrate-co-3-hydroxyvalerate) (PHBV), poly(3-hydroxybutyrate-co-4-hydroxybutyrate) or poly(3-hydroxybutyrate-co-3-hydroxy- hexanoate) may overcome some of these shortcomings [5,9,10]. Different flexibilities and strengths may be obtained by variation of the co-monomer unit or composition in these copolyesters [10,11]. The addition of different modifiers, plasticizers, micro- and nanofillers was also tested, aiming to improve PHB properties [4,5,8,9,12]. Cellulose nanofibers are mostly used in PHB and other PHAs due to their important effect of increasing mechanical and barrier properties [1–4,6,8,13]. Bacterial cellulose nanofibers (BC) are preferred for packaging and biomedical applications due to their high purity. For example, a remarkable improvement of PHB properties was obtained by mixing PHB with medium chain length PHAs and BC as "soft" and "stiff" modifiers [8,14]. Therefore, PHB and copolymers modified by BC nanofibers are promising materials for food packaging and biomedical applications.

Several studies have shown that PHB could replace conventional plastics that are used in juices and dressings [15] or fat-rich products' packaging [16] if its melt processability is improved. Still, the high rigidity and low flexibility of PHB leads to the cracking of jars and caps during injection molding [16]. Post-processing of PHB electrospun fiber mats by different physical treatments (annealing and cooling) resulted in PHB films with higher elongation at break and toughness compared to common compression-molded films [17]. The addition of high molecular weight natural rubber or poly(ε-caprolactone) (PCL) may also improve the elongation to break of PHB/PHBV and lead to higher thermal stability [18] or toughness [19] with the increase of rubber/PCL content in the blends. Blends of poly(lactic acid) (PLA) with 25 wt % PHB were intensively studied for food packaging applications [20–22]. A large amount of plasticizers must be used in these blends to ensure the high flexibility required by packaging processes [20–22], which is often detrimental to the mechanical properties.

However, PHB-based materials are susceptible to microbial attack, which is undesirable for food packaging applications. Different methods were proposed to obtain antimicrobial active packaging (AP) for maintaining the quality and safety of foods [23,24]. Generally, migratory and non-migratory systems were developed for this purpose [23]. In a migratory packaging system, volatile or non-volatile active components are released from the polymer matrix to the surface of the food and, in the case of non-migratory systems, the antimicrobial agents or oxygen/moisture absorbers are bound to the polymer packaging.

Several agents were tested to obtain PHB-based biomaterials with antimicrobial activity for AP [25–31]. Natural antimicrobial agents such as vanillin [25], sophorolipid [26] or eugenol [27] were incorporated in PHB in different proportions using a solvent casting-evaporation method with chloroform as a solvent. Vanillin and eugenol in small concentrations (80 μg/g PHB) [25,27], and sophorolipid in a much higher concentration (9–29 wt %) [26] were found to be effective against a large number of bacterial species. Carvacrol, a constituent of several essential oils, was also incorporated in PLA/PHB blends and improved their antioxidant and antibacterial activity [27].

However, the incorporation of these natural antimicrobial agents in PHB frequently leads to the decrease of thermal stability and mechanical properties [25,28].

Synthetic antibacterial agents were also tested in PHB formulations; however, their concentration should be limited in the case of food packaging application [29,30]. Triclosan showed good bactericidal activity against *Escherichia coli* and *Staphylococcus aureus* after its incorporation in PHBV, and increased its flexibility, but significantly reduced its elastic modulus and tensile strength [29]. PHB modified by chlorhexidine digluconate showed a controlled release of the antibacterial agent into a food area [30]. Similarly, PHB/PCL/organo-clay nanocomposites activated with nisin, an antibacterial peptide, showed good inhibiting effect against *Lactobacillus plantarum* and extended the shelf life of processed meat [31].

In recent years, metal nanoparticles have been intensively studied for their ability to induce antimicrobial activity in PHB films for AP applications [32–38]. For example, zinc oxide (ZnO) nanoparticles, which are efficient UV blockers, were incorporated in PHB using a solution casting technique [32]. The nanocomposite films showed good antibacterial activity against both Gram-positive and Gram-negative bacteria, depending on the ZnO concentration within the nanocomposite. Furthermore, PHB/ZnO films showed better mechanical properties and thermal stability compared to the PHB matrix [32]. Similar improvement was reported for (3-hydroxybutyrate-co-3-hydroxyvalerate) (PHBV) reinforced with ZnO [33]. An optimal concentration of 4.0 wt % ZnO determined the maximum tensile strength, Young's and storage moduli, crystallinity and barrier properties, and antibacterial activity. Another route to obtain PHB films with antimicrobial activity for AP was proposed by Castro-Mayorga et al. [34]. They obtained PHB nanocomposites containing silver nanoparticles using a biological synthesis without the addition of a reducing agent to prepare these nanoparticles. The nanocomposite films showed strong antimicrobial activity against *Salmonella enteric* and *Listeria monocytogenes*, known food-borne pathogens. Moreover, no significant change of thermal stability or biodegradability of PHB was caused by these nanoparticles [34]. Another "green" method to obtain silver nanoparticles and PHB-Ag nanocomposites consists of the grafting of PHB onto the chitosan biguanidine, which reduced silver nitrate to silver nanoparticles [35]. These nanocomposites showed good antibacterial activity against Gram-positive and Gram-negative bacteria. PHBV films coated with PHBV/ZnO ultrathin fiber mats and annealed at 160 °C showed effective and prolonged antibacterial activity against *Listeria monocytogenes* [36].

In most of the methods proposed so far, large amounts of chloroform or other solvents are used to prepare the PHB films with antimicrobial activity, which may cause environmental or health problems. Melt processing is environmentally friendly and cost-effective and, thus, a preferred choice for industrial scale-up. Moreover, the incorporation of metal nanoparticles in the whole PHB matrix leads to the use of a large amount of nanoparticles for the same effect, with only the surface ones being efficient as antibacterial agents. Moreover, the migration of nanoparticles from the bulk to the surface is a difficult process in PHB due to its high crystallinity and additional additives like lubricants, plasticizers, surfactants being necessary to increase migration. Surface treatment of PHB melt-processed films using plasma is an innovative and environmentally friendly approach to obtain antibacterial activity with a minimal change in bulk properties.

In this work, PHB was modified with different amounts of BC nanofibers as a reinforcing agent [4]; PHB-BC nanocomposite films were produced by melt compounding and compression molding and were plasma-treated to improve their properties and enhance antibacterial activity. Moreover, a PHB–BC–ZnO nanocomposite film was prepared using a new plasma coating process. ZnO nanoparticles were selected as an affordable and safe solution for food packaging, ZnO being recognized as safe by the U.S. Food and Drug Administration [36]. Earlier works have reported on the effect of low-pressure plasma treatment in the modification of PHB by grafting, polymerization or functionalization for biomedical applications [39–43]. Low-pressure plasma treatments were mostly used to improve the surface hydrophilicity of PHB/PHBV films for enhancing cell compatibility [44–52]. The treatments were applied to increase cell adhesion and proliferation for scaffolding and other biomedical purposes. In a study, PHB was plasma-activated and subsequently silver-coated for being used in packaging [52].

However, low-pressure plasma processes have limitations due to the complicated installations and high costs inherent to vacuum technology. Moreover, the influence of plasma treatment on PHB-nanocellulose composites was not yet studied. In this work we propose a simple atmospheric pressure plasma treatment and coating process, which ensure the surface modification of PHB and ZnO nanoparticles deposition. The effect of the plasma treatment and ZnO plasma coating on the thermal, mechanical and antimicrobial properties of PHB and its nanocomposites with 2 and 5 wt % bacterial cellulose nanofibers was investigated here for the first time. Plasma treatments modified to a different extent the thermal and mechanical properties of PHB and its antibacterial activity depending on composition and treatment conditions. Plasma-treated PHB-BC nanocomposites are new materials that are proposed as a "green" solution for food packaging. This work is an important step for the design of new biomaterials obtained by eco-friendly processes, as alternatives to non-biodegradable plastics commonly used in the food packaging industry.

2. Materials and Methods

2.1. Materials

Bacterial cellulose pellicles consist of a network of entangled cellulose nanofibers, with a width of 40–110 nm [53]. Individual bacterial cellulose nanofibers were produced by the mechanical disintegration of bacterial cellulose pellicles using a high-speed blender for 30 min and a vertical colloid mill at about 2000 min^{-1} for 180 min. Cold water (5–10 °C) was added to prevent temperature rising above 40 °C. BC nanofibers with widths mostly between 50 and 90 nm and a thickness of 8–10 nm, as determined by atomic force microscopy (AFM, JPK Instruments, Berlin, Germany), peak force quantitative nanomechanical mapping (QNM) mode, were obtained by defibrillation. The suspension containing BC nanofibers was concentrated in a rotary evaporator (Heidolph Instruments, Schwabach, Germany) at 40 °C and then freeze-dried using FreeZone 2.5 L (Labconco, Kansas City, MO, USA). Pelletized PHB from Goodfellow (Huntingdon, UK) with a density of 1.25 g cm^{-3} was used to prepare the nanocomposites. A rough proportion of 10% tributyl citrate (TBC) plasticizer was estimated in the commercial PHB granules [12]. The argon (Ar) gas used to generate plasma was 99.999% purity. The ZnO nanoparticles with a mean diameter of 50 nm and ethanol with a purity higher than 99.8% were purchased from Sigma-Aldrich (St. Louis, MI, USA) and used as received.

2.2. Preparation of PHB Nanocomposite Films

PHB pellets and BC were dried in vacuum ovens at 60 °C for 2 h and 50 °C for 4 h. Cellulose nanofibers (2 and 5 wt %) were added in melted PHB using a Brabender LabStation (Duisburg, Germany) with a mixing chamber of 30 cm^3. Melt mixing was carried out at 170 °C for 8 min starting from the melting of PHB at a rotor speed of 40 min^{-1}. Films with a thickness of 200 μm were obtained by compression molding using an electrically heated press (P200E, Dr. Collin, Ebersberg, Germany) at 175 °C, with 120 s preheating (5 bar), 75 s under pressure (100 bar), and cooling for 1 min in a cooling cassette.

2.3. Plasma Treatments of PHB Nanocomposite Films

The experimental system designed for the dielectric barrier discharge (DBD) plasma treatment of PHB nanocomposite films is shown in Figure 1a. In the first setup, the nanocomposite film was placed on a CNC XYZ router table (StepCraft 300, Stepcraft, Iserlohn, Germany) and swept by a longitudinal jet of atmospheric pressure Ar plasma. The distance between the polymer film and the longitudinal nozzle was 1.8 mm and the length and thickness of the plasma jet were 10 cm and 1 mm, respectively. The plasma source consists of a flat ceramic body that allows the gas flow and two metal electrodes, one connected to the ground (GND) and the other to the radiofrequency (RF) generator (Cesar 136

13.56 MHz—Advanced Energy Industries, Metzingen, Germany). The discharge impedance was matched to the generator via an automatic impedance matching network (matching box Advanced Energy Navio). The Ar feeding through the top of the plasma source was assured by a mass flow controller (EL-FLOW 10 slm—Bronkhorst, AK Ruurlo, The Netherlands).

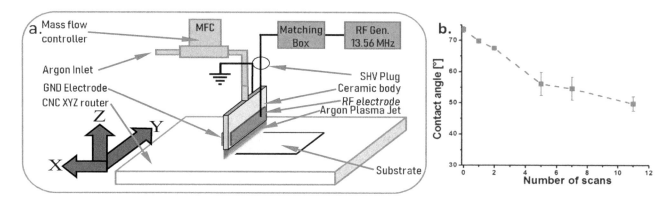

Figure 1. Experimental setup for the DBD plasma treatment of PHB nanocomposite films (substrate) (**a**); contact angle of the nanocomposite films vs. the number of scans (**b**).

The number of scans was determined based on contact angle measurements. Different number of scans, from 1 to 11, was carried out with the DBD plasma source on the surface of PHB nanocomposites and the contact angle was measured with a CAM 200 (KSV Instruments, Helsinki Finland) equipped with a high resolution camera (Basler A602f, Basler, Ahrensburg, Germany) and an auto-dispenser, at room temperature and ambient humidity. Five scans was considered optimal (Figure 1b). All the plasma-treated nanocomposites were denoted with a "p" after their name. In the second setup, the nanocomposite films were plasma-treated in the same conditions as mentioned in the first setup for increasing their hydrophilicity and then coated with ZnO nanoparticles dispersed in an alcohol solution using an ultrasonic spraying device. In both experimental setups, the control of the devices involved in the experiments was carried out by a computer using our own software.

2.4. Characterization

2.4.1. Scanning Electron Microscopy Coupled with Energy-Dispersive X-ray Analysis (SEM-EDX)

SEM images were captured with a Quanta Inspect F scanning electron microscope (FEI-Philips, Hillsboro, OR, USA), equipped with a field emission gun, working at an accelerating voltage of 30 kV with a resolution of 1.2 nm. The surface composition was analyzed with an energy dispersive X-ray (EDX) spectrometer coupled to SEM, with a resolution of 133 eV at MnKα. The blends and nanocomposites films were sputter-coated with gold before analysis.

2.4.2. Thermal Characterization

Thermogravimetric analysis (TGA) was used to characterize the thermal stability of PHB nanocomposites before and after treatments. TGA was carried out on TA-Q5000 V3.13 system (TA Instruments Inc., New Castle, DE, USA) with signal resolution of 0.01 μg, using nitrogen as the purge gas at a flow rate of 40 mL/min. Samples were heated from 25 °C to 700 °C at a heating rate of 10 °C/min.

Differential scanning calorimetry (DSC) was carried out with a DSC Q2000 from TA Instruments (New Castle, DE, USA) under a helium flow (100 mL/min). Samples weighing around 8 mg were cooled to −65 °C, allowed to stabilize for several minutes, heated to 200 °C and held at that temperature for 3 min to delete the thermal history, then cooled to −65 °C, kept isothermal at that temperature for 3 min and heated again to 200 °C at a constant heating/cooling rate of 10 °C/min. The instrument was calibrated using indium standard (TA Instruments) and has a temperature precision of ±0.01 °C and a calorimetric precision of $\pm 0.05\%$. The melting temperature (T_m) was taken as the peak temperature of the melting endotherm. The degree of crystallinity (C) was calculated from the second melting scan as in Equation (1):

$$C = \frac{\Delta H}{\Delta H_0 \, w_{PHB}} \cdot 100 \qquad (1)$$

where ΔH_m and ΔH_0 are the apparent melting enthalpy of the nanocomposite and of 100% crystalline PHB, respectively, while w_{PHB} is the weight fraction of PHB in nanocomposite. ΔH_0 was taken as 146 J/g [32].

2.4.3. Dynamic Mechanical Analysis (DMA)

The thermomechanical properties of blends and nanocomposites were analyzed using a DMA Q800 (TA Instruments, New Castle, DE, USA), which has a force resolution of 0.00001 N and a strain resolution of 1 nm, operating in multi-frequency–strain mode. The films were maintained at room temperature for four weeks before characterization. The bar specimens with the length × width of 20 mm × 6 mm were obtained by cutting. Samples in duplicate were cooled to −15 °C with 10 °C/min, equilibrated for 5 min at this temperature and heated to 160 °C with a heating rate of 3 °C/min.

2.4.4. Fourier Transform Infrared Spectroscopy (FTIR)

FTIR spectrometer with ATR setup (Tensor 37) from Bruker Optics (Ettlingen, Germany) was used for the analysis of plasma-treated nanocomposite films. The spectra were collected at room temperature, from 4000 to 400 cm^{-1}, at a resolution of 4 cm^{-1} using 16 scans. Spectra were baseline-corrected using the OPUS software (Bruker Optics, Ettlingen, Germany).

2.4.5. Peak Force Quantitative Nanomechanical Mapping

AFM images of PHB and nanocomposite with 5 wt % BC (PHB-5BC) were captured using a MultiMode 8 AFM instrument (Bruker, Santa Barbara, CA, USA) in peak force QNM mode. Measurements were carried out at room temperature with a scan rate of 0.7–0.8 Hz and a scan angle of 90° using an etched silicon tip (nominal radius 8–10 nm) with a cantilever length of 225 μm and a resonance frequency of about 75 kHz. The data and images were processed with NanoScope software version 1.20.

2.4.6. Tensile Characterization

The tensile properties of PHB and PHB-BC nanocomposites were measured at room temperature using an Instron 3382 universal testing machine (Instron, Norwood, MA, USA) with a crosshead speed of 2 mm/min. Five specimens according to ISO 527 were tested for each nanocomposite and the results were analyzed using the Bluehill 2 software (Instron, Norwood, MA, USA).

2.4.7. Antibacterial Activity

The antibacterial activity of plasma-treated nanocomposite films was evaluated against *Staphylococcus aureus* by the colony counting method [54]. Square shaped fragments of 5 × 5 mm^2 were cut from the nanocomposite films and were placed in sterile Petri dishes. The plasma-treated samples were placed with the impregnated face up. The Petri dishes with the nanocomposite fragments were exposed to a UV source for 15 min. Ten 10-mm square fragments of reference and nanocomposite films were placed in a sterile 25/25 mm tube. An aliquot of 1 mL bacterial culture (*Staphylococcus aureus* ATCC 6538 or *Escherichia coli* ATCC 35218) was added over the nanocomposite fragments to a concentration of about 10^2 UFC / mL. The tubes were then incubated with stirring at 36 ± 1 °C for 24 h. A set of 10 fragments from the untreated reference was seeded with 1 mL of non-inoculated nutrient broth and served as a sterility control. The concentration of the bacterial suspensions used in the experiment was checked by inoculating 50 μL of each suspension, in triplicate, onto a blood agar plate. After incubation, 25 mL of physiological saline solution was added to each tube, the tubes were vortexed and then 100 µL were transferred into Corning tubes with 9.9 mL of physiological saline solution, diluted at 10^{-2} and then seeded on blood agar plates. After the incubation of seeded plates, the colonies were counted for each of the analyzed samples. The percentage reduction in bacterial growth, X (%) was calculated as follows [55]:

$$X\ (\%) = (\text{viable count at 0 h} - \text{viable count at 24 h}) \times 100/\text{viable count at 0 h} \qquad (2)$$

3. Results and Discussion

3.1. Thermogravimetric Analysis

Plasma surface treatment may affect the thermal stability of PHB nanocomposites films due to the reactive species, radicals, excited atoms or charged particles that are generated during this treatment. Therefore, the influence of the plasma treatment on the thermal behavior of PHB modified with cellulose nanofibers was investigated by thermogravimetric analysis.

The thermal degradation of PHB and nanocomposites consists in a main degradation step between 230 °C and 280 °C (Figure 2a). Two small shoulders were observed in the derivative thermogravimetric (DTG) curves (Figure 2b), one at about 195 °C, probably determined by the release of TBC plasticizer from the PHB matrix [12], and the second at about 330 °C. This second shoulder is absent in PHB films and occurs only in nanocomposites, increasing with the amount of BC (see Figure S1). Therefore, it may be attributed to cellulose decomposition.

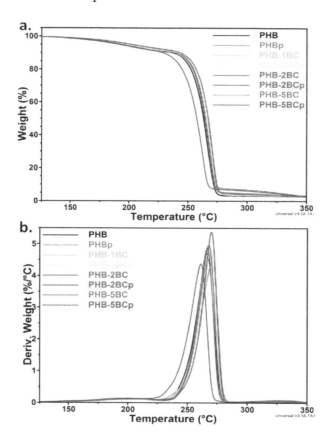

Figure 2. TGA (**a**) and DTG (**b**) curves of PHB nanocomposites with different BC content before and after the plasma treatment.

The thermal stability is slightly influenced by the plasma treatment or composition (Figure 2). An increase in the maximum degradation temperature (T_{max}) of 1.5–4.4 °C was observed following plasma exposure for pristine PHB and nanocomposites with 1–2 wt % BC; however, a decrease of 5.2 °C was noticed for PHB-5BC (Table 1). This suggests a higher plasma sensitivity of cellulose, also noticed in our previous works [56,57]. The differences in the onset degradation temperature (T_{on}) values before and after the plasma treatment were similar to those of T_{max}. However, $T_{5\%}$ (temperature at 5% weight loss) values showed different changes following the treatment (Table 1). This characteristic temperature may be considered as a measure of the thermal stability during melt processing, being close to the melt temperature range. An important increase of the $T_{5\%}$ value, more than 10 °C, was observed for PHB and PHB-2BC nanocomposite after the plasma treatment. This may be due to the chemical changes of PHB induced by the plasma exposure.

Table 1. TGA results for PHB nanocomposites before and after the plasma treatment.

Nanocomposites	$T_{5\%}$ (°C)	T_{on} (°C)	T_{max} (°C)	R_{600} (%)
PHB	191.9	254.1	266.3	2.23
PHBp	202.4	255.5	270.7	2.26
PHB-1BC	195.3	255.2	268.4	2.22
PHB-1BCp	196.7	256.6	269.9	2.29
PHB-2BC	192.4	256.4	267.9	2.29
PHB-2BCp	203.0	259.1	270.5	2.43
PHB-5BC	192.8	255.4	266.7	2.48
PHB-5BCp	194.4	247.7	261.5	2.50
PHB-2BC-ZnO	183.5	251.0	265.0	3.43

Plasma deposition of ZnO nanoparticles decreased the T_{max} of PHB-2BC nanocomposite by 2.9 °C. Similarly, the T_{on} of the nanocomposite decreased by about 5 °C (Figure 3, Table 1).

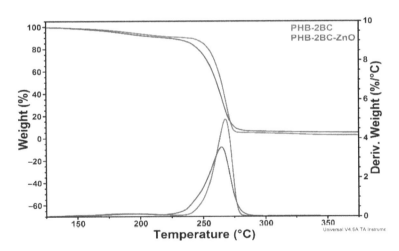

Figure 3. TGA-DTG curves of PHB nanocomposite plasma-coated with ZnO nanoparticles.

A decrease of the T_{max} of about 12 °C was observed after the incorporation of ZnO nanoparticles (6 wt %) in PHBV using a melt mixing procedure [36]. However, a strong increase of the thermal stability, with about 20 °C, was reported for PHB and PHBV nanocomposites with only 2 wt % ZnO prepared by solution casting [32,33]. The different effect of ZnO nanoparticles on the thermal stability of PHAs may be due to the different procedure or degree of dispersion and other factors [58]. However, the slight decrease in thermal stability observed for ZnO plasma-coated PHB may be rather an effect of the plasma treatment and of the active species that appear during this treatment. Nanocomposite films were analyzed by ATR-FTIR to observe the surface changes induced by the treatments.

3.2. FTIR Analysis

FTIR spectra of PHB and PHB nanocomposites with 2 wt % and 5 wt % BC are given in Figure S2. The addition of BC in PHB leads to no spectral differences because of the small amount of cellulose.

Several differences were observed after the plasma treatment of PHB and nanocomposites (Figure S3). Most of the differences occur in the range from 3100 to 2800 cm^{-1}, assigned to C-H stretching modes (Figure 4).

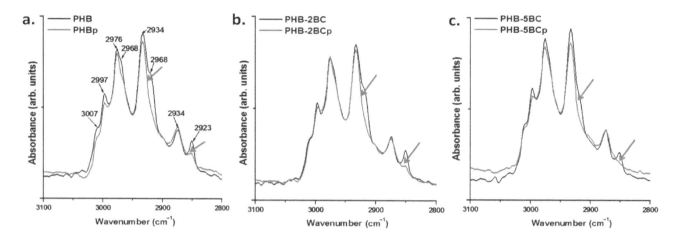

Figure 4. FTIR spectra of untreated and plasma-treated PHB and nanocomposites in the C–H stretching vibration region.

The bands in the range 3015–2955 cm^{-1} are assigned to CH_3 asymmetric stretching vibrations and those in the range 2940–2915 cm^{-1} to CH_2 asymmetric stretching vibrations [59]. The range from 2885 to 2845 cm^{-1} is characteristic of symmetric stretching modes of CH_3 and CH_2 [59,60]. The peaks at 2976 cm^{-1}, 2934 cm^{-1} and 2874 cm^{-1} and the shoulder at 2923 cm^{-1} arise from the crystalline state and that at 2997 cm^{-1} from the amorphous state [61,62]. The shoulder at 3007 cm^{-1}, related to CH_3 asymmetric stretching, indicates the presence of intermolecular CH ... O hydrogen bonds in PHB crystals [61]. Assignments of the peaks in Figure 4 (3050–2800 cm^{-1} region) are given in Table S1. PHB and nanocomposites show similar response to the plasma treatment, i.e., disappearance of the shoulder at 2923 cm^{-1} and of the peak at 2851 cm^{-1}. Previous works have shown that the two peaks at 2934 and 2923 cm^{-1}, which are related to the CH_2 asymmetric stretching [62], result from the crystal field splitting, caused by inter- or intramolecular interactions [61]. Therefore, the disappearance of the peak at 2923 cm^{-1} may be related to changes in the molecular packing and interactions caused by the plasma treatment. The same reasons may cause the reduction or disappearance of the peak at 2851 cm^{-1}, which is attributed to the symmetric CH_2 stretching vibrations. The cleavage of chemical bonds and release of small molecular products and the occurring of cross-linking reactions may also be reflected in this FTIR region. The disappearance of several peaks in the range from 800 to 400 cm^{-1} was also observed in all the samples after the plasma treatment and it may be explained by the cleaning effect of Ar plasma [56,57].

In contrast, a significant effect of ZnO plasma coating on PHB nanocomposite film with 2 wt % BC was observed (Figure 5). ATR-FTIR is surface-sensitive and it is a very suitable method to investigate the effects of plasma coating with ZnO nanoparticles. The main changes in the FTIR spectra are summarized below:

(i) the appearance of a new wide band at 3600–3100 cm^{-1}, which is ascribed to free and hydrogen bounded OH groups in alcohol and water; this probably comes from the alcoholic suspension of ZnO nanoparticles used for the treatment;

(ii) the shift of the peak ascribed to CH_3 asymmetric stretching vibrations, from 2976 to 2961cm^{-1}; this shift may be caused by several factors; however, the participation of methyl groups on the surface of PHB in reactions with the active species from ethanol and ZnO under the plasma bombardment is, probably, the most important [63]; the catalytic effect of Zn should have a remarkable enhancing role; moreover the absorption of ethanol, acetaldehyde, ethylene, water or other reaction products could also take place;

(iii) the appearance of a new large peak at 1587 cm^{-1}, which may be due to the C=C stretching vibrations of polymer-ZnO complex ion, similar to other observations [64]; ethylene active species result from ethanol under plasma and catalytic effect of Zn [63];

(iv) the appearance of a new band at 1430 cm^{-1}, in the range characteristic to the CH$_3$ and CH$_2$ deformation vibrations, which may be ascribed to the metal complexes; similarly, the asymmetric and symmetric stretching vibrations of metal carboxylate complexes occur at 1590 and 1430 cm^{-1} [65].

(v) the appearance of a shoulder at 547 cm^{-1}, which is characteristic to ZnO nanoparticles [66].

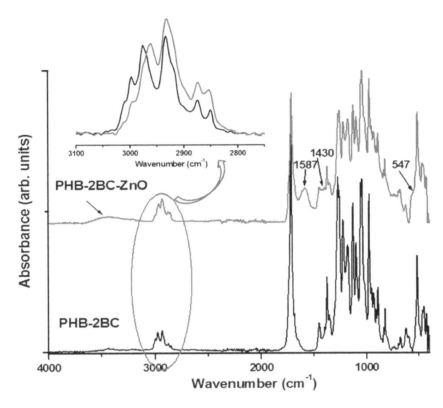

Figure 5. FTIR spectra of PHB-2BC and ZnO plasma-coated nanocomposite.

Overall, the peaks observed in the FTIR spectra of PHB nanocomposite after ZnO coating support the formation of polymer-ZnO complex species, i.e., strong chemical bonding of the metal nanoparticles on the surface of PHB nanocomposite. The role of cellulose nanofillers in the bonding of Zn nanoparticles could be also an active one.

3.3. Differential Scanning Calorimetry

DSC analysis may highlight the changes induced by the plasma treatments in the crystalline and amorphous phases. The heating and cooling DSC thermograms of nanocomposites before and after the plasma treatment are shown in Figure 6a–d. PHB shows double melting peaks due to melting, re-crystallization and re-melting during heating with the higher temperature peak arising from the re-crystallized polymer [67]. Different lamellar thickness and perfection may be another cause of the double melting peak besides the different molecular weight species and orientations [12]. Moreover, much attention has been paid to the influence of the compositional heterogeneity as a source of the multiple melting behavior of PHB compositions [8,68].

Minor changes in the melting temperature and enthalpy of nanocomposites containing cellulose nanofibers vs. pristine PHB were observed (Table 2), in agreement with other reports [69,70]. In particular, the addition of 2–10% cellulose crystals in PHB did not change its melting temperature or crystallinity [70]. However, all the nanocomposites showed higher crystallization temperature (T_c) compared to pristine PHB (Table 2), which is a result of the nucleating effect of BC [71].

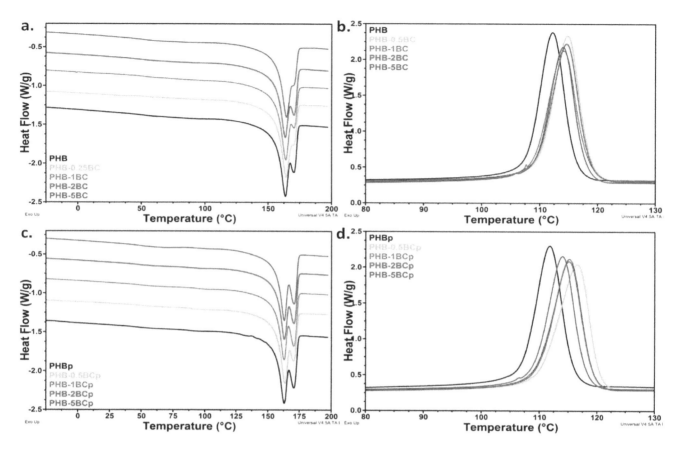

Figure 6. DSC melting (**a**) and cooling (**b**) scans of PHB nanocomposites; DSC melting (**c**) and cooling (**d**) scans of PHB nanocomposites after the plasma treatment.

Table 2. Data obtained from the analysis of DSC melting and cooling thermograms.

Nanocomposites	T_{m1}/T_{m2} °C	ΔH_m J/g	$\Delta H_{m1}/\Delta H_{m2}$ J/g	T_c °C	ΔH_c J/g	C^* %
PHB	163.8/170.5	76.1	57.7/18.4	112.2	73.6	57.9
PHBp	163.0/170.7	75.5	53.2/22.3	111.8	73.8	57.5
PHB-0.5BC	164.2/169.0	75.1	63.1/12.0	114.8	75.2	57.4
PHB-0.5BCp	163.8/170.4	74.6	55.3/19.3	116.5	74.6	57.1
PHB-1BC	163.6/170.5	75.8	61.3/14.5	114.1	73.7	58.3
PHB-1BCp	162.9/170.8	74.2	50.5/23.7	115.2	74.0	57.0
PHB-2BC	163.6/170.5	74.7	56.5/18.2	114.6	73.5	58.0
PHB-2BCp	162.8/170.6	72.4	50.7/21.7	115.1	72.1	56.2
PHB-5BC	164.4/169.0	71.2	60.3/10.9	114.1	69.8	57.0
PHB-5BCp	163.0/170.8	71.5	49.7/21.8	114.0	71.0	57.3
PHB-2BC-ZnO	162.7/170.6	76.5	51.7/24.8	113.7	77.5	59.4

* calculated with Equation (1); the amount of TBC plasticizer (10 wt %) was extracted from the w_{PHB} value. T_{m1}/T_{m2}—melting temperature of PHB corresponding to the double melting peaks; T_c—crystallization temperature; $\Delta H_{m1}/\Delta H_{m2}$—melting enthalpy of PHB corresponding to the double melting peaks; ΔH_m—total melting enthalpy.

Furthermore, the melting temperature of PHB and nanocomposites did not change significantly after the plasma treatments. However, a slight but systematic increase of T_c (by 1–2 °C) and decrease of crystallinity (with less than 2%) were observed after the plasma treatment of nanocomposites with less than 5 wt % BC (Figure 6b,d; Table 2). In general, the changes in the bulk properties following the plasma treatments are small due to the surface character of these treatments. The transformations induced by plasma in polymers occur in a small depth from the surface, keeping unmodified the bulk material and, therefore, its properties [72]. The slight increase of T_c and decrease of crystallinity after plasma exposure may be caused by several factors: (i) the slight increase of temperature at the surface of nanocomposites, which induces faster aging of PHB [71]; (ii) new active species and functions generated by plasma, which act as nucleating sites; and (iii) crosslinking reactions involving surface PHB chains.

An important aspect observed after the plasma treatment is the increased proportion of crystalline PHB with higher melting temperature (higher melting peak), coming from recrystallization. The proportion of crystalline PHB corresponding to the higher and lower melting peaks was calculated by dividing the enthalpy of the peak to the total enthalpy ($\Delta H_{m1}/\Delta H_m$ and $\Delta H_{m2}/\Delta H_m$) (Figure 7).

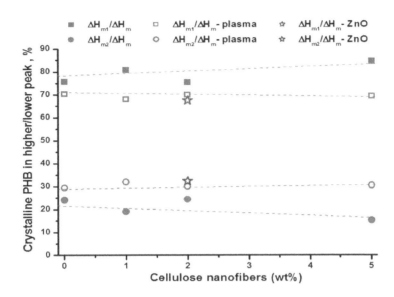

Figure 7. Influence of cellulose nanofibers concentration and plasma treatments on the proportion of crystalline PHB with lower (upper squares) and higher melting temperature (lower circles).

Before the plasma treatment, a decreasing tendency of $\Delta H_{m2}/\Delta H_m$ proportion with the concentration of cellulose nanofibers was observed and almost no variation after plasma curing. Therefore, it may be assumed that cellulose hinders the recrystallization; however the plasma treatment favors this process. Thus, plasma bombardment increased with a couple degrees the local temperature and the polymer chains mobility is also increased favoring the formation of larger, more stable crystals during melting. Moreover, the small amount of new functions induced by plasma may act as seeds for the melting recrystallization process. Higher surface crystallinity was observed after oxygen and carbon dioxide vacuum plasma treatment of PHB, which was explained by the preferential etching of the softer amorphous parts on the surface of PHB in plasma treatment [51].

These effects are more pronounced in the case of ZnO plasma-coated PHB-2BC nanocomposite (Figure S4): higher crystallinity and higher proportion of more perfect crystallites. ZnO acts probably as a nucleating agent enhancing the crystallization. Previous works have shown the effectiveness of ZnO nanoparticles as nucleating agents, raising the crystallinity of PHB or PHBV in PHAs-ZnO nanocomposites [32,33].

3.4. Dynamic Mechanical Analysis

The glass transition temperature (T_g) cannot be detected in the DSC diagrams (Figure 6). Moreover, the mechanical properties might be modified by the plasma treatment [41,51]. Therefore, dynamic mechanical analysis (DMA) was used to characterize the viscoelastic behavior and transitions (Figure 8). The values of the T_g and storage modulus (E') at different temperatures were collected in Table 3.

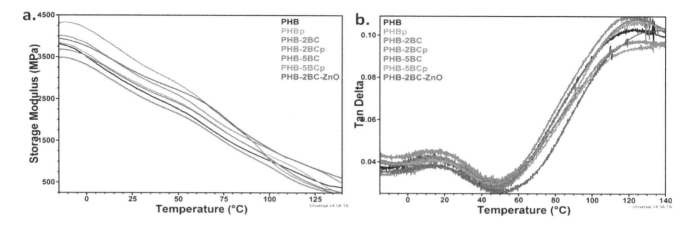

Figure 8. Storage Module (**a**) and tangent of mechanical loss angle (loss factor, tan δ) (**b**) of PHB nanocomposites with 2 and 5 wt % BC before (-) and after plasma treatment (p).

Table 3. DMA results: glass transition of PHB (T_g), $T_α{}^*$ transition (crystal-crystal slip) and storage modulus (E') values at different temperatures.

Nanocomposites	T_g [°C]	$T_α{}^*$ [°C]	E' [MPa] −15 °C	E' [MPa] 30 °C	E' [MPa] 100 °C
PHB	14.2	123	3807	2687	1024
PHBp	16.2	126	3833	2827	1141
PHB-2BC	17.1	122	3456	2532	901
PHB-2BCp	18.0	125	3980	3000	1290
PHB-5BC	14.9	125	3652	2768	1075
PHB-5BCp	17.8	128	4289	3324	1058
PHB-2BC-ZnO	15.5	136	3950	3064	1311

The plasma treatment has as result a slight increase of the storage modulus of PHB, between 0.7% and 11% depending on the temperature; however, a strong increase of the storage modulus was observed for nanocomposites after the plasma treatment. In particular, an increase between 15% and 43% was noticed in the case of PHB-2BC and an increase of 17–20% for PHB-5BC. ZnO plasma coating led to a further increase in the storage modulus of PHB-2BC of 21–46%. Data available in the literature on the influence of plasma treatment on PHA mechanical properties are scarce [41,51]. PHB films modified on their surface with poly(acrylic acid) using a vacuum plasma process showed increased tensile strength and Young's modulus indicating recrystallization during plasma irradiation [41]. Small change of the storage modulus was also reported in the case of O_2 plasma-treated PHB [51].

Correlating DMA, DSC and FTIR results, it can be assumed that the plasma treatment led to both physical (different crystal thicknesses and perfection, accelerated aging, heterogeneous nucleation) and chemical changes (active species, functionalization and/or crosslinking) that decrease the chain mobility in the surface layers and increase the stiffness, leading to this remarkable increase of the storage modulus. The dynamic mechanical properties are also sensitive to the concentration of cellulose in nanocomposites, an increase of the storage modulus with the increase of BC concentration being noticed after the treatment. The largest increase of stiffness (27%) compared to pristine PHB was observed for the nanocomposites with 5 wt % BC. The highest increase of stiffness at 100 °C was noticed in the case of PHB-2BCp and PHB-2BC-ZnO, with 26% and 28%, showing a synergistic effect of BC and plasma on the mechanical properties of PHB. It can be assumed that the plasma modification of nanocellulose is easier compared to PHB due to its more polar character; however, the addition of cellulose nanofibers may also induce a higher porosity, which may favor the diffusion of the active species generated during the plasma treatment. To elucidate this aspect, plasma-treated films were investigated by SEM.

Two relaxations were observed in the tan δ versus temperature curves, the first at 14 . . . 18 °C, depending on the treatment and composition, which corresponds to the glass transition of PHB and the second at 122 . . . 136 °C, which is a transition associated to the slippage between crystallites [8,73] A slight increase of T_g, with 1–3 °C, was noticed for the plasma-treated films, showing a restriction of PHB chain mobility in the amorphous regions induced by plasma. Besides the influence of the larger crystals on the surrounding amorphous phase, the chemical changes and, especially crosslinking, may decrease the chain mobility leading to higher T_g values after the treatments. These changes are supported by increased values of the T_α^* transition after the plasma treatment for all the samples with about 3 °C, showing a more difficult slippage between crystallites after the treatment. The highest T_α^* (136 °C) was noticed for ZnO plasma-coated PHB-2BC nanocomposite, showing a high influence of ZnO plasma treatment; it can be assumed that ZnO nanoparticles are strongly bond to the surface of the films, which is consistent with the FTIR results.

3.5. Tensile Characterization

Tensile tests were carried out to determine the influence of cellulose nanofibers on the strength and flexibility of the prepared films (Figure 9a). An increase in the tensile strength of nanocomposites with BC concentration was observed, although no chemical modification was applied, and no significant variation of the elongation at break. Thus, an increase of tensile strength by 24% and 28%, from 11.1 MPa for PHB to 13.8 MPa for the nanocomposite with 1 wt % BC and 14.2 MPa for the nanocomposite with 2 wt % BC was noticed. A concentration of 5 wt % BC nanofibers led to a less significant increase in tensile strength. Similar results were reported for PHBV with 12 mol % hydroxyvalerate (HV) reinforced with cellulose nanowhiskers (CNW), with 2.3 wt % the optimum concentration of CNW [74]; however, no variation in mechanical properties was observed in the case of PHBV (low and high HV content) incorporating bacterial cellulose nanowhiskers [75], both obtained using a solution casting technique. On the other hand, a slight increase of tensile strength was reported for PHBV nanocomposites with nanofibrillated cellulose prepared by melt mixing [76]. Therefore, the increase in tensile strength (Figure 9a) may be caused by stronger interfacial bonding between PHB and BC nanofibers, favored by melt compounding, and by a good dispersion of nanofibers. The leveling off observed for 5 wt % BC may be attributed to the PHB embrittlement caused by BC nanofiber agglomeration. The most favorable mechanical behavior in terms of improvement of properties vs. concentration of nanofiller was observed for the nanocomposite with 1 wt % BC. This was also characterized after treatment with plasma.

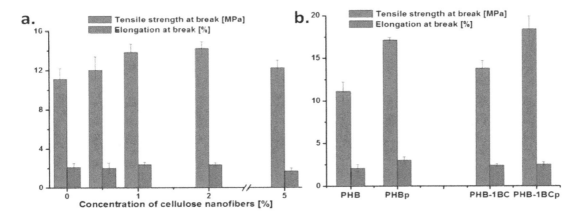

Figure 9. Tensile tests results for (**a**) nanocomposites with different amount of cellulose nanofibers and (**b**) PHB and PHB-1BC before and after plasma treatment.

A remarkable increase of tensile strength was observed after the plasma treatment for both PHB and the PHB-1BC nanocomposite, with 54% and 21%, respectively. This is in line with the important increase of storage modulus observed after the plasma treatment (Figure 8).

3.6. Morphological Analysis by SEM and AFM

The SEM image of the PHB film (Figure 10a) shows an ordered spherulitic structure (circled areas), which is less obvious after the plasma treatment (Figure 10b).

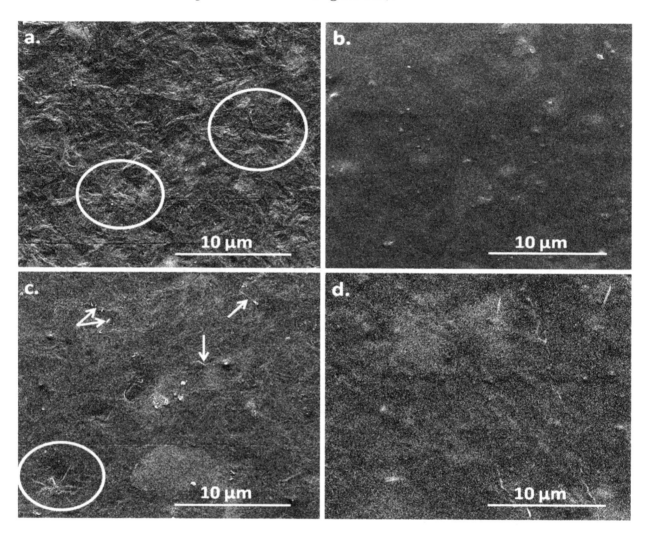

Figure 10. SEM images of PHB and PHB-5BC before (**a**,**c**) and after (**b**,**d**) the plasma treatment.

This image was captured one week after the plasma treatment of PHB; however, an embossed surface was observed 18 h after the treatment (Figure S5). This pattern is similar to an "egg box" structure. The embossing of the film surface may be due to the different effect of the plasma exposure on the crystalline and amorphous regions in PHB. The areas with higher crystallinity or the spherulites are more resistant to plasma exposure than the amorphous ones, leading to different degree of deformation. It is worth to mention that several particles mostly in nanometric range were observed on the surface of PHB and nanocomposite, before and after the treatment. They may be inorganic nanoparticles usually found in commercial PHB samples.

The spherulitic morphology is hardly seen on the surface of PHB-5BC nanocomposite (Figure 10c) because it is masked on certain areas by the presence of cellulose nanofibers. They are well dispersed on the surface of nanocomposite and they have a thickness of less than 100 nm and lengths from 500 nm to several microns. Similarly with PHB, a ordered structure was not observed on the surface of PHB-5BC nanocomposite after the plasma treatment.

Atomic force microscopy is a sensitive tool to analyze nanostructures. Figure 11 shows that the organized structure observed in PHB and PHB-5BC was maintained after the plasma treatment, but was attenuated.

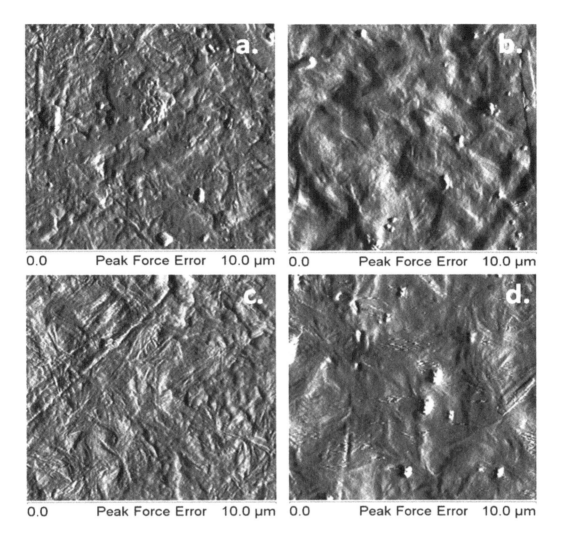

Figure 11. AFM peak force error images of PHB and PHB-5BC before (**a,c**) and after the plasma treatment (**b,d**).

The lamellar morphology of PHB may be seen in Figure 11a and, especially, in Figure 11c, in the case of the nanocomposite. Although DSC did not show an increase of crystallinity in PHB-5BC compared to pristine PHB, being a bulk analysis, the surface morphology of the nanocomposite showed a higher degree of order and better organization. Although spherulites may not be seen at this magnitude, the twisted lamellar morphology of PHB was visible in Figure 11c [77]. This may be an influence of BC nanofibers. Previous works have shown that cellulose nanofibers, regardless the source or fabrication, act as nucleating agents, enhancing the crystallization and the degree of order in PHB [8,70,76]. Plasma treatment induced visible changes in the surface morphology. The crystalline organization was less visible as compared to untreated samples, similar to SEM observations. The analysis of topographic images (Figure S6) reveals new details. The crystalline lamellar structure is also visible after the plasma treatment; however, the structure is more regular and the surface roughness is lower. Thus, root mean square roughness (RRMS) was determined using the AFM topographic images. RRMS of PHB surface was 19.0 ± 2.3 nm before and 16.6 ± 0.5 nm after the plasma treatment; for the nanocomposite, RRMS was 16.1 ± 1.9 nm before and 13.3 ± 1.1 nm after the plasma treatment. This decrease of surface roughness may result from the high energy and active species, which allow a higher mobility of the PHB chains in plasma-treated samples. Similar observations were reported for PHB films treated with oxygen and carbon dioxide plasma in a vacuum chamber [51].

PHB nanocomposite with 2 wt % bacterial cellulose nanofibers, which led to the best thermal and mechanical properties, was thoroughly investigated by SEM-EDX before and after plasma treatments, with and without ZnO deposition. EDX results (Figure S6 and Table S2) show the presence of less than 1% Si in all the samples and traces of different other elements (Na, K, Mg, Al), which suggests an aluminosilicate filler. Therefore, the particles observed in the SEM images from Figure 10 may be associated to aluminosilicate filler or nucleating agent. The EDX results for ZnO plasma-treated PHB-2BC clearly show the presence of about 10% ZnO on the surface of nanocomposite. Moreover, the O/C weight ratio is 0.60, so almost identical to the theoretical value for PHB, and slightly decreases after the plasma treatment, from 0.60 to 0.57; however, it strongly increases (to 1.16) after ZnO deposition due to the O atoms in ZnO.

The SEM images of the PHB-2BC nanocomposite film before and after the plasma treatment are shown in Figure 12. Bacterial cellulose nanofibers are well dispersed on the surface of nanocomposite film and embedded into the PHB matrix. A couple of nanofibers with the top part outside (Figure 12b) were used to determine their size. Nanofibers have a thickness between 30 and 70 nm. After the plasma treatment the surface seems smoother (Figure 12c) and the nanofiber size is in the same range (Figure 12d).

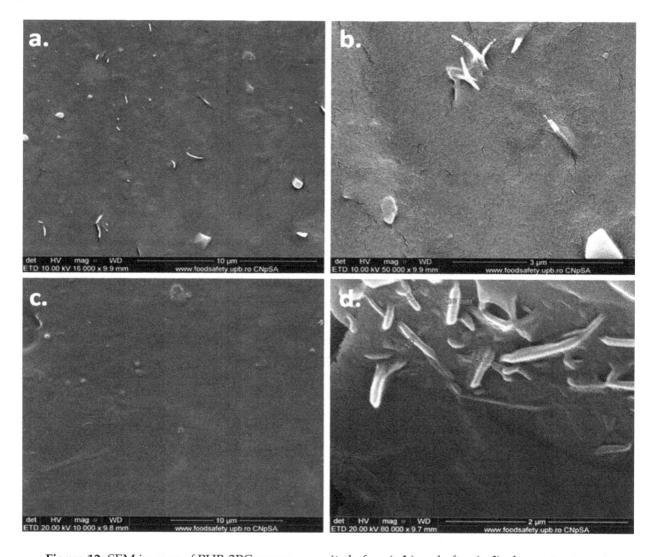

Figure 12. SEM images of PHB-2BC nanocomposite before (**a,b**) and after (**c,d**) plasma treatment.

SEM images of PHB-2BC nanocomposite after ZnO plasma coating are shown in Figure 13. The ZnO nanoparticles continuously cover the surface of PHB-2BC nanocomposite; however, they self-aggregate similar to a coral reef structure (Figure 13a,b). This pattern results from the conjugated influence of ZnO and polymeric substrate. SEM image of PHB film right after the treatment (Figure S5) shows a "egg box" structure. We may presume that the holes of this structure will allow the deposition of more ZnO nanoparticles, which will aggregate, forming the coral reef pattern. The plasma coating process also causes the detachment of cellulose fibers and polymer strips, as observed in Figure 13c. Figure 13d shows the aggregated ZnO nanoparticles with a granular shape and an average diameter of only 18 nm. Similar shape and size of ZnO nanoparticles were observed by SEM on the surface of paper impregnated with an ammoniacal suspension of ZnO [78].

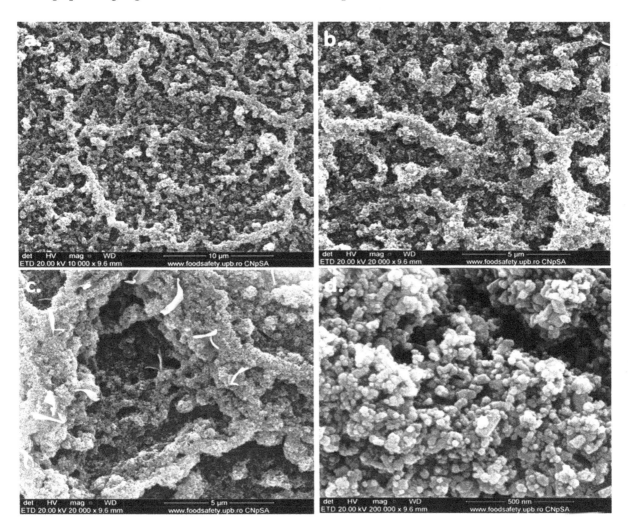

Figure 13. SEM images of PHB-2BC nanocomposite after ZnO plasma coating.

3.7. Antibacterial Properties

The antibacterial activity of ZnO was demonstrated by previous works [32,33,36]; however, the influence of a simple Ar plasma exposure on the antibacterial activity of PHAs had not yet been studied.

Previous work has shown better adhesion and proliferation of mouse adipose-derived stem cells on the surface of plasma-treated PHBV film [39]. Moreover, the presence of cellulose in the nanocomposite favors cells proliferation [14,71]. Better compatibility with L929 cells was reported for poly(3-hydroxyoctanoate) (PHO) reinforced with 3 wt % BC compared to pristine PHO due to a more hydrophilic surface [71], similarly for PHB modified with a polyhydroxyalkanoate and 2 wt % BC compared to PHB [14]. Therefore, the number of colonies obtained after incubation of the bacterial suspensions in contact with the untreated and plasma-treated PHB-2BC nanocomposite films was determined. A decrease by 44% of the *Staphylococcus aureus* bacteria colonies' mean number, from 100 to 56, was obtained after the plasma treatment. Similarly, 63% inhibition of *Escherichia coli* growth after 24 h was obtained in the case of plasma-treated PHB-2BC compared to the untreated film (Figure 14a,b). For comparison, a PHB-2BC-ZnO nanocomposite was characterized in the same conditions and almost complete inhibition of *Staphylococcus aureus* growth was obtained (Figure 14c,d). This significant decrease of the number of colonies, by 44% in the case of *Staphylococcus aureus* and 63% in the case of *Escherichia coli*, shows that the plasma treatment is able to induce antibacterial activity. Similar antibacterial efficiency against *Escherichia coli* was reported in the case of a PHB blended with a quaternarized N-halamine polymer obtained using a complicate chemical route [79]. Besides biodegradability and eco-friendliness, the new nanocomposite films show good thermal and mechanical properties and increased antibacterial activity. Thus, plasma-treated PHB-2BC nanocomposite films are proposed for food packaging to protect fresh foods from decay.

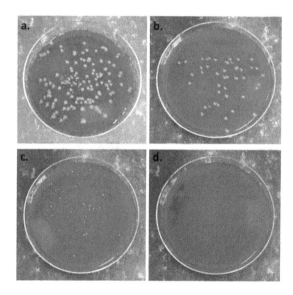

Figure 14. Antibacterial activity of PHB-2BC nanocomposite films against *Escherichia coli* before (**a**) and after the plasma treatment (**b**) and against *Staphylococcus aureus* before (**c**) and after ZnO plasma coating (**d**).

4. Conclusions

Biodegradable and eco-friendly films of poly(3-hydroxybutyrate) modified with bacterial cellulose nanofibers were prepared using a melt compounding-compression molding technique and surface-treated by a plasma or ZnO nanoparticle plasma coating. The thermal, mechanical and antibacterial properties of these films were studied for possible applications in food packaging. The thermal stability, crystallinity and melting behavior of nanocomposites were not changed by the plasma treatment; however, a remarkable increase in stiffness and strength was obtained. ZnO plasma coating consisted of a continuous layer of self-aggregated ZnO nanoparticles and led to the improvement of the crystallinity and mechanical properties of the PHB-2BC nanocomposite. Higher antibacterial activity was noted after the plasma treatment and complete inhibition of *Staphylococcus aureus* growth after ZnO nanoparticle plasma coating. Plasma-treated PHB nanocomposites containing bacterial cellulose nanofibers are proposed as a "green" solution for food packaging.

Supplementary Materials: Figure S1: DTG curves (300–350 °C) of PHB nanocomposites before and after the plasma treatment; Figure S2: FTIR spectra of PHB and PHB nanocomposites with 2 wt % and 5 wt % BC; Figure S3: FTIR spectra of PHB (A), PHB-2BC (B) and PHB-5BC (C) before and after the plasma treatments; Figure S4: DSC first melting and cooling scans for ZnO plasma-coated PHB-2BC compared to the untreated nanocomposite; Figure S5: SEM image of PHB right after the plasma treatment; Figure S6: AFM topographic images of PHB and PHB-5BC before (a,c) and after the plasma treatment (b,d); Figure S7: EDX results for PHB-2BC and PHB-2BC-Zn; Table S1: EDX data for PHB-2BC nanocomposite before and after plasma and ZnO plasma coating; Table S1: Peak assignments in the FTIR spectra of PHB and nanocomposites in the 3050–2800 cm^{-1} region.

Author Contributions: D.M.P. conceived and designed the experiments; G.D. conceived and designed the plasma experiments, E.R.I. designed and developed the plasma source and developed the software running on the coating device; C.-A.N. and A.R.G. carried out the thermal and thermomechanical characterization and R.T. the morphological analysis by SEM; M.D.I. carried out the plasma experiments and contact angle measurements; B.-E.L. and I.C. performed the antimicrobial tests; all the authors contributed to the interpretation of the results; D.M.P. wrote and reviewed the paper. The manuscript was read and approved by all the authors.

Acknowledgments: The financial support from Romanian Ministry of Research and Innovation is greatly appreciated.

References

1. Rhim, J.W.; Park, H.M.; Ha, C.S. Bio-nanocomposites for food packaging applications. *Prog. Polym. Sci.* **2013**, *38*, 1629–1652. [CrossRef]

2. Youssef, A.M.; El-Sayed, S.M. Bionanocomposites materials for food packaging applications: Concepts and future outlook. *Carbohydr. Polym.* **2018**, *193*, 19–27. [CrossRef] [PubMed]

3. Peelman, N.; Ragaert, P.; De Meulenaer, B.; Adons, D.; Peeters, R.; Cardon, L.; van Impe, F.; Devlieghere, F. Application of bioplastics for food packaging. *Trends Food Sci. Technol.* **2013**, *32*, 128–141. [CrossRef]

4. Panaitescu, D.M.; Frone, A.N.; Chiulan, I. Nanostructured biocomposites from aliphatic polyesters and bacterial cellulose. *Ind. Crops Prod.* **2016**, *93*, 251–266. [CrossRef]

5. Plackett, D.; Siró, I. Polyhydroxyalkanoates (PHAs) for food packaging. In *Multifunctional and Nanoreinforced Polymers Food Packaging*; Lagaron, J.M., Ed.; Woodhead Publishing Ltd.: Cambridge, UK, 2011; pp. 498–526.

6. Yu, H.; Yan, C.; Yao, J. Fully biodegradable food packaging materials based on functionalized cellulose nanocrystals/poly(3-hydroxybutyrate-*co*-3-hydroxyvalerate) nanocomposites. *RSC Adv.* **2014**, *4*, 59792–59802. [CrossRef]

7. Doi, Y.; Kanesawa, Y.; Kawaguchi, Y.; Kunioka, M. Hydrolytic degradation of microbial poly(hydroxya1kanoates). *Macromol. Chem. Rapid Commun.* **1989**, *10*, 227–230. [CrossRef]

8. Panaitescu, D.M.; Frone, A.N.; Chiulan, I.; Nicolae, C.A.; Trusca, R.; Ghiurea, M.; Gabor, A.R.; Mihailescu, M.; Casarica, A.; Lupescu, I. Role of bacterial cellulose and poly (3-hydroxyhexanoate-*co*-3-hydroxyoctanoate) in poly (3-hydroxybutyrate) blends and composites. *Cellulose* **2018**, *25*, 5569–5591. [CrossRef]

9. Koller, M. Poly(hydroxyalkanoates) for food packaging: Application and attempts towards implementation. *Appl. Food Biotechnol.* **2014**, *1*, 3–15.

10. Kunioka, M.; Tamaki, A.; Doi, Y. Crystalline and thermal properties of bacterial copolyesters: Poly (3-hydroxybutyrate-*co*-3-hydroxyvalerate) and poly (3-hydroxybutyrate-*co*-4-hydroxybutyrate). *Macromolecules* **1989**, *22*, 694–697. [CrossRef]

11. Zhila, N.; Shishatskaya, E. Properties of PHA bi-, ter-, and quarter-polymers containing 4-hydroxybutyrate monomer units. *Int. J. Biol. Macromol.* **2018**, *111*, 1019–1026. [CrossRef] [PubMed]

12. Panaitescu, D.M.; Nicolae, C.A.; Frone, A.N.; Chiulan, I.; Stanescu, P.O.; Draghici, C.; Iorga, M.; Mihailescu, M. Plasticized poly(3-hydroxybutyrate) with improved melt processing and balanced properties. *J. Appl. Polym. Sci.* **2017**, *134*, 44810. [CrossRef]

13. Seoane, I.T.; Manfredi, L.B.; Cyras, V.P.; Torre, L.; Fortunati, E.; Puglia, D. Effect of cellulose nanocrystals and bacterial cellulose on disintegrability in composting conditions of plasticized PHB nanocomposites. *Polymers* **2017**, *9*, 561. [CrossRef]

14. Chiulan, I.; Panaitescu, D.M.; Frone, A.N.; Teodorescu, M.; Nicolae, C.A.; Casarica, A.; Tofan, V.; Salageanu, A. Biocompatible polyhydroxyalkanoates/bacterial cellulose composites: Preparation, characterization, and in vitro evaluation. *J. Biomed. Mater. Res. A* **2016**, *104*, 2576–2584. [CrossRef] [PubMed]

15. Haugaard, V.K.; Danielsen, B.; Bertelsen, G. Impact of polylactate and poly(hydroxybutyrate) on food quality. *Eur. Food Res. Technol.* **2003**, *216*, 233–240. [CrossRef]

16. Bucci, D.Z.; Tavares, L.B.B.; Sell, I. PHB packaging for the storage of food products. *Polym. Test.* **2005**, *24*, 564–571. [CrossRef]

17. Cherpinski, A.; Torres-Giner, S.; Cabedo, L.; Lagaron, J.M. Postprocessing optimization of electrospun sub-micron poly(3-hydroxybutyrate) fibers to obtain continuous films of interest in food packaging applications. *Food Addit. Contam. Part A* **2017** *34*, 1817–1830. [CrossRef] [PubMed]

18. Modi, S.J.; Cornish, K.; Koelling, K.; Vodovotz, Y. Fabrication and improved performance of poly(3-hydroxybutyrate-*co*-3-hydroxyvalerate) for packaging by addition of high molecular weight natural rubber. *J. Appl. Polym. Sci.* **2016**, *133*, 43937. [CrossRef]

19. Garcia-Garcia, D.; Ferri, J.M.; Boronat, T.; Lopez-Martinez, J.; Balart, R. Processing and characterization of binary poly(hydroxybutyrate) (PHB) and poly(caprolactone) (PCL) blends with improved impact properties. *Polym. Bull.* **2016**, *73*, 3333–3350. [CrossRef]

20. Arrieta, M.P.; Samper, M.D.; Aldas, M.; López, J. On the use of PLA-PHB blends for sustainable food packaging applications. *Materials* **2017**, *10*, 1008. [CrossRef] [PubMed]

21. Armentano, I.; Fortunati, E.; Burgos, N.; Dominici, F.; Luzi, F.; Fiori, S.; Jiménez, A.; Yoon, K.; Ahn, J.; Kang, S.; Kenny, J.M. Processing and characterization of plasticized PLA/PHB blends for biodegradable multiphase systems. *eXPRESS Polym. Lett.* **2015**, *9*, 583–596. [CrossRef]

22. Arrieta, M.P.; Samper, M.D.; Lopez, J.; Jimenez, A. Combined Effect of Poly(hydroxybutyrate) and Plasticizers on Polylactic acid Properties for Film Intended for Food Packaging. *J. Polym. Environ.* **2014**, *22*, 460–470. [CrossRef]

23. Khaneghah, A.M.; Hashemi, S.M.B.; Limbo, S. Antimicrobial agents and packaging systems in antimicrobial active food packaging: An overview of approaches and interactions. *Food Bioprod. Process.* **2018**, *111*, 1–19. [CrossRef]

24. Mlalila, N.; Hilong, A.; Swai, H.; Devlieghere, F.; Ragaert, P. Antimicrobial packaging based on starch, poly(3-hydroxybutyrate) and poly (lactic-*co*-glycolide) materials and application challenges. *Trends Food Sci. Technol.* **2018**, *74*, 1–11. [CrossRef]

25. Janifer, R.X.; Sudalaimuthu, T.B.; Johnsy, G.; Karna, V.R. Material Properties and Antimicrobial Activity of Polyhydroxybutyrate (PHB) Films Incorporated with Vanillin. *Appl. Biochem. Biotechnol.* **2015**, *176*, 1498–1510.

26. Solaiman, D.K.Y.; Ashby, R.D.; Zerkowski, J.A.; Krishnama, A.; Vasanthan, N. Control-release of antimicrobial sophorolipid employing different biopolymer matrices. *Biocatal. Agric. Biotechnol.* **2015**, *4*, 342–348. [CrossRef]

27. Burgos, N.; Armentano, I.; Fortunati, E.; Dominici, F.; Luzi, F.; Fiori, S.; Cristofaro, F.; Visai, L.; Jiménez, A.; Kenny, J.M. Functional Properties of Plasticized Bio-Based Poly(Lactic Acid)-Poly(Hydroxybutyrate) (PLA-PHB) Films for Active Food Packaging. *Food Bioprocess Technol.* **2017**, *10*, 770–780. [CrossRef]

28. Narayanan, A.; Mallesha, N.; Ramana, K.V. Synergized Antimicrobial Activity of Eugenol Incorporated Polyhydroxybutyrate Films against Food Spoilage Microorganisms in Conjunction with Pediocin. *Appl. Biochem. Biotechnol.* **2013**, *170*, 1379–1388. [CrossRef] [PubMed]

29. Sabharwal, P.K.; Chattopadhyay, S.; Singh, H. Preparation and characterization of antimicrobial, biodegradable, triclosan-incorporated polyhydroxybutyrate-co-valerate films for packaging applications. *J. Appl. Polym. Sci.* **2018**, *135*, 46862. [CrossRef]

30. Iordanskii, A.; Zhulkina, A.; Olkhov, A.; Fomin, S.; Burkov, A.; Stilman, M. Characterization and Evaluation of Controlled Antimicrobial Release from Petrochemical (PU) and Biodegradable (PHB) Packaging. *Polymers* **2018**, *10*, 817. [CrossRef]

31. Correa, J.P.; Molina, V.; Sanchez, M.; Kainz, C.; Eisenberg, P.; Massani, M.B. Improving ham shelf life with a polyhydroxybutyrate/polycaprolactone biodegradable film activated with nisin. *Food Packag. Shelf Life* **2017**, *11*, 31–39. [CrossRef]

32. Díez-Pascual, A.M.; Díez-Vicente, A.L. Poly(3-hydroxybutyrate)/ZnO Bionanocomposites with Improved Mechanical, Barrier and Antibacterial Properties. *Int. J. Mol. Sci.* **2014**, *15*, 10950–10973. [CrossRef] [PubMed]

33. Díez-Pascual, A.M.; Díez-Vicente, A.L. ZnO-Reinforced Poly(3-hydroxybutyrate-*co*-3-hydroxyvalerate) Bionanocomposites with Antimicrobial Function for Food Packaging. *ACS Appl. Mater. Interfaces* **2014**, *6*, 9822–9834. [CrossRef] [PubMed]

34. Castro-Mayorga, J.L.; Freitas, F.; Reis, M.A.M.; Prieto, M.A.; Lagaron, J.M. Biosynthesis of silver nanoparticles and polyhydroxybutyrate nanocomposites of interest in antimicrobial applications. *Int. J. Biol. Macromol.* **2018**, *108*, 426–435. [CrossRef] [PubMed]

35. Salama, H.E.; Aziz, M.S.A.; Saad, G.R. Thermal properties, crystallization and antimicrobial activity of chitosan biguanidine grafted poly(3-hydroxybutyrate) containing silver nanoparticles. *Int. J. Biol. Macromol.* **2018**, *111*, 19–27. [CrossRef] [PubMed]

36. Castro-Mayorga, J.L.; Fabra, M.J.; Pourrahimi, A.M.; Olsson, R.T.; Lagaron, J.M. The impact of zinc oxide particle morphology as an antimicrobial and when incorporated in poly(3-hydroxybutyrate-co-3-hydroxyvalerate) films for food packaging and food contact surfaces applications. *Food Bioprod. Process.* **2017**, *101*, 32–44. [CrossRef]

37. Castro-Mayorga, J.L.; Fabra Rovira, M.J.; Mas, L.C.; Moragas, G.S.; Lagaron, J.M. Antimicrobial nanocomposites and electrospun coatings based on poly(3-hydroxybutyrate-co-3-hydroxyvalerate) and copper oxide nanoparticles for active packaging and coating applications. *J. Appl. Polym. Sci.* **2018**, *135*, 45673. [CrossRef]

38. Cherpinski, A.; Gozutok, M.; Sasmazel, H.T.; Torres-Giner, S.; Lagaron, J.M. Electrospun oxygen scavenging films of poly(3-hydroxybutyrate) containing palladium nanoparticles for active packaging applications. *Nanomaterials* **2018**, *8*, 469. [CrossRef] [PubMed]

39. Chang, C.-K.; Wang, H.-M.D.; Lan, J.C.W. Investigation and characterization of plasma-treated poly(3-hydroxybutyrate) and poly(3-hydroxybutyrate-co-3-hydroxyvalerate) biopolymers for an in vitro cellular study of mouse adipose-derived stem cells. *Polymers* **2018**, *10*, 355. [CrossRef]

40. Karahaliloglu, Z.; Demirbilek, M.; Sam, M.; Erol-Demirbilek, M.; Saglam, N.; Denkbas, E.B. Plasma polymerization-modified bacterial polyhydroxybutyrate nanofibrillar scaffolds. *J. Appl. Polym. Sci.* **2013**, *128*, 1904–1912. [CrossRef]

41. Zhang, J.; Kasuya, K.; Takemura, A.; Isogai, A.; Iwata, T. Properties and enzymatic degradation of poly(acrylic acid) grafted polyhydroxyalkanoate films by plasma-initiated polymerization. *Polym. Degrad. Stab.* **2013**, *98*, 1458–1464. [CrossRef]

42. Qu, X.H.; Wu, Q.; Liang, J.; Qu, X.; Wang, S.G.; Chen, G.Q. Enhanced vascular-related cellular affinity on surface modified copolyesters of 3-hydroxybutyrate and 3-hydroxyhexanoate (PHBHHx). *Biomaterials* **2005**, *26*, 6991–7001. [CrossRef] [PubMed]

43. Slepickova Kasalkova, N.; Slepicka, P.; Sajdl, P.; Svorcík, V. Surface changes of biopolymers PHB and PLLA induced by Ar+ plasma treatment and wet etching. *Nucl. Instrum. Methods Phys. Res. B* **2014**, *332*, 63–67. [CrossRef]

44. Morent, R.; De Geyter, N.; Desmet, T.; Dubruel, P.; Leys, C. Plasma Surface Modification of Biodegradable Polymers: A Review. *Plasma Process. Polym.* **2011**, *8*, 171–190. [CrossRef]

45. Aflori, M. Chitosan-based silver nanoparticles incorporated at the surface of plasma-treated PHB films. *Chem. Lett.* **2017**, *46*, 65–67. [CrossRef]

46. Wang, Y.; Lu, L.; Zheng, Y.; Chen, X. Improvement in hydrophilicity of PHBV films by plasma treatment. *J. Biomed. Mater. Res. A* **2006**, *76*, 589–595. [CrossRef] [PubMed]

47. Pompe, T.; Keller, K.; Mothes, G.; Nitschke, M.; Teese, M.; Zimmermann, R.; Werner, C. Surface modification of poly(hydroxybutyrate) films to control cell–matrix adhesion. *Biomaterials* **2007**, *28*, 28–37. [CrossRef] [PubMed]

48. Tezcaner, A.; Bugra, K.; Hasýrcý, V. Retinal pigment epithelium cell culture on surface modified poly(hydroxybutyrate-co-hydroxyvalerate) thin films. *Biomaterials* **2003**, *24*, 4573–4583. [CrossRef]

49. Ferreira, B.M.P.; Pinheiro, L.M.P.; Nascente, P.A.P.; Ferreira, M.J.; Duek, E.A.R. Plasma surface treatments of poly(L-lactic acid) (PLLA) and poly(hydroxybutyrate-co-hydroxyvalerate) (PHBV). *Mater. Sci. Eng. C* **2009**, *29*, 806–813. [CrossRef]

50. Slepicka, P.; Malá, Z.; Rimpelová, S.; Svorcík, V. Antibacterial properties of modified biodegradable PHB non-woven fabric. *Mater. Sci. Eng. C* **2016**, *65*, 364–368. [CrossRef] [PubMed]

51. Mirmohammadi, S.A.; Khorasani, M.T.; Mirzadeh, H.; Irani, S. Investigation of plasma treatment on poly(3-hydroxybutyrate) film surface: Characterization and in vitro assay. *Polym.-Plast. Technol. Eng.* **2012**, *51*, 1319–1326. [CrossRef]

52. Aflori, M. Embedding silver nanoparticles at PHB surfaces by means of combined plasma and chemical treatments. *Rev. Roum. Chim.* **2016**, *61*, 405–409.

53. Panaitescu, D.M.; Frone, A.N.; Chiulan, I.; Casarica, A.; Nicolae, C.A.; Ghiurea, M.; Trusca, R.; Damian, C.M. Structural and morphological characterization of bacterial cellulose nano-reinforcements prepared by mechanical route. *Mater. Des.* **2016**, *110*, 790–801. [CrossRef]

54. Watthanaphanit, A.; Supaphol, P.; Tamura, H.; Tokura, S.; Rujiravanit, R. Wet-spun alginate/chitosan whiskers nanocomposite fibers: Preparation, characterization and release characteristic of the whiskers. *Carbohydr. Polym.* **2010**, *79*, 738–746. [CrossRef]

55. Li, Y.; Leung, P.; Yao, L.; Song, Q.W.; Newton, E. Antimicrobial effect of surgical masks coated with nanoparticles. *J. Hosp. Infect.* **2006**, *62*, 58–63. [CrossRef] [PubMed]

56. Panaitescu, D.M.; Vizireanu, S.; Nicolae, C.A.; Frone, A.N.; Casarica, A.; Carpen, L.G.; Dinescu, G. Treatment of Nanocellulose by Submerged Liquid Plasma for Surface Functionalization. *Nanomaterials* **2018**, *8*, 467. [CrossRef] [PubMed]

57. Vizireanu, S.; Panaitescu, D.M.; Nicolae, C.A.; Frone, A.N.; Chiulan, I.; Ionita, M.D.; Satulu, V.; Carpen, L.G.; Petrescu, S.; Birjega, R.; Dinescu, G. Cellulose defibrillation and functionalization by plasma in liquid treatment. *Sci. Rep.* **2018**, *8*, 15473. [CrossRef] [PubMed]

58. Castro Mayorga, J.L.; Fabra, M.J.; Lagaron, J.M. Stabilized nanosilver based antimicrobial poly(3-hydroxybutyrate-co-3- hydroxyvalerate) nanocomposites of interest in active food packaging. *Innov. Food Sci. Emerg. Technol.* **2016**, *33*, 524–533. [CrossRef]

59. Sato, H.; Murakami, R.; Padermshoke, A.; Hirose, F.; Senda, K.; Noda, I.; Ozaki, Y. Infrared spectroscopy studies of CH . . . O hydrogen bondings and thermal behavior of biodegradable poly(hydroxyalkanoate). *Macromolecules* **2004**, *37*, 7203–7213. [CrossRef]

60. Socrates, G. *Infrared and Raman Characteristic Group Frequencies*, 3rd ed.; John Wiley & sons Ltd.: Chichester, UK, 2001; pp. 50–67.

61. Zhang, J.; Sato, H.; Noda, I.; Ozaki, Y. Conformation rearrangement and molecular dynamics of poly(3-hydroxybutyrate) during the melt-crystallization process investigated by infrared and two-dimensional infrared correlation spectroscopy. *Macromolecules* **2005**, *38*, 4274–4281. [CrossRef]

62. Padermshoke, A.; Katsumoto, Y.; Sato, H.; Ekgasit, S.; Noda, I.; Ozaki, Y. Melting behavior of poly(3-hydroxybutyrate) investigated by two-dimensional infrared correlation spectroscopy. *Spectrochim. Acta A* **2005**, *61*, 541–550. [CrossRef] [PubMed]

63. Reinicker, A.; Miller, J.B.; Kim, W.; Yong, K.; Gellman, J.A. CH3CH2OH, CD3CD2OD, and CF3CH2OH decomposition on ZnO(1ī00). *Top. Catal.* **2015**, *58*, 613–622. [CrossRef]

64. Sunderrajan, S.; Freeman, B.D.; Hall, C.K. Fourier transform infrared spectroscopic characterization of olefin complexation by silver salts in solution. *Ind. Eng. Chem. Res.* **1999**, *38*, 4051–4059. [CrossRef]

65. Norazzizi, N.; Wan Zurina, S.; Muhammad, R.Y.; Rozali, M. Othman Synthesis and characterization of copper(II) carboxylate with palm-based oleic acid by electrochemical technique. *MJAS* **2015**, *19*, 236–243.

66. Handore, K.; Bhavsar, S.; Horne, A.; Chhattise, P.; Mohite, K.; Ambekar, J.; Pande, N.; Chabukswar, V. Novel green route of synthesis of ZnO nanoparticles by using natural biodegradable polymer and its application as a catalyst for oxidation of aldehydes. *J. Macromol. Sci. A: Pure Appl. Chem.* **2014**, *51*, 941–947. [CrossRef]

67. Saito, M.; Inoue, Y.; Yoshie, N. Cocrystallization and phase segregation of blends of poly(3-hydroxybutyrate) and poly(3-hydroxybutyrate-co-3-hydroxyvalerate). *Polymers* **2001**, *42*, 5573–5580. [CrossRef]

68. Zini, E.; Focarete, M.L.; Noda, I.; Scandola, M. Bio-composite of bacterial poly(3-hydroxybutyrate-co-3-hydroxyhexanoate) reinforced with vegetable fibers. *Compos. Sci. Technol.* **2007**, *67*, 2085–2094. [CrossRef]

69. Xie, Y.; Kohls, D.; Noda, I.; Schaefer, D.W.; Akpalu, A.Y. Poly(3-hydroxybutyrate-co-3-hydroxyhexanoate) nanocomposites with optimal mechanical properties. *Polymers* **2009**, *50*, 4656–4670. [CrossRef]

70. Jianxiang, C.; Chunjiang, X.; Defeng, W.; Keren, P.; Aiwen, Q.; Yulu, S.; Li, W.; Wei, T. Insights into the nucleation role of cellulose crystals during crystallization of poly(3-hydroxybutyrate). *Carbohydr. Polym.* **2015**, *134*, 508–515.

71. Panaitescu, D.M.; Lupescu, I.; Frone, A.N.; Chiulan, I.; Nicolae, C.A.; Tofan, V.; Stefaniu, A.; Somoghi, R.; Trusca, R. Medium chain-length polyhydroxyalkanoate copolymer modified by bacterial cellulose for medical devices. *Biomacromolecules* **2017**, *18*, 3222–3232. [CrossRef] [PubMed]

72. Nedela, O.; Slepicka, P.; Svorcík, V. Surface modification of polymer substrates for biomedical applications. *Materials* **2017**, *10*, 1115. [CrossRef] [PubMed]

73. Wei, L.; Liang, S.; McDonald, A.G. Thermophysical properties and biodegradation behavior of green composites made from polyhydroxybutyrate and potato peel waste fermentation residue. *Ind. Crops Prod.* **2015**, *69*, 91–103. [CrossRef]

74. Ten, E.; Bahr, D.F.; Li, B.; Jiang, L.; Wolcott, M.P. Effects of cellulose nanowhiskers on mechanical, dielectric, and rheological properties of poly(3-hydroxybutyrate-*co*-3-hydroxyvalerate)/cellulose nanowhisker composites. *Ind. Eng. Chem. Res.* **2012**, *51*, 2941–2951. [CrossRef]

75. Martínez-Sanz, M.; Villano, M.; Oliveira, C.; Albuquerque, M.G.; Majone, M.; Reis, M.; Lopez-Rubio, A.; Lagaron, J.M. Characterization of Polyhydroxyalkanoates Synthesized from Microbial Mixed Cultures and of Their Nanobiocomposites with Bacterial Cellulose Nanowhiskers. *New Biotechnol.* **2014**, *31*, 364–376. [CrossRef] [PubMed]

76. Srithep, Y.; Ellingham, T.; Peng, J.; Sabo, R.; Clemons, C.; Turng, L.-S.; Pilla, S. Melt Compounding of Poly (3-hydroxybutyrate-*co*-3-hydroxyvalerate)/Nanofibrillated Cellulose Nanocomposites. *Polym. Degrad. Stab.* **2013**, *98*, 1439–1449. [CrossRef]

77. Xu, J.; Ye, H.; Zhang, S.; Guo, B. Organization of Twisting Lamellar Crystals in Birefringent Banded Polymer Spherulites: A Mini-Review. *Crystals* **2017**, *7*, 241.

78. Ghule, K.; Ghule, A.V.; Chen, B.-J.; Ling, Y.-C. Preparation and characterization of ZnO nanoparticles coated paper and its antibacterial activity study. *Green Chem.* **2006**, *8*, 1034–1041. [CrossRef]

79. Fan, X.; Ren, X.; Huang, T.-S.; Sun, Y. Cytocompatible antibacterial fibrous membranes based on poly(3-hydroxybutyrate-*co*-4-hydroxybutyrate) and quaternarized N-halamine polymer. *RSC Adv.* **2016**, *6*, 42600–42610. [CrossRef]

Antimicrobial LDPE/EVOH Layered Films Containing Carvacrol Fabricated by Multiplication Extrusion

Max Krepker [1], Cong Zhang [2], Nadav Nitzan [1], Ofer Prinz-Setter [1], Naama Massad-Ivanir [1], Andrew Olah [2], Eric Baer [2] and Ester Segal [1,3,*]

[1] Department of Biotechnology and Food Engineering, Technion-Israel Institute of Technology, Haifa 3200003, Israel; makskrepker@gmail.com (M.K.); nitzannadav@technion.ac.il (N.N.); oferp@campus.technion.ac.il (O.P.-S.); naamam@bfe.technion.ac.il (N.M.-I.)

[2] Center for Layered Polymeric Systems, Department of Macromolecular Science and Engineering, Case Western Reserve University, Cleveland, OH 44106-7202, USA; cxz177@case.edu (C.Z.); amo5@case.edu (A.O.); exb6@case.edu (E.B.)

[3] The Russell Berrie Nanotechnology Institute, Technion-Israel Institute of Technology, Haifa 3200003, Israel

* Correspondence: esegal@technion.ac.il

Abstract: This work describes the fabrication of antimicrobial multilayered polymeric films containing carvacrol (used as a model essential oil) by co-extrusion and multiplication technique. The microlayering process was utilized to produce films, with up to 65 alternating layers, of carvacrol-containing low-density polyethylene (LDPE) and ethylene vinyl alcohol copolymer (EVOH). Carvacrol was melt compounded with LDPE or loaded into halloysite nanotubes (HNTs) in a pre-compounding step prior film production. The detailed nanostructure and composition (in terms of carvacrol content) of the films were characterized and correlated to their barrier properties, carvacrol release rate, and antibacterial and antifungal activity. The resulting films exhibit high carvacrol content despite the harsh processing conditions (temperature of 200 °C and long processing time), regardless of the number of layers or the presence of HNTs. The multilayered films exhibit superior oxygen transmission rates and carvacrol diffusivity values that are more than two orders of magnitude lower in comparison to single-layered carvacrol-containing films (i.e., LDPE/carvacrol and LDPE/(HNTs/carvacrol)) produced by conventional cast extrusion. The (LDPE/carvacrol)/EVOH and (LDPE/[HNTs/carvacrol])/EVOH films demonstrated excellent antimicrobial efficacy against *E. coli* and *Alternaria alternata* in *in vitro* micro-atmosphere assays and against *A. alternata* and *Rhizopus* in cherry tomatoes, used as the food model. The results presented here suggest that sensitive essential oils, such as carvacrol, can be incorporated into plastic polymers constructed of tailored multiple layers, without losing their antimicrobial efficacy.

Keywords: antimicrobial; EVOH; essential oils; carvacrol; halloysite nanotubes; multilayered films

1. Introduction

Polymeric materials with antimicrobial activity have gained significant interest over the past decade. These polymers hold immense potential in combating bacterial and fungal contaminations, including those involving pathogenic microorganisms [1–3]. Among these polymeric systems, biocide-releasing polymers, in which molecules or nanoparticles with biocidical activity are released from the polymer matrix, have been increasingly studied [4,5]. However, potential health and safety risks associated with the release of synthetic antimicrobials (e.g., antibiotics) and nanoparticles could limit the application of these materials [2]. Essential oils (EOs), which are natural and highly-effective antimicrobials against both bacteria and fungi [6], are categorized as GRAS (generally recognized as safe) by the Food and Drug Administration (FDA). Numerous studies have highlighted the immense potential of these antimicrobial compounds in food packaging, pharmaceuticals and hygiene [7–10]. The incorporation of EOs into polymers offers significant advantages in terms of their broad spectrum activity, safety, and as they can be released as a vapor, their antimicrobial function does not require direct contact with the target microorganism [10–14]. Yet, the integration of these sensitive compounds with commodity polymers using conventional manufacturing techniques is challenging, due to their loss during high-temperature processing and diminished antimicrobial efficacy [15–20].

In our recent studies, we have demonstrated that Halloysite nanotubes (HNTs), which are naturally-occurring clays with a characteristic tubular structure and chemical composition similar to kaolin [21,22], can serve as active carriers for carvacrol or its synergistic mixture with thymol [23,24]. Entrapment of these model EOs in HNTs was demonstrated to enhance the thermal stability of these volatile compounds and allowing their high-temperature melt compounding (up to 250 °C) with various polymers, including low-density polyethylene (LDPE) [23,24], polyamide [25] and polypropylene [26]. The resulting polymer nanocomposite films exhibit high EOs content, a sustained release profile of the volatile compounds, and excellent antimicrobial properties, both in synthetic media and in complex food systems [23–26]. The capability of processing EOs at high temperatures opens new opportunities to produce more sophisticated materials with superior properties and advanced functionalities. It should be noted that several other studies have reported on polymeric films containing EOs encapsulated in HNTs [27–31]; however, the majority of these films were prepared by solution casting.

In this study, we investigate the feasibility of fabricating complex antimicrobial EOs-containing multilayered films by layer-multiplication coextrusion [32]. In the latter technique, coextrusion through a series of layer multiplying dies allows for producing sophisticated films containing numerous layers with a tailored structure and thickness from the micro- to the nanoscale [32,33]. Herein, we produce multilayered films (number of layers varies from 9 to 65) comprising of alternating layers of carvacrol-containing LDPE and ethylene vinyl alcohol copolymer (EVOH) by layer-multiplication coextrusion. EVOH copolymers are well known for their outstanding oxygen and odor barrier properties [34,35], including organic compounds such as EOs [36]. However, due to its polar nature, EVOH tends to adsorb moisture which in turn impedes with its barrier properties. This moisture-dependent behavior was exploited for controlling the release of EOs from EVOH coating on different polymeric substrates [37–39].

First, carvacrol, a model EO, was loaded into HNTs in a pre-compounding step. Next, the resulting HNTs/carvacrol hybrids were melt-compounded with LDPE, followed by a high-temperature multilayer coextrusion process in which the LDPE/(HNTs/carvacrol) compound and EVOH were coextruded to produce layered (LDPE/[HNTs/carvacrol])/EVOH films. The detailed nanostructure and composition (in terms of carvacrol content) of the films were characterized and correlated to their barrier properties, carvacrol release rate, as well as antibacterial and antifungal activity, both *in vitro* and in real food systems.

2. Materials and Methods

2.1. Materials

Halloysite Nanotubes (HNTs) were supplied by NaturalNano (Rochester, NY, USA) and dried at 150 °C for 3 h prior to use. Low-density polyethylene (LDPE), Ipethene 320, is supplied by Carmel Olefins Ltd. (Haifa, Israel) with melt flow rate of 2 g/10 min. Carvacrol (1-methyl-4-(1-methylethylidene) cyclohexene, >97%, CAS 586-62-9), Yeast extract and Tryptone for Lysogeny broth (LB) medium, NaCl, Bacto agar, Nutrient Broth (NB) medium, Potato Dextrose Agar (PDA) are purchased from Sigma Aldrich (Rehovot, Israel). NB bacto-agar is purchased from Becton Dickinson (Sparks, MD, USA).

Ethylene vinyl alcohol (EVAL E105B), (EVOH) containing 44 mole percent of ethylene was purchased from Kuraray, Tokyo Japan. The properties reported by the manufacturer of the EVOH are a density of 1.11 g cm^{-3}, a melting temperature of 165 °C, a glass transition temperature of 53 °C, and an oxygen transmission rate at 20 °C and a 65% relative humidity of 0.03 cc·mm/(m^2·day).

2.2. Preparation of HNTs Loaded with Carvacrol

HNTs were dried at 150 °C for 3 h prior to use to remove adsorbed moisture. HNTs/carvacrol hybrids were prepared by shear mixing of carvacrol with HNTs followed by ultrasonication (Vibra cell VCX 750 instrument, Sonics & Materials Inc., Newtown, CT, USA) at a constant amplitude of 40% for 20 min on ice [23–26]. NOTATION: This procedure should be performed in a fume hood to minimize inhalation of HNTs [40].

2.3. Preparation of Films

The forced-assembly multilayer coextrusion process was utilized to fabricate several different multilayer films having alternating layers of LDPE/EVOH, (LDPE/carvacrol)/EVOH, (LDPE/HNTs)/EVOH and (LDPE/[HNTs/carvacrol])/EVOH. This process, illustrated in Figure 1, comprises two single screw extruders, two melt pumps, a feed-block and a series of multipliers. During this process two coextruded polymer melts form three vertical layers within the feed-block. This three-layer melt structure then proceeds through a series of multipliers. Within each multiplier the melt is sectioned vertically and reoriented to double the number of layers as shown in Figure 1. The coextrusion temperature was selected on the rheological compatibility of the LDPE and the EVOH. The rheological properties of the polymer melts were initially determined as a function of temperature by using a Kayeness Galaxy 1 melt flow indexer (MFI) at a shear rate of 10 s^{-1} in order to simulate the flow conditions during multilayer coextrusion. This analysis identified 200 °C as the optimum multilayer coextrusion temperature. Films having 9, 17, 33 and 65 layers were produced with LDPE as the outer layers. The film thickness was maintained at a constant 100 µm (note that the actual film thickness was found to vary across the machine direction from 95 to 105 µm). The composition ratio between LDPE and EVOH within each film was kept as 80% to 20% as shown in Table 1. For comparison, LDPE was melt compounded at 150 °C with, pure carvacrol and HNTs/carvacrol hybrids, as shown in Table 1, using a 25-mm twin-screw extruder (Berstorff, Munich, Germany) with L/D ratio of 25:1 at a screw speed of 300 rpm and feeding rate of 5 kg h^{-1}. Following the melt compounding process, ~100 µm thick films were prepared by cast extrusion using a 45-mm screw diameter extruder (Dr. Collin, Ebersberg, Germany) at 150 °C.

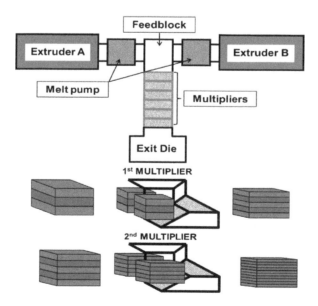

Figure 1. Schematic illustration of the coextrusion and multiplication line used for the fabrication of the multilayered films.

Table 1. Composition of the studied multilayered films. The thickness of all films was maintained constant at 100 μm.

System	LDPE/EVOH Ratio	No. of Layers	Carvacrol Content * (w/w %)	HNTs Content (w/w %)
LDPE/EVOH	80/20	9, 17, 33, 65	0	0
(LDPE/carvacrol)/EVOH	80/20	9, 17, 33, 65	6	0
(LDPE/HNTs)/EVOH	80/20	9, 17, 33, 65	0	3
(LDPE/[HNTs/carvacrol])/EVOH	80/20	9, 17, 33, 65	6	3
Neat LDPE	N/A	1	0	0
LDPE/carvacrol	N/A	1	4	0
LDPE/(HNTs/carvacrol)	N/A	1	4	2

* Refers to initial carvacrol content in LDPE.

2.4. Characterization of Films

2.4.1. High-Resolution Scanning Electron Microscopy

For cross-sectional high-resolution scanning electron microscope (HR-SEM) imaging, a polymer film was embedded in epoxy (Agar 100 resin, Agar Scientific, Stansted, UK) cured at 60 °C for 24 h. The cured epoxy block was then sectioned (along the transverse direction) at −120 °C in liquid nitrogen by a diamond knife mounted on Reichert Ultracut Ultra-microtome. The morphology of the cross-sectioned films was studied using a Carl Zeiss Ultra Plus instrument operated at 1 keV accelerating voltage.

2.4.2. Thermal Gravimetric Analysis (TGA)

The post-processing content of carvacrol in the films was investigated by thermal gravimetric analysis (TGA) using a TGA-Q5000 system (TA Instruments, Newcastle, DE, USA). The films were heated under nitrogen atmosphere from room temperature to 600 °C at a heating rate of 20 °C min^{-1}. The results were analyzed using Universal Analysis 200 version 4.5A build 4.5.0.5 software. The measurements were performed in triplicates, at least. The mass loss at 225 °C was attributed to the total volatile content. The carvacrol content within films was calculated by subtracting total volatile content within corresponding reference films (attributed to moisture) from the total volatile content within carvacrol-containing films.

2.4.3. Carvacrol Release Studies

The accelerated release of carvacrol from the films was studied by isothermal gravimetric analysis using a TGA-Q5000 system at a constant temperature of 60 °C under nitrogen atmosphere, until a constant mass was attained.

Several methods have been reported to characterize the diffusion of active agents and other additives in polymers [41–45]. In this study, we express the diffusion coefficient from the initial linear slope of the fractional mass release ratio vs. $t^{1/2}$ according to Equation (1) [41,45].

$$\frac{m_t}{m_\infty} = 4\left(\frac{Dt}{\pi l^2}\right)^{1/2} \tag{1}$$

where m_t and m_∞ are the amounts of additive (carvacrol) released from the film at time t and at equilibrium $t = \infty$, respectively, D (m^2 s^{-1}) is the diffusion coefficient and l is the overall film thickness. The thickness values of the specimens were determined with an accuracy of ±1 um using a hand-held micrometer (Hahn & Kolb, Stuttgart, Germany). Five readings were taken for each sample and the results were analyzed for their statistical significance using ANOVA with significance reported at the 0.05 probability level.

2.4.4. Oxygen Transmission Rate

Oxygen flux rates $J(t)$ for the extruded multilayer films comprising LDPE/EVOH, (LDPE/HNTs)/EVOH, (LDPE/carvacrol)/EVOH, (LDPE/[HNTs/carvacrol])/EVOH and the control neat LDPE film were all determined using a MOCON (Minneapolis, MN, USA) OxTran 2/20 unit. The measurements were carried out at 1 atm pressure at 23 °C (+0.1) and a 65% relative humidity. The instrument was calibrated at 23 °C using a NIST-certified Mylar film with known oxygen transport characteristics. The oxygen transmission rate (OTR) value for each film was calculated according to Equation (2) from steady-state flux, J, rates,

$$OTR = J\frac{l}{\Delta p} \tag{2}$$

where, l is the overall film thickness and Δp is partial pressure difference of the oxygen across the film. The average reported $P(O_2)$ values were taken from at least two samples. The OTR values are reported as $[cc\cdot mm/(m^2\cdot day)]$.

2.4.5. Antimicrobial In Vitro Assays

Micro-atmosphere diffusion antimicrobial assays: The antibacterial activity of the films was evaluated by measuring the inhibition of *Escherichia coli* (*E. coli*, ATCC 8739) growth by a modified micro-atmosphere diffusion method on LB agar [46]. *E. coli* was maintained on polystyrene beads at −80 °C and a bacterial culture was prepared by incubating one polystyrene bead in 5 mL of LB medium for 16 h at 37 °C under shaking (250 rpm). Subsequently, the culture was diluted with 0.85% w/w NaCl in distilled water to obtain a bacterial stock solution at a concentration of 10^3 colony forming units (CFU) mL^{-1}. Petri dishes containing LB agar were seeded with 0.1 mL of 10^3 CFU mL^{-1} *E. coli* stock culture, and a film sample (with an area of ~36 cm²) was attached using a double-sided tape to the center of the Petri dish lid, assuring no direct contact between the film and the agar. The plates were tightly sealed with Parafilm® and incubated for 12 h at 37 °C. The antibacterial potency of the films was estimated by measuring the observed inhibition zone below the films. All measurements were performed in triplicates Neat LDPE, LDPE/EVOH and (LDPE/HNTs)/EVOH films were used as reference materials.

Micro-atmosphere diffusion antifungal assay: *Alternaria alternata* (*A. alternata*; source: tomato) a phytopathogenic, clinical and food spoilage fungi was cultured on 1% potato dextrose agar (PDA: 10 g L^{-1}; Bacto-agar: 15 g L^{-1} in deionized water) at 25 °C in the dark. Film efficacy was tested as follows: agar plugs (diameter = 6 mm) were removed from the edges of a 5 days old growing colony with a cork-borer. The plugs were placed at the center of a 9-cm Petri dish onto 1% PDA. Then, a film sample (6 × 6 cm; 36 cm²) was attached with a double-sided masking-tape to the center of the Petri dish lid (see previous section). The plates were tightly sealed with Parafilm™ and incubated inverted for 5 days at 25 °C in the dark. Neat LDPE, LDPE/EVOH and (LDPE/HNTs)/EVOH films were used as references. Following 5 days incubation, the diameter of the growing colonies was recorded, subtracting the diameter of the mycelial agar plug. Growth rate reduction was calculated as per-cent of Neat LDPE control.

Micro-atmosphere diffusion antifungal in vivo bioassay: The efficacy of the films to extend the shelf life of products was examined using cherry tomato as a model. Fresh cultures of the decay causing molds *A. alternata* and *Rhizopus* spp. (source: tomato) were prepared by transferring mycelial fragments onto 1% PDA in 9-cm Petri dish and incubating inverted in the dark for 5 days at 25 °C.

Fresh cherry tomatoes were purchased at the local grocery store. The tomatoes were surface sanitized with by thoroughly washing in tap water to remove dust followed by immersion in 10% bleach for 10 min. Then the tomatoes were rewashed with tap water and let dry on clean paper towels on the bench. Inoculation was carried out by puncturing the tomato with a sanitized scalpel, creating a 1 mm long × 1 mm deep wound. Mycelial plugs (1 × 1 mm) of *A. alternata* and of *Rhizopus* spp. (each tested separately) were placed directly on the fresh wound. The bioassays were carried out in 6-wells plate. Film samples were situated at the bottom (2 × 2 cm) and circumference (10 × 1 cm) of each well. An additional film piece (2 × 2 cm) was attached to a 'SealPlate' cover (Excel Scientific Inc., Victorville, CA, USA), which was used to seal the entire plate at the top, creating a packaging simulation. The test system was incubated at 23 °C in the dark for 4 days. Each film sample was tested in triplicates and fungal growth was quantified as a binomial outcome (growth vs. no growth).

3. Results

3.1. Films Preparation and Morphology

The production of the nanocomposite multilayered films consists of several processing steps. First, carvacrol was loaded onto dry HNTs by shear mixing and ultrasonication, producing HNTs/carvacrol hybrids [23,25,26]. The latter were melt compounded with LDPE in a single screw extruder and LDPE/(HNTs/carvacrol) pellets were obtained. Finally, multilayered nanocomposite films were prepared by forced-assembly layer-multiplying coextrusion with alternating layers of the LDPE/(HNTs/carvacrol) and EVOH, termed as (LDPE/[HNTs/carvacrol])/EVOH films. The films were processed through 2, 3, 4 and 5 multipliers in a multiplication extrusion system, depicted in Figure 1, resulting in films with 9, 17, 33 and 65 layers. The total film thickness was maintained constant at 100 μm, while the weight ratio of LDPE/(HNTs/carvacrol) to EVOH was 80 to 20. Multilayered films of LDPE/EVOH, (LDPE/HNTs)/EVOH and (LDPE/carvacrol)/EVOH were produced using the same procedure. For comparison, single-layered neat LDPE, LDPE/carvacrol and LDPE/(HNTs/carvacrol) films were also produced by conventional blown extrusion. All carvacrol-containing films were transparent and continuous, with a slight yellow tint due to carvacrol presence (see Figure S1 in Supplementary Materials).

The films' structure and morphology were studied by HR-SEM. Figure 2 presents characteristic micrographs of the cross-sectioned multilayered films. The films exhibit intact and continuous alternating layers throughout their cross-section. The calculated periodicity of the 100-μm thick 65-layer LDPE/EVOH 80/20 films is approximately 1.5 μm, consisting of alternating layers of ~1.3 μm LDPE and ~0.2 μm EVOH. Nevertheless, our microscopy studies reveal that the thickness of the individual layers is not consistent throughout the film's cross-section, see Figure 2b,e. This behavior was also observed for all studied films, regardless of the number of layers. The HNTs are observed to be well dispersed in the LDPE matrix, and aggregates or percolating networks are not detected (see Figure 2d,e). In Figure 2f the individual HNTs are clearly observed to protrude from the film's surface, as they are oriented in the machine direction. These results coincide well with our previous studies [23,25,26]. In general, HNTs are reported to be easily dispersed in many polymers owing to their weak cohesion, lacking strong primary interactions, mainly relying on secondary hydrogen bonds or van der Waals forces [47]. Some of the HNTs are observed to penetrate into the thin EVOH layers, however, it is impossible to determine whether it occurred during film production or was introduced during film cryo-cross sectioning.

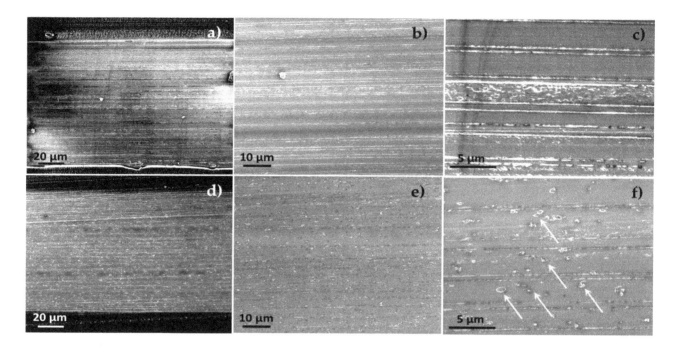

Figure 2. HR-SEM images of cryo-fractured cross-sections of (a–c) a 65-layer (low-density polyethylene (LDPE)/carvacrol)/ethylene vinyl alcohol copolymer (EVOH) film; and (d–f) a 65-layer (LDPE/[halloysite nanotubes (HNTs)/carvacrol])/EVOH film, at a low magnification demonstrating the entire film thickness, medium and higher magnifications. HNTs appear to be finely dispersed in the LDPE layer, marked by arrows for clarity. Note that the films were cross-sectioned in the transverse direction.

3.2. Thermal Gravimetric Analysis (TGA)

As the multiplication extrusion technique process involves high exerted temperatures and long processing time (10–15 min), it may not be suitable for the processing of blends containing sensitive and bioactive ingredients. Specifically, highly volatile compounds such as carvacrol. Therefore, carvacrol retention in the multilayer films is highly important as it will determine their resulting antimicrobial functionality. The pre-processing carvacrol content in the films was 4.8% w/w, while the total carvacrol content in the films (i.e., post melt-compounding and film production) was determined by TGA and the results are summarized in Table 2. For most films, the carvacrol concentration is found to vary between 3.1–3.4% w/w, indicating that >60–70% of the initial carvacrol content is retained during the high-temperature processing steps. It should be noted that in this case, the entrapment of carvacrol within HNTs (prior to melt compounding) was not found to significantly affect its residual content within the films. Our previous studies have demonstrated that carvacrol loading into HNTs was crucial to achieve high EO content in LDPE (~70%) and prolonged antimicrobial performance [23,24]. It should be noted that in these studies the LDPE processing temperature was only 140 °C and films were prepared by compression molding and blown extrusion.

We suggest that the unique layered microstructure of the films, incorporating EVOH, which is known for its outstanding low permeability to organic vapors (including flavors and aroma compounds) [48], delays the loss of carvacrol during the extended high-temperature processing and results in a high final carvacrol content within film [36]. Thus, our results demonstrate that carvacrol can withstand the harsh conditions (high temperature and shearing) applied during the layer-multiplying coextrusion forced-assembly, resulting in uniform and transparent carvacrol-containing multilayered LDPE/EVOH films.

Table 2. Carvacrol content in the different films as measured by Thermal Gravimetric Analysis (TGA) (the number in parenthesis indicate the standard deviation obtained from at least three independent experiments).

Polymer Films	No. of Layers	Pre-Processing Content of Carvacrol (% *w/w*)	Post-Processing Content of Carvacrol by TGA (% *w/w*)
(LDPE/carvacrol)/EVOH	9	4.8	3.2 (0.1)
(LDPE/carvacrol)/EVOH	17	4.8	3.1 (0.1)
(LDPE/carvacrol)/EVOH	33	4.8	3.4 (0.1)
(LDPE/carvacrol)/EVOH	65	4.8	2.9 (0.2)
(LDPE/[HNTs/carvacrol])/EVOH	9	4.8	3.2 (0.1)
(LDPE/[HNTs/carvacrol])/EVOH	17	4.8	3.4 (0.2)
(LDPE/[HNTs/carvacrol])/EVOH	33	4.8	3.2 (0.1)
(LDPE/[HNTs/carvacrol])/EVOH	65	4.8	3.3 (0.1)

3.3. Carvacrol Release Studies

The release kinetics of the volatile antimicrobial agent from the film is crucial for determining its antimicrobial performance and potential applicability as a packaging material. Moreover, the ability to control the release kinetics is desired for tailoring the material's properties for a specific application or condition [49]. TGA was used to characterize the carvacrol release from the different films by monitoring their weight loss over time at a constant temperature (60 °C) [26,50], and Table 3 presents the calculated effective diffusion coefficient values. The highest value of effective carvacrol diffusivity was obtained for LDPE/carvacrol films, while the addition of HNTs slows down the out-diffusion of carvacrol from the LDPE/(HNTs/carvacrol) films. This effect, was already described in our earlier work and is ascribed to the effective role of HNTs as nano-carrier for carvacrol in LDPE systems [23]. Incorporation of EVOH in the layered films has significantly lowered the effective carvacrol's diffusivity by more than 2 orders of magnitude in comparison to LDPE/carvacrol and LDPE/(HNTs/carvacrol) films. This profound effect may be attributed to the high-barrier EVOH layered structure of the films. EVOH is well-known for its outstanding barrier properties and was shown to profoundly reduce fragrance and aroma migration [36]. The number of layers was not found to affect the effective carvacrol's diffusivity, while the addition of HNTs induced a slight increase in diffusivity. The latter may be ascribed to the penetration of HNTs (which are 0.2 to 2 μm length) into the EVOH layers, as was observed in Figure 2, disrupting the layer integrity and enhancing carvacrol diffusion. It should be noted that in order to eliminate the plasticizing effect of water on the EVOH properties, carvacrol diffusivity was measured in dry conditions. Previous studies revealed that humidity significantly facilitates diffusion of carvacrol in EVOH systems [39,51,52].

Table 3. Calculated effective diffusivity values for carvacrol from various films at 60 °C. Determined by fitting the mathematical model for short times diffusion-limited desorption from a polymer film surface to the release profile of carvacrol from the films measured by TGA (the number in parenthesis indicate the standard deviation obtained from at least three independent experiments).

Polymer Film	No of Layers	Carvacrol Effective Diffusivity $\times 10^{14}$, m^2 s^{-1}
(LDPE/carvacrol)/EVOH	9	1.11 [a] (0.19)
(LDPE/carvacrol)/EVOH	17	1.57 [b] (0.23)
(LDPE/carvacrol)/EVOH	33	2.22 [c] (0.37)
(LDPE/carvacrol)/EVOH	65	1.35 [a,b] (0.22)
(LDPE/[HNTs/carvacrol])/EVOH	9	2.45 [e] (0.08)
(LDPE/[HNTs/carvacrol])/EVOH	17	2.39 [e,d] (0.33)
(LDPE/[HNTs/carvacrol])/EVOH	33	2.54 [e,d] (0.39)
(LDPE/[HNTs/carvacrol])/EVOH	65	2.76 [d] (0.02)
LDPE/carvacrol	1	418 [f] (72)
LDPE/(HNTs/carvacrol)	1	285 [g] (24)

[a,b,c,d,e,f,g]—note that values with different superscripts are statistically different.

3.4. Oxygen Transmission Rate

The oxygen transmission rate (OTR) values of the multilayered films and the reference single-layers were measured and the results are summarized in Table 4. The OTR values for the carvacrol-containing multilayered LDPE/EVOH films are at least two orders of magnitude lower in comparison to those of LDPE/carvacrol and LDPE/(HNTs/carvacrol) films, demonstrating the role of EVOH as a barrier layer. In general, all carvacrol-containing films exhibit higher OTR values than their corresponding references. This effect may be attributed to the plasticizing effect of carvacrol on the polymer matrix [26,53], In the multilayered films, diffusion of carvacrol from LDPE layers to the EVOH layers, where it plasticized the EVOH matrix triggering increase in OTR.

Table 4. Oxygen transmission rate $(OTR)(O_2)$ [cc·mm/(m^2·day)], of the films (the number in parenthesis indicate the standard deviation obtained from at least three independent experiments).

Systems	No. of Layers			
	9	17	33	65
LDPE/EVOH	0.26 (0.03)	0.29 (0.02)	0.35 (0.02)	0.41 (0.02)
(LDPE/carvacrol)/EVOH	1.49 (0.04)	1.67 (0.04)	1.94 (0.04)	2.14 (0.06)
(LDPE/HNTs)/EVOH	0.26 (0.02)	0.28 (0.01)	0.30 (0.02)	0.33 (0.02)
(LDPE/[HNTs/carvacrol])/EVOH	1.36 (0.19)	2.31 (0.06)	2.52 (0.09)	3.24 (0.11)
LDPE/carvacrol	237.0 (6.5)			
LDPE/(HNTs/carvacrol)	254.2 (7.0)			
Neat LDPE	181.1 (5.1)			

3.5. Antimicrobial Assays

To evaluate the biofunctionality of the carvacrol in the films in terms of its antimicrobial activity, we studied the effect of the carvacrol-containing films on the growth of common bacterial and fungal microorganisms. The Gram-negative *Escherichia coli* (*E. coli*) was chosen as a model bacterial species as some of its strains are widely-spread foodborne pathogens. The fungus used was *Alternaria alternate* (*A. alternate*), a phytopathogen with clinical and food spoilage relevance [54]. Figure 3 depicts characteristic images of Petri dishes after the exposure of the two model microorganisms to the different films. All carvacrol-containing films were found to exhibit full inhibition of *E. coli* growth. Figure 3a shows a colony-free dish after 16-h incubation with the 9-layer (LDPE/[HNTs/carvacrol])/EVOH film, while for films without carvacrol the development of *E. coli* colonies was unhindered (see Figure 3b). All carvacrol-containing films also display a strong antifungal activity, as presented in Figure 3 and Table 5. The films completely arrest hyphal growth and sporulation of *A. alternata* as can be observed in Figure 3c. Whereas, for all carvacrol-free films no effect on fungal development is observed (see Table 5 and Figure 3d).

So far, all carvacrol-containing films have demonstrated excellent *in vitro* antimicrobial activity in micro-atmosphere diffusion assay against *E. coli* and *A. alternata* on synthetic LB and PDA agar, respectively. However, EOs often exhibit inferior antimicrobial performance in real foods comparing to antimicrobial assays *in vitro*. The latter is ascribed to greater availability of nutrients in real foods comparing to synthetic growth media, which allows microorganisms to repair damage faster, and to possible interactions of EOs with various food components (e.g., fats, proteins and sugars) and other extrinsic factors (such as temperature, pH etc.) [55]. Thus, assessing antimicrobial activity of the films in real food systems is important for assessing their potential application of these films as antimicrobial packaging. Herein, the antifungal efficacy of the films was investigated using cherry tomatoes as a relevant model and *A. alternata* and *Rhizopus* spp. as model microorganisms for postharvest decay causing molds. Figure 4a,c show the outstanding antifungal activity of a 9-layer (LDPE/[HNTs/carvacrol])/EVOH film; a complete fungal growth inhibition is evident by the lack of mycelial growth and sporulation on cherry tomatoes. Table 5 summarize the *in vivo* efficacy of the studied films and the results demonstrate that carvacrol-containing films exhibit a complete fungal growth inhibition, while the reference films (no carvacrol) showed insignificant effect (see Figure 4b,d).

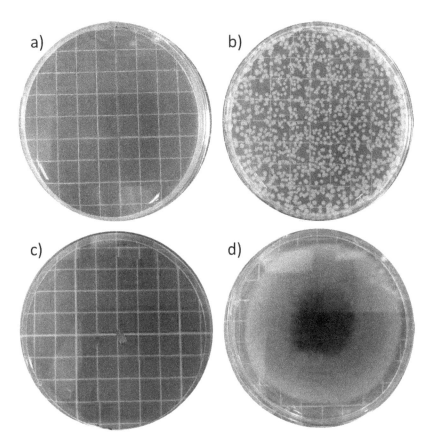

Figure 3. Antimicrobial and antifungal effects of the carvacrol-containing multilayered films exhibited in the micro-atmosphere diffusion *in vitro* assays, i.e., without direct contact between the studied films and the microbial cultures. (**a**) full inhibition of *E. coli* growth after incubation with a 9-layer (LDPE/[HNTs/carvacrol])/EVOH film; (**b**) unhindered *E. coli* growth after incubation with a 9-layer LDPE/EVOH film; (**c**) full inhibition of *A. alternata* development after incubation with a 9-layer (LDPE/[HNTs/carvacrol])/EVOH film; and (**d**) unhindered *A. alternata* development after incubation with a 9-layer LDPE/EVOH film. The *E. coli* were incubated with the films for 16 h at 37 °C and the *A. alternata* for 5 days at 25 °C in the dark. The results presented in (**a,c**) are characteristic for all carvacrol-containing films. The results presented in (**b,d**) are characteristic for all carvacrol-free films. All photographs were taken after the films were removed.

Table 5. Growth reduction of *A. alternata* and *Rhizopus spp.* exposed to multilayered films. Studies were performed *in vitro* and using cherry tomatoes as a food model.

Polymer Composition	No. of Layers	*A. alternata In Vitro* Growth Reduction (%) [x]	*A. alternata* Growth on Cherry Tomato [y]	*Rhisopus* spp. Growth on Cherry Tomato [y]
LDPE control 100%	1	0 ± 0	3/3	3/3
LDPE/EVOH control	9	1.1 ± 2.3	3/3	3/3
LDPE/EVOH control	17	2.9 ± 6.8	3/3	3/3
LDPE/EVOH control	33	7.2 ± 1.2	3/3	3/3
LDPE/EVOH control	65	7.9 ± 3.1	3/3	3/3
LDPE/carvacrol/EVOH	9	100 ± 0.0	0/3	0/3
LDPE/carvacrol/EVOH	17	100 ± 0.0	0/3	0/3
LDPE/carvacrol/EVOH	33	100 ± 0.0	0/3	0/3
LDPE/carvacrol/EVOH	65	100 ± 0.0	0/3	0/3
LDPE/[HNT/carvacrol]/EVOH	9	100 ± 0.0	0/3	0/3
LDPE/[HNT/carvacrol]/EVOH	17	100 ± 0.0	0/3	0/3
LDPE/[HNT/carvacrol]/EVOH	33	100 ± 0.0	0/3	0/3
LDPE/[HNT/carvacrol]/EVOH	65	100 ± 0.0	0/3	0/3

[x] Values are mean ± standard error of the mean. [y] Frequency of fungal growth on cherry tomato. All tests were carried out with 3 replications per film.

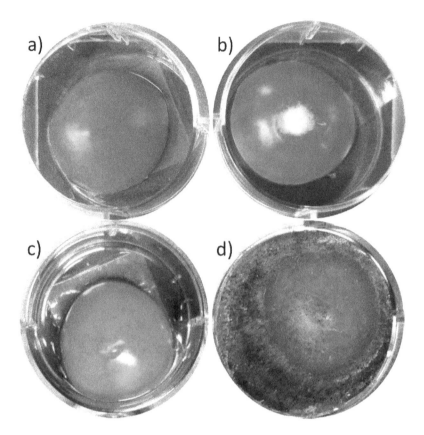

Figure 4. Antifungal effects of the carvacrol-containing films exhibited in the micro-atmosphere bioassays, i.e. without direct contact between the studied films and the microbial cultures using cherry tomatoes as the food model. (**a**) full inhibition of *A. alternata* development after incubation with a 9-layer (LDPE/[HNTs/carvacrol])/EVOH film; (**b**) *A. alternata* development after incubation with a 9-layer LDPE/EVOH film; (**c**) full inhibition of *Rhizopus* spp. development after incubation with a 9-layer (LDPE/[HNTs/carvacrol])/EVOH film; (**d**) *Rhizopus* spp. development after incubation with a 9-layer LDPE/EVOH film. The cherries were incubated with the films at 23 °C in the dark for 4 days. The results presented in (**a,c**) are characteristic for all carvacrol-containing films. The results presented in (**b,d**) are characteristic for all carvacrol-free films. All photographs were taken after the top and circumference films were removed.

4. Conclusions

The present work shows that multilayered LDPE/EVOH films produced by forced assembly coextrusion process exhibit high carvacrol content despite the harsh processing conditions (temperature of 200 °C and long processing time). The post-processing carvacrol content was similar in all films and was not affected by the number of layers or HNTs presence at the studied concentrations. This is ascribed to the unique layered structure of the films and the barrier properties of EVOH, which makes the entrapment of carvacrol within HNTs (prior to melt compounding) not necessary for obtaining high-quality films with high carvacrol content. The multilayered carvacrol-containing films exhibit OTR and carvacrol diffusivity values that are >2 orders of magnitude lower in comparison to single-layered films (LDPE/carvacrol and LDPE/(HNTs/carvacrol)) produced by conventional cast extrusion. The (LDPE/carvacrol)/EVOH and (LDPE/[HNTs/carvacrol])/EVOH films demonstrated excellent antimicrobial efficacy against *E. coli* and *A. alternata* in *in vitro* micro-atmosphere assays and against *A. alternata* and *Rhizopus* spp. in in vivo micro-atmosphere assays using cherry tomatoes as food model. The results presented here suggest that sensitive essential oils, such as carvacrol, can be incorporated into plastic polymers constructed of tailored multiple layers without losing their antimicrobial efficacy. This is a novel technology provides the ability to customize packaging for various food products that are affected by bacteria, such as meat and fish, or by molds such as cheese, bread and fresh produce. Hence, providing food producers, farmers, and cold chain logistics personnel a versatile tool-box to overcome public health issues in situations of product abuse throughout the supply chain.

Author Contributions: E.S., N.M-I. and E.B. conceived the research concepts. M.K. has prepared HNTs/carvacrol hybrids and carried out subsequent melt compounding. C.Z. and A.O. produced the multilayered films and characterized their barrier properties. M.K. and O.S. conducted TGA studies. N.N. has designed and performed the antimicrobial studies. O.S. and N.M-I. carried out the electron microscopy studies including sample preparation. M.K. and E.S. analyzed the data and wrote the paper.

Acknowledgments: This research was partially supported by the European Union's Horizon 2020 research and innovation programme under grant agreement No 720815. Any dissemination of results must indicate that it reflects only the author's view and that the Commission is not responsible for any use that may be made of the information it contains.

References

1. Muñoz-Bonilla, A.; Fernández-García, M.; Mu, A. Polymeric materials with antimicrobial activity. *Prog. Polym. Sci.* **2012**, *37*, 281–339. [CrossRef]
2. Santos, M.R.E.; Fonseca, A.C.; Mendonça, P.V.; Branco, R.; Serra, A.C.; Morais, P.V.; Coelho, J.F.J. Recent developments in antimicrobial polymers: A review. *Materials.* **2016**, *9*, 599. [CrossRef] [PubMed]
3. Ravensdale, J.T.; Coorey, R.; Dykes, G.A. Integration of Emerging Biomedical Technologies in Meat Processing to Improve Meat Safety and Quality. *Compr. Rev. Food Sci. Food Saf.* **2018**, *17*, 615–632. [CrossRef]
4. Siedenbiedel, F.; Tiller, J.C. Antimicrobial polymers in solution and on surfaces: Overview and functional principles. *Polymers* **2012**, *4*, 46–71. [CrossRef]
5. Huang, K.-S.; Yang, C.-H.; Huang, S.-L.; Chen, C.-Y.; Lu, Y.-Y.; Lin, Y.-S. Recent Advances in Antimicrobial Polymers: A Mini-Review. *Int. J. Mol. Sci.* **2016**, *17*, 1578. [CrossRef] [PubMed]
6. Puškárová, A.; Bučková, M.; Kraková, L.; Pangallo, D.; Kozics, K. The antibacterial and antifungal activity of six essential oils and their cyto/genotoxicity to human HEL 12469 cells. *Sci. Rep.* **2017**, *7*, 8211. [CrossRef] [PubMed]
7. Calo, J.R.; Crandall, P.G.; O'Bryan, C.A.; Ricke, S.C. Essential oils as antimicrobials in food systems—A review. *Food Control* **2015**, *54*, 111–119. [CrossRef]
8. Hyldgaard, M.; Mygind, T.; Meyer, R.L. Essential oils in food preservation: Mode of action, synergies, and interactions with food matrix components. *Front. Microbiol.* **2012**, *3*, 12. [CrossRef] [PubMed]
9. Bassolé, I.H.N.; Juliani, H.R. Essential oils in combination and their antimicrobial properties. *Molecules* **2012**, *17*, 3989–4006. [CrossRef] [PubMed]
10. Ribeiro-Santos, R.; Andrade, M.; de Melo, N.R.; Sanches-Silva, A. Use of essential oils in active food packaging: recent advances and future trends. *Trends Food Sci. Technol.* **2017**, *61*, 132–140. [CrossRef]
11. Ramos, M.; Jiménez, A.; Garrigós, M.C. *Carvacrol-Based Films: Usage and Potential in In Antimicrobial Food Packaging*; Academic Press: San Diego, CA, USA, 2016; pp. 329–338.
12. Han, J.-W.; Ruiz-Garcia, L.; Qian, J.-P.; Yang, X.-T. Food Packaging: A Comprehensive Review and Future Trends. *Compr. Rev. Food Sci. Food Saf.* **2018**, *17*, 860–877. [CrossRef]
13. Kumar, S.; Sarita; Nehra, M.; Dilbaghi, N.; Tankeshwar, K.; Kim, K.-H. Recent advances and remaining challenges for polymeric nanocomposites in healthcare applications. *Prog. Polym. Sci.* **2018**, *80*, 1–38. [CrossRef]
14. Chouhan, S.; Sharma, K.; Guleria, S. Antimicrobial Activity of Some Essential Oils—Present Status and Future Perspectives. *Medicines* **2017**, *4*, 58. [CrossRef] [PubMed]
15. Suppakul, P.; Miltz, J.; Sonneveld, K.; Bigger, S.W. Characterization of antimicrobial films containing basil extracts. *Packag. Technol. Sci.* **2006**, *19*, 259–268. [CrossRef]
16. Ramos, M.; Jiménez, A.; Peltzer, M.; Garrigós, M.C. Characterization and antimicrobial activity studies of polypropylene films with carvacrol and thymol for active packaging. *J. Food Eng.* **2012**, *109*, 513–519. [CrossRef]

17.   Sung, S.-Y.; Sin, L.T.; Tee, T.-T.; Bee, S.-T.; Rahmat, A.R.; Rahman, W.A.; Tan, A.-C.; Vikhraman, M. Antimicrobial agents for food packaging applications. *Trends Food Sci. Technol.* **2013**, *33*, 110–123. [CrossRef]

18.   Malhotra, B.; Keshwani, A.; Kharkwal, H. Antimicrobial food packaging: Potential and pitfalls. *Front. Microbiol.* **2015**, *6*, 611. [CrossRef] [PubMed]

19.   Valdés, A.; Mellinas, A.C.; Ramos, M.; Burgos, N.; Jiménez, A.; Garrigós, M.C. Use of herbs, spices and their bioactive compounds in active food packaging. *RSC Adv.* **2015**, *5*, 40324–40335. [CrossRef]

20.   Campos-Requena, V.H.; Rivas, B.L.; Pérez, M.A.; Figueroa, C.R.; Figueroa, N.E.; Sanfuentes, E.A. Thermoplastic starch/clay nanocomposites loaded with essential oil constituents as packaging for strawberries—In vivo antimicrobial synergy over Botrytis cinerea. *Postharvest Biol. Technol.* **2017**, *129*, 29–36. [CrossRef]

21.   Santos, A.C.; Ferreira, C.; Veiga, F.; Ribeiro, A.J.; Panchal, A.; Lvov, Y.; Agarwal, A. Halloysite clay nanotubes for life sciences applications: From drug encapsulation to bioscaffold. *Adv. Colloid Interface Sci.* **2018**, *258*, 58–70. [CrossRef] [PubMed]

22.   Yuan, P.; Tan, D.; Annabi-Bergaya, F. Properties and applications of halloysite nanotubes: Recent research advances and future prospects. *Appl. Clay Sci.* **2015**, *112–113*, 75–93. [CrossRef]

23.   Shemesh, R.; Krepker, M.; Natan, M.; Danin-Poleg, Y.; Banin, E.; Kashi, Y.; Nitzan, N.; Vaxman, A.; Segal, E. Novel LDPE/halloysite nanotube films with sustained carvacrol release for broad-spectrum antimicrobial activity. *Rsc Adv* **2015**, *5*, 87108–87117. [CrossRef]

24.   Krepker, M.; Shemesh, R.; Danin Poleg, Y.; Kashi, Y.; Vaxman, A.; Segal, E. Active food packaging films with synergistic antimicrobial activity. *Food Control* **2017**, *76*, 117–126. [CrossRef]

25.   Shemesh, R.; Krepker, M.; Nitzan, N.; Vaxman, A.; Segal, E. Active packaging containing encapsulated carvacrol for control of postharvest decay. *Postharvest Biol. Technol.* **2016**, *118*, 175–182. [CrossRef]

26.   Krepker, M.; Prinz-Setter, O.; Shemesh, R.; Vaxman, A.; Alperstein, D.; Segal, E. Antimicrobial carvacrol-containing polypropylene films: Composition, structure and function. *Polymers* **2018**, *10*, 79. [CrossRef]

27.   Biddeci, G.; Cavallaro, G.; Di Blasi, F.; Lazzara, G.; Massaro, M.; Milioto, S.; Parisi, F.; Riela, S.; Spinelli, G. Halloysite nanotubes loaded with peppermint essential oil as filler for functional biopolymer film. *Carbohydr. Polym.* **2016**, *152*, 548–557. [CrossRef] [PubMed]

28.   Gorrasi, G. Dispersion of halloysite loaded with natural antimicrobials into pectins: Characterization and controlled release analysis. *Carbohydr. Polym.* **2015**, *127*, 47–53. [CrossRef] [PubMed]

29.   Lee, M.H.; Kim, S.Y.; Park, H.J. Effect of halloysite nanoclay on the physical, mechanical, and antioxidant properties of chitosan films incorporated with clove essential oil. *Food Hydrocoll.* **2018**, *84*, 58–67. [CrossRef]

30.   Hendessi, S.; Sevinis, E.B.; Unal, S.; Cebeci, F.C.; Menceloglu, Y.Z.; Unal, H. Antibacterial sustained-release coatings from halloysite nanotubes/waterborne polyurethanes. *Prog. Org. Coat.* **2016**, *101*, 253–261. [CrossRef]

31.   Tenci, M.; Rossi, S.; Aguzzi, C.; Carazo, E.; Sandri, G.; Bonferoni, M.C.; Grisoli, P.; Viseras, C.; Caramella, C.M.; Ferrari, F. Carvacrol/clay hybrids loaded into in situ gelling films. *Int. J. Pharm.* **2017**, *531*, 676–688. [CrossRef] [PubMed]

32.   Ponting, M.; Hiltner, A.; Baer, E. Polymer Nanostructures by Forced Assembly: Process, Structure, and Properties. *Macromol. Symp.* **2010**, *294*, 19–32. [CrossRef]

33.   Ponting, M.; Burt, T.M.; Korley, L.T.J.; Andrews, J.; Hiltner, A.; Baer, E. Gradient multilayer films by forced assembly coextrusion. *Ind. Eng. Chem. Res.* **2010**, *49*, 12111–12118. [CrossRef]

34.   Bhunia, K.; Sablani, S.S.; Tang, J.; Rasco, B. Migration of chemical compounds from packaging polymers during microwave, conventional heat treatment, and storage. *Compr. Rev. Food Sci. Food Saf.* **2013**, *12*, 523–545. [CrossRef]

35.   Yam, K.L. *The Wiley Encyclopedia of Packaging Technology*; Yam, K.L., Ed.; John Wiley & Sons, Inc.: Hoboken, NJ, USA, 2009.

36.   Catalá, R.; Muriel-Galet, V.; Cerisuelo, J.P.; Domínguez, I.; Carballo, G.L.; Hernández-Muñoz, P.; Gavara, R. Antimicrobial Active Packaging Systems Based on EVOH Copolymers. In *Antimicrobial Food Packaging*; Academic Press: San Diego, CA, USA, 2016; pp. 297–303.

37.   Cerisuelo, J.P.; Gavara, R.; Hernández-Muñoz, P. Natural antimicrobial-Containing EVOH coatings on PP and PET films: Functional and active property characterization. *Packag. Technol. Sci.* **2014**, *27*, 901–920. [CrossRef]

38. Muriel-Galet, V.; Cerisuelo, J.P.; López-Carballo, G.; Aucejo, S.; Gavara, R.; Hernández-Muñoz, P. Evaluation of EVOH-coated PP films with oregano essential oil and citral to improve the shelf-life of packaged salad. *Food Control* **2013**, *30*, 137–143. [CrossRef]

39. Cerisuelo, J.P.; Bermúdez, J.M.; Aucejo, S.; Catalá, R.; Gavara, R.; Hernández-Muñoz, P. Describing and modeling the release of an antimicrobial agent from an active PP/EVOH/PP package for salmon. *J. Food Eng.* **2013**, *116*, 352–361. [CrossRef]

40. Koivisto, A.J.; Bluhme, A.B.; Kling, K.I.; Fonseca, A.S.; Redant, E.; Andrade, F.; Hougaard, K.S.; Krepker, M.; Prinz, O.S.; Segal, E.; et al. Occupational exposure during handling and loading of halloysite nanotubes–A case study of counting nanofibers. *NanoImpact* **2018**, *10*, 153–160. [CrossRef]

41. Crank, J. *The Mathematics of Diffusion*, 2nd ed.; Oxford University Press: Oxford, UK, 1975.

42. Cran, M.J.; Rupika, L.A.S.; Sonneveld, K.; Miltz, J.; Bigger, S.W. Release of Naturally Derived Antimicrobial Agents from LDPE Films. *J. Food Sci.* **2010**, *75*, E126–E133. [CrossRef] [PubMed]

43. Suppakul, P.; Sonneveld, K.; Bigger, S.W.; Miltz, J. Loss of AM additives from antimicrobial films during storage. *J. Food Eng.* **2011**, *105*, 270–276. [CrossRef]

44. Lim, L.-T.; Tung, M.A. Vapor Pressure of Allyl Isothiocyanate and Its Transport in PVDC/PVC Copolymer Packaging Film. *J. Food Sci.* **1997**, *62*, 1061–1062. [CrossRef]

45. Miltz, J. Migration of low molecular weight species from packaging materials: Theoretical and practical considerations. In *Food Product-Package Compatibility*; Technomic Publishing Company: Lancaster, PA, USA, 1987; pp. 30–43.

46. Cardiet, G.; Fuzeau, B.; Barreau, C.; Fleurat-Lessard, F. Contact and fumigant toxicity of some essential oil constituents against a grain insect pest Sitophilus oryzae and two fungi, *Aspergillus westerdijkiae* and *Fusarium graminearum*. *J. Pest Sci.* **2012**, *85*, 351–358. [CrossRef]

47. Du, M.; Guo, B.; Cai, X.; Jia, Z.; Liu, M.; Jia, D. Morphology and properties of halloysite nanotubes reinforced polypropylene nanocomposites. *e-Polymers* **2008**, *8*, 130. [CrossRef]

48. Cabedo, L.; Giménez, E.; Lagaron, J.M.; Gavara, R.; Saura, J.J. Development of EVOH-kaolinite nanocomposites. *Polymer* **2004**, *45*, 5233–5238. [CrossRef]

49. Han, J.H. Antimicrobial packaging systems. In *Innovations in Food Packaging*; Elsevier: New York, NY, USA, 2005; pp. 80–107.

50. Ouali, L.; Léon, G.; Normand, V.; Johnsen, H.; Dyrli, A.; Schmid, R.; Benczedi, D. Mechanism of Romascone® release from hydrolyzed vinyl acetate nanoparticles: Thermogravimetric method. *Polym. Adv. Technol.* **2006**, *17*, 45–52. [CrossRef]

51. Cerisuelo, J.P.; Alonso, J.; Aucejo, S.; Gavara, R.; Hernández-Muñoz, P. Modifications induced by the addition of a nanoclay in the functional and active properties of an EVOH film containing carvacrol for food packaging. *J. Membr. Sci.* **2012**, *423–424*, 247–256. [CrossRef]

52. Cerisuelo, J.P.; Muriel-Galet, V.; Bermúdez, J.M.; Aucejo, S.; Catalá, R.; Gavara, R.; Hernández-Muñoz, P. Mathematical model to describe the release of an antimicrobial agent from an active package constituted by carvacrol in a hydrophilic EVOH coating on a PP film. *J. Food Eng.* **2012**, *110*, 26–37. [CrossRef]

53. Persico, P.; Ambrogi, V.; Carfagna, C.; Cerruti, P.; Ferrocino, I.; Mauriello, G. Nanocomposite polymer films containing carvacrol for antimicrobial active packaging. *Polym. Eng. Sci.* **2009**, *49*, 1447–1455. [CrossRef]

54. Pitt, J.I.; Hocking, A.D. Fungi and food spoilage. In *Fungi and Food Spoilage*; Springer: Berlin, Germany, 2009; pp. 1–519.

55. Burt, S. Essential oils: Their antibacterial properties and potential applications in wound—A review. *Microbiol. Int. J. Food* **2004**, *94*, 223–254. [CrossRef] [PubMed]

The Influence of Accelerated UV-A and Q-SUN Irradiation on the Antibacterial Properties of Hydrophobic Coatings Containing *Eucomis comosa* Extract

Małgorzata Mizielińska [1,*], Urszula Kowalska [1], Piotr Salachna [2], Łukasz Łopusiewicz [1] and Michał Jarosz [1]

[1] Center of Bioimmobilisation and Innovative Packaging Materials, Faculty of Food Sciences and Fisheries, West Pomeranian University of Technology Szczecin, Janickiego 35, 71-270 Szczecin, Poland; urszula.kowalska@zut.edu.pl (U.K); lukasz.lopusiewicz@zut.edu.pl (Ł.Ł.); michal.jarosz@zut.edu.pl (M.J.)

[2] Department of Horticulture, West Pomeranian University of Technology, 3 Papieża Pawła VI Str., 71-434 Szczecin, Poland; piotr.salachna@zut.edu.pl

* Correspondence: malgorzata.mizielinska@zut.edu.pl.

Abstract: The purpose of this research was to examine the antimicrobial properties against Gram-positive bacteria, as well as the water vapour characteristic of polylactic acid (PLA) films covered with a methyl–hydroxypropyl–cellulose (MHPC)/cocoa butter carrier containing *Eucomis comosa* extract as an active substance. The second purpose of the study was to evaluate the influence of accelerated UV-A and Q-SUN irradiation (UV-aging) on the antimicrobial properties and the barrier characteristic of the coatings. The results of the study revealed that MHPC/cocoa butter coatings had no influence on the growth of *Staphylococcus aureus*, *Bacillus cereus*, and *Bacillus atrophaeus*. MHPC/cocoa butter coatings containing *E. comosa* extract reduced the number of bacterial strains. MHPC/cocoa butter coatings also decreased the water vapour permeability of PLA. It was shown that accelerated UV-A and Q-SUN irradiations altered the chemical composition of the coatings containing cocoa butter. Despite the alteration of the chemical composition of the layers, the accelerated Q-SUN and UV-A irradiation had no influence on the antimicrobial properties of *E. comosa* extract coatings against *S. aureus* and *B. cereus*. It was found that only Q-SUN irradiation decreased the coating activity with an extract against *B. atrophaeus*, though this was to a small degree.

Keywords: coatings; *Eucomis comosa* extract; antibacterial; antimicrobial properties

1. Introduction

The food packaging industry is seeking to replace synthetic polymers with natural and biodegradable materials that present grease, water vapour, and gas barrier properties. Polylactic acid (PLA) is a compostable and renewable biopolymer that has been commercialized as a good alternative substitute for synthetic polymers [1–6]. It can be produced from the bacterial fermentation of renewable resources, such as sugar beet or corn starch. PLA has been approved as safe by the United States Food and Drug Administration (FDA). Polylactic acid has an advantage, as its physical and mechanical properties can be easily changed and tailored by simply varying its chemical composition (quantities of D- and L-isomer), as well as altering any processing conditions. It is also important that this biopolymer is commercially available, and its price is low [6]. As a consequence, these advantages give PLA the greatest potential for packaging and medical applications.

With respect to barrier behaviour, PLA shows high gas transmission and low water vapour barrier that renders this polymer unsuitable for several food packaging applications in the case of synthetic polymers. To improve barrier characteristics, the surface of the polymer should be modified. A covering of biopolymer with coatings might be a solution [2–7]. For example, paraffin wax emulsions and polyurethane- or styrene-based copolymers are typical hydrophobic sizing agents that are applied in molten form to the surface of biopolymers and offer an improvement in water vapour barrier property. The purpose of the wax is to provide a moisture barrier. However, paraffin wax has relatively poor durability and flexibility as a surface-coating material. The addition of synthetic polymers and modifiers, such as cellulose derivatives, rubber derivatives, vinyl copolymers, polyamides, polyesters, and butadiene–styrene copolymers, compatible with wax, may overcome its defects. Often, special resins or plastic polymers can be added to the wax to improve adhesion and low-temperature performance and to prevent cracking [8]. Commercial products such as Eurocryl 2080 (Cebra Chemie, GmbH, Bramsche, Germany), Exceval HR 3010 (Kuraray, Europe GmBH, Hattersheim am Main, Germany), Ecroprint RA 112 (Michelman, Ecronova Polymer GmbH, Recklinghausen, Germany), Ultralub (Keim Additec Surface, GmbH, Kirchberg, Germany), Aquacer 2650 (Byk, Wesel, Germany) or cocoa butter (which is used in the food industry) are used as hydrophobic carriers to cover packaging materials. This butter is considered the most important cocoa by-product, due to its physical and chemical characteristics, which offer highly valued functional properties in the food industry. The amount of cocoa butter and the fatty acid profiles depend on the growing conditions of the cocoa beans. In cocoa butter, fatty acids are organised as triacylglycerol (TAG), the majority of these TAGs being 2-oleyl glycerides (O) of palmitic (P) and stearic (S) acids (POP, POS, SOS). Due to its hydrophobic properties, a cocoa butter carrier can be used to improve the water vapour barrier properties of PLA [9]. Unfortunately, cocoa butter is not compatible with PLA, which is a polyester. The addition of methyl–hydroxypropyl–cellulose, cellulose derivative, may improve in coating adhesion to the PLA film.

In recent years, a great deal of effort has been devoted not only to barrier properties, but also to the development of coatings with antimicrobial activity that decreases bacterial growth. Attention has been concentrated on the development of coatings offering the highest possible eradication activity in the shortest possible time. For antibacterial coatings, long lifetime is a key requirement in many practical applications when these coatings are deposited on the contact surfaces of substrates, and must simultaneously perform two functions: antibacterial and protective [4–7,10–13]. The contact of active materials with foodstuffs, offering the ability to change composition or the atmosphere around it, is an active packaging system that inhibits or decreases the growth of microorganisms present on the surface of these perishable goods. Coatings with antimicrobial properties may contain an active substance e.g., zinc oxide (ZnO) nanoparticles, essential oils, organic acids, bacteriocins, exopolysaccharides, or plant extracts [4–7,10–17].

Eucomis (L.) L'Hér (Asparagaceae, formerly Hyacinthaceae) is a small genus consisting of bulbous geophytes extensively used in southern African traditional medicine [18–22]. Various plant parts and extract solvents of various *Eucomis* species have been tested for in vitro and in vivo antimicrobial screening [18,22–25]. It has been shown that *Eucomis* extracts inhibited *Bacillus subtilis*, *Escherichia coli*, and *Staphylococcus aureus* [22–25], as well as fungal cells, such as *Candida albicans* [24]. It was proven that *E. comosa* water extracts had a marked influence on the viability of the *S. aureus* strain in a previous study. A medium containing the *E. comosa* water extract caused a 2-log decrease in the number of bacterial cells on average. A decrease in the number of *S. aureus* was dependent on the extract concentration and was confirmed in tests, with the highest results being found in 25% extracts. It was also noted that *E. comosa* water extracts triggered a decrease in the viability of the *B. atrophaeus* strain [25,26].

The coating, which contained cocoa butter as a hydrophobic carrier, could be used to cover PLA films. The cocoa butter was able to increase the water vapour barrier of PLA. In addition, *E. comosa* extract as an active substance could be introduced into an MHPC/cocoa butter carrier to create antimicrobial coating activity. Boxes covered with Methocel™ containing polylysine and boxes containing polyethylene (PE) films covered with MHPC with ZnO nanoparticles were used as packaging material for fresh cod fillets [27]. It was demonstrated that the number of bacterial cells stored in boxes covered with Methocel™ containing polylysine or in the boxes containing the PE films coated with MHPC with ZnO nanoparticles did not go over 10^7 cfu/g. Quite opposite results were obtained for boxes that were not covered with the active coatings (control samples). These results suggested that the active coatings improved the quality of cod fillets after storage. In marine fish stored in refrigerated aerobic conditions, *Pseudomonas* sp. and *Shewanella* spp. (Gram-negative bacteria) were observed as dominant. It was proved that MHPC coatings with ZnO nanoparticles [28] and polylysine [27] were active against Gram-negative bacteria, and they could be chosen as a packaging material to extend the quality and freshness of cod fillets after storage. The *Eucomis* extracts were active against *Bacillus* sp. and *Staphylococcus aureus* [25,26]. The polymer films or boxes covered with active coatings containing *E. comosa* extract could be used as packaging material for vacuum packed raw meat, fish, or cheese. The active coatings could extend the shelf life, quality, and freshness of food products.

In general, an active packaging material should function during storage to inhibit microorganism growth and extend the shelf life of any given food product. This means that coatings should offer sufficient resistance against UV radiation [28]. Ultraviolet (UV) radiation is a section of the non-ionizing region of the electromagnetic spectrum that comprises of approximately 8–9% of total solar radiation. It can lead to a degradation in the physical–mechanical, optical, and antimicrobial properties of materials. Introducing an active substance that is sensitive to UV in a coating carrier can lead to coating inactivation after UV-aging. Introducing an active substance that is resistant to UV in a coating carrier can prevent the inactivation of this coating after UV-aging [28–32].

The aim of this research was to examine the antimicrobial properties against Gram-positive bacteria, as well as the water vapour characteristic of PLA films covered with an MHPC/cocoa butter carrier containing *E. comosa* extract as an active substance. The second aim of the study was to evaluate the influence of accelerated UV-A and Q-SUN irradiation (UV-aging) on the antimicrobial properties and the barrier characteristic of the coatings.

2. Materials and Methods

2.1. Materials

The test microorganisms used in this study were obtained from a collection from the Leibniz Institute DSMZ (Deutsche Sammlung von Mikroorganismen und Zellkulturen, Braunschweig, Germany). The strains were supplied from an American Type Culture Collection (ATCC, Manassas, VA, USA). The organisms used in this study were *S. aureus* DSMZ 346, *B. cereus* ATCC 14579, and *B. atrophaeus* DSM 675 IZT.

Polylactide films, (A4, 20 μm) (CBIMO—Center of Bioimmobilisation and Innovative Packaging Materials, Szczecin, Poland) were used in this research. MHPC (Chempur, Piekary Śląskie, Poland) was used as a coating carrier. *E. comosa* bulbs (Department of Horticulture, Szczecin, Poland) were used to prepare a water extract (as an active substance). To verify the antimicrobial properties of any coatings, agar-agar, TSB, and TSA mediums (Merck, Darmstadt, Germany) were used. All mediums were prepared according to the Merck protocol (each medium was weighed according to the manufacturer's instructions, then suspended in 1000 mL of distilled water and autoclaved at 121 °C for 15 min).

2.2. Extract Preparation

E. comosa dried bulbs were ground to powder and a sample of 5 g was extracted with 50 g of 70% aqueous acetone. The sample was then kept in a sonication water bath for one hour. The temperature of the bath was maintained at 15 °C by adding ice. The acetone extract was concentrated at 40 °C. After the evaporation of the acetone, the sample was filtered through a 0.2 μm filter. The resulting 15 g of extract was used in further analyses.

2.3. Coating Preparation and Antimicrobial Properties Analysis

(1)   MHPC (2 g) was introduced into 48 g of water, this was mixed for 1 h using a magnetic stirrer (Ika, Staufen im Breisgau, Germany) at 40 °C at 1500 rpm. Cocoa butter (10 g) was heated to 40 °C. Then, 40 g of MHPC was mixed with 10 g of cocoa butter and homogenized (1000 rpm) (Heidolph, Sigma-Aldrich, Poznań, Poland). The hydrophobic mixture was used to cover the PLA films to obtain coatings devoid of any active substances.

(2)   *E. comosa* extract (12.5 g) was mixed with 25.5 g of water. Next, 2 g of MHPC was introduced into 38 g of *E. comosa* solution. The mixture was mixed for 1 h using a magnetic stirrer (Ika, Staufen im Breisgau, Germany) at 40 °C at 1000 rpm. Cocoa butter (10 g) was heated to 40 °C and 40 g of MHPC containing *E. comosa* extract was then mixed with the cocoa butter and homogenized (1000 rpm) (Heidolph, Sigma-Aldrich, Poznań, Poland). The hydrophobic mixture was used to cover the PLA films to obtain 25% active substance coatings.

(3)   MHPC (2 g) was introduced into 48 g of water. The mixture was mixed for 1 h using a magnetic stirrer (Ika, Staufen im Breisgau, Germany) at 40 °C at 1500 rpm. The hydrophilic MHPC was used to cover PLA films to obtain coatings devoid of any active substances and hydrophobic substances. The films covered with MHPC were used for the analyses of Fourier transform infrared (FT-IR) and water vapour transmission rate (WVTR).

PLA films were covered using Unicoater 409 (Erichsen, Hemer, Germany) at a temperature of 40 °C with a 40 μm diameter roller. The coatings were dried for 2 h at a temperature of 25 °C. Layers (1.6 g) of MHPC/cocoa butter (1.28 g of MHPC and 0.32 g of cocoa butter) per 1 m^2 of PLA were obtained. PLA films that were not covered were used as control samples (K). PLA films with MHPC/cocoa butter coatings were also used as control samples (MHPC/CB).

The film samples were cut into square shapes (3 cm × 3 cm). The antimicrobial properties of non-covered and covered films were carried out according to the ASTM E 2180-01 standard [33]. As the first step of the experiments, *S. aureus*, *B. cereus*, and *B. atrophaeus* cultures originated from 24 h growth (coming from stock cultures) were prepared. The concentrations of the cultures were standardized to 1.5×10^8 cfu/mL. The concentration of each culture was measured using Cell Density Meter (WPA, Cambridge, UK. CB4 OF J). The agar slurry was prepared by dissolving 0.85 g of NaCl and 0.3 g of agar-agar in 100 mL of deionized water and autoclaved for 15 min at 121 °C and equilibrated at 45 °C (one agar slurry was prepared for each strain). One millilitre of the culture (separately) was placed into the 100 mL of agar slurry. The final concentration of each culture was 1.5×10^6 cfu/mL in molten agar slurry. The square samples of each film (not covered PLA films, covered PLA film, irradiated, and not irradiated) were introduced (separately) into the sterile Petri dishes with a diameter of 55 mm. Inoculated agar slurry (1.0 mL) was pipetted onto each square sample. The samples were incubated 24 h at 30 °C with relative humidity at 90%. After incubation, the samples were aseptically removed from the Petri dishes and introduced into the 100 mL of TSB. The samples were sonicated 1 min in the Bag Mixer® CC (Interscience, St Nom la Brètech, France). The sonication facilitated the complete release the agar slurry from the samples. Then serial dilutions of the initial inoculum were performed. Each dilution was spread into the TSA and incubated at 30 °C for 48 h. The results were presented as an average value with standard deviations.

2.4. Accelerated Irradiation

The non-covered and covered film samples were cut into rectangle shapes (23.5 cm × 7.0 cm and 26.0 cm × 2.5 cm) respectively. The samples were introduced into a UV-A accelerated weathering tester with 1.55 W/m^2 (QUV/spray, Q-LAB, Homestead, FL, USA) and into Q-SUN accelerated Xenon Test Chamber with 1.5 W/m^2 (Model Xe-2, Q-LAB, Homestead, FL, USA) and irradiated for 24 h [34].

2.5. FT-IR

Fourier transform infrared (FT-IR) spectrum of the non-covered and covered film samples was measured using a FT-IR spectroscopy (Perkin Elmer Spectrophotometer, Spectrum 100, Waltham, MA, USA), operated at a resolution of 4 cm^{-1} and four scans. Film samples were cut into square shapes (2 cm × 2 cm) and placed directly at the ray-exposing stage. The spectrum was recorded at a wavelength of 650–4000 cm^{-1}.

2.6. Barrier Characteristic

Water vapour transmission rate (WVTR) was performed according to DIN 53122-1 [35] and ISO 2528:1995 [36]. WVTR was measured by means of a gravimetric method that is based on the sorption of humidity by calcium chloride and a comparison of sample weight gain. Initially, the amount of dry $CaCl_2$ inside the container was 9 g. The area of PLA film (covered or not covered) was 8.86 cm^2. Measurement was carried out for a period of 4 days, and each day, the containers were weighed to determine the amount of absorbed water vapour through the films. The results were expressed as average values from each day of measurement and each container. Analyses were carried out at 6 independent containers (6 repetitions) for each type of PLA films (covered with the coatings and not covered), calculated as a standard unit g/m^2·h, and presented as a mean ± standard deviation.

2.7. Statistical Analysis

The statistical significance was determined using an analysis of variance (ANOVA) followed by Duncan's test. This test was used to determine significant differences between numbers of the bacterial cells. The values were considered as significantly different when $p < 0.05$. All analysis was performed with Statistica version 10 (StatSoft, Kraków, Poland).

3. Results

3.1. Antimicrobial Properties

Study results indicated that MHPC coatings containing cocoa butter as hydrophobic additive had no influence on the growth of S. aureus cells. It was demonstrated that the number of S. aureus cells for the PLA film (control sample) was 2.1×10^5 cfu/mL. The amount of the bacterial cells for the MHPC coating containing cocoa butter was similar compared to the control sample (4.56×10^5 cfu/mL). The accelerated Q-SUN and UV-A irradiation did not influence the antimicrobial properties of the coatings devoid of bulb extract, while the MHPC coatings containing E. comosa extract inhibited the growth of S. aureus. The 3-log reduction of the number of S. aureus was noticed for the samples containing bulb extract (the number of the cells was 3.60×10^2 cfu/mL). The E. comosa bulb extract demonstrated antimicrobial activity against Gram-positive bacteria, a point confirmed in a previous study [25,26]. Statistical analysis showed that the decrease in the number of bacterial cells was significant ($p < 0.05$). The accelerated UV-A irradiation did not change the antimicrobial properties of the coatings with bulb extract (3.2×10^2 cfu/mL). In the case of Q-SUN irradiation (Figure 1), the number of bacterial cells marginally increased (5.86×10^2 cfu/mL), later confirmed by a Duncan test ($p > 0.05$).

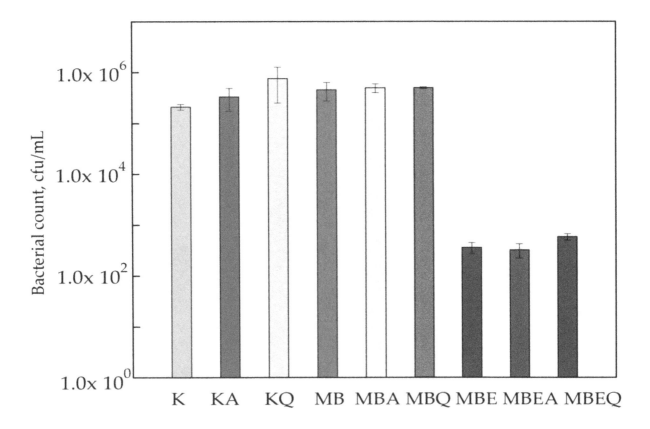

Figure 1. The influence of coatings on *S. aureus* growth. K—polylactic acid (PLA) film; KA—UV-A irradiated PLA film; KQ—Q-SUN irradiated PLA film; MB—PLA film, covered with methyl–hydroxypropyl–cellulose (MHPC)/cocoa butter coating; MBA—UV-A irradiated PLA film, covered with MHPC/cocoa butter coating; MBQ—Q-SUN irradiated PE film, covered with MHPC/cocoa butter coating; MBE—PLA film, covered with MHPC/cocoa butter coating, containing 25% of *E. comosa* extract; MBEA—UV-A irradiated PLA film, covered with MHPC/cocoa butter coating, containing 25% of *E. comosa* extract; MBEQ—Q-SUN irradiated PLA film, covered with MHPC/cocoa butter coating, containing 25% of *E. comosa* extract.

The susceptibility assay of *B. cereus* with respect to the active coatings containing *E. comosa* extract is shown in Figure 2. Comparing the numbers of *B. cereus* cells for PLA films (1.01×10^4 cfu/mL) to the numbers the *B. cereus* for PLA films covered with MHPC (1.14×10^4 cfu/mL), it should be said that the numbers were almost the same. The results of this research determined that MHPC coatings containing cocoa butter were not found to be active against bacteria. Accelerated Q-SUN and UV-A irradiation did not influence the antimicrobial properties of the coatings devoid of any bulb extract, later confirmed by a Duncan test ($p > 0.05$). The marginally low change in the numbers of bacteria (lower than 1-log reduction) was noticed. The *B. cereus* cells exhibited sensitivity towards coatings containing *E. comosa* extract. The number of *S. aureus* decreased from 1.14×10^4 to 1.84×10^2 cfu/mL (2-log reduction). Statistical analysis showed that the decrease in the number of bacterial cells was significant ($p < 0.05$). Q-SUN and UV-A irradiation only marginally influenced the antimicrobial properties of coatings. UV-A irradiation deactivated the antimicrobial properties of the coatings with *E. comosa* extract. An increase in the number of bacterial cells (2.76×10^2 cfu/mL) for these coatings irradiated with UV-A was also observed. In contrast to UV-A aging, Q-SUN improved the antimicrobial activity of the coatings (1.10×10^2 cfu/mL). The differences between the numbers of viable cells were not significant, as later confirmed by Duncan's test ($p > 0.05$).

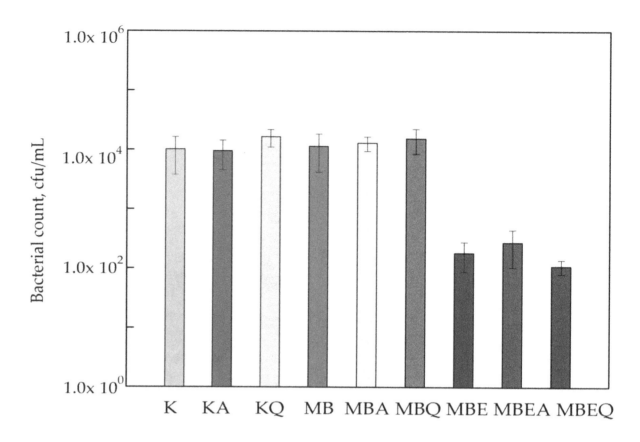

Figure 2. The influence of coatings on *B. cereus* growth. K—PLA film; KA—UV-A irradiated PLA film; KQ—Q-SUN irradiated PLA film; MB—PLA film, covered with MHPC/cocoa butter coating; MBA—UV-A irradiated PLA film, covered with MHPC/cocoa butter coating; MBQ—Q-SUN irradiated PE film, covered with MHPC/cocoa butter coating; MBE—PLA film, covered with MHPC/cocoa butter coating, containing 25% of *E. comosa* extract; MBEA—UV-A irradiated PLA film, covered with MHPC/cocoa butter coating, containing 25% of *E. comosa* extract; MBEQ—Q-SUN irradiated PLA film, covered with MHPC/cocoa butter coating, containing 25% of *E. comosa* extract.

The results of this research demonstrated that MHPC coatings with cocoa butter had no influence on the decrease in *B. atrophaeus* cell growth. It was also observed that the number of bacterial cells (5.44×10^5 cfu/mL) increased when compared to PLA films (4.34×10^4 cfu/mL) that were not covered with coatings. It is tempting to suggest that a coating devoid of an active substance may be used by *B. atrophaeus* as a carbon source. It should be added that the increase in viable cells was significant, later confirmed by statistical analysis ($p < 0.05$). The growth of bacterial cells decreased after 24 h contact with MHPC coatings containing *E. comosa* extract from 5.44×10^5 to 1.73×10^3 cfu/mL. As seen below (Figure 3), the influence of accelerated UV-A irradiation on the antimicrobial properties of coatings with *E. comosa* extract was also not noted. In the case of Q-SUN irradiation, it was observed that the number of viable cells increased from 1.73×10^3 to 1.24×10^4 cfu/mL when compared to the non-irradiated samples. A statistical analysis demonstrated that the differences between numbers of *B. atrophaeus* cells were not significant ($p > 0.05$).

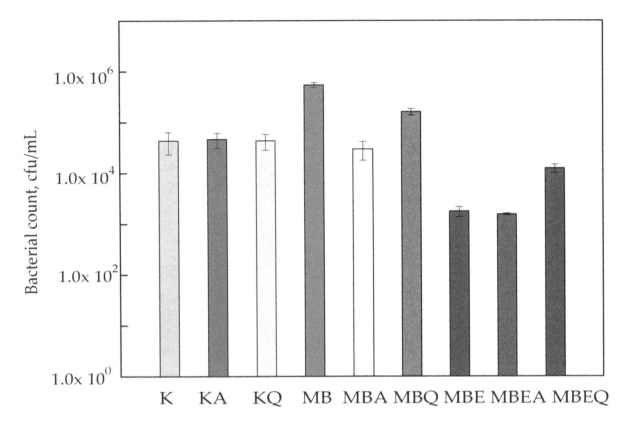

Figure 3. The influence of coatings on *B. atrophaeus* growth. K—PLA film; KA—UV-A irradiated PLA film; KQ—Q-SUN irradiated PLA film; MB—PLA film, covered with MHPC/cocoa butter coating; MBA—UV-A irradiated PLA film, covered with MHPC/cocoa butter coating; MBQ—Q-SUN irradiated PE film, covered with MHPC/cocoa butter coating; MBE—PLA film, covered with MHPC/cocoa butter coating, containing 25% of *E. comosa* extract; MBEA—UV-A irradiated PLA film, covered with MHPC/cocoa butter coating, containing 25% of *E. comosa* extract; MBEQ—Q-SUN irradiated PLA film, covered with MHPC/cocoa butter coating, containing 25% of *E. comosa* extract.

3.2. FT-IR Analysis

PLA belonging to the polyester family has characteristic peaks. The infrared (IR) spectra at 2996.3 and 2946.49 cm^{-1} were assigned to the asymmetric and symmetric –CH stretching region of the –CH$_3$ mode, respectively. The C=O stretching of the ester group was attributed as a broad and strong absorption bend at 1747.91 cm^{-1}. The –CH$_3$ bend was characterized by a peak of 1451.85 cm^{-1}. –CH deformation and asymmetric bends were observed at 1381.82 and 1360.06 cm^{-1}, respectively. The C=O stretching mode of the ester group were noted at 1266.41 cm^{-1} and an asymmetric C–O–C stretching mode were observed at 1180.81, 1127.28, and 1083 cm^{-1}. In the region of 1000 and 800 cm^{-1}, the bend at 955.57 cm^{-1} was attributed to the characteristic vibration of a helical backbone with CH$_3$ rocking mode. Two bends related to the crystalline and amorphous phases of PLA were found at 868.06 (assigned to the amorphous phase), 766.57 and 755.32 cm^{-1} (crystalline phase). The results were confirmed by Chu Z. et al. [37] and Seda Tığlı Aydın R. et al. [37].

The influence of UV irradiation and Q-SUN irradiation on coatings can be clearly noted through the use of FT-IR spectroscopy. Properties that were found to influence the absorption peak and bend positions were; structure, chemical composition, as well as the morphology of thin films or coating [28,29]. Study results demonstrated that differences in chemical composition and morphology of the PLA films (K—PLA) after UV-A irradiation (K—UV-A) and Q-SUN irradiation (K—Q-UV) were not noted (Figure 4). As shown in Figure 4, the curves of irradiated PLA films were similar to the PLA film that was not irradiated. However, with the UV-aging of the samples, the bend, at 1381.82, 1360.06, and 868.06 cm^{-1}, decreased in intensity. The results were confirmed by Yingfeng Z. et al. [38]. The authors demonstrated that the position of characteristic absorption peak of PLA did not change after aging treatment for different length of time, but significant change of the intensity of absorption peak was observed. The accelerated weather testing was performed also by Van Cong D. et al. [39] to evaluate the effects of TiO$_2$ crystal forms on the degradation behaviour of an EVA/PLA/TiO$_2$ nanocomposites compared to poly (ethylene-*co*-vinyl acetate) (EVA)/polylactic acid (PLA) blend. The results of FT-IR analysis, and thermal–mechanical properties confirmed the degradation of samples under accelerated weather testing. The degradation level of samples depended on TiO$_2$ crystal forms present in samples. The TiO$_2$ nanoparticles promoted the photodegradation of EVA/PLA/TiO$_2$ nanocomposites, in which mixed crystals of TiO$_2$ nanoparticles showed the highest photocatalytic activity.

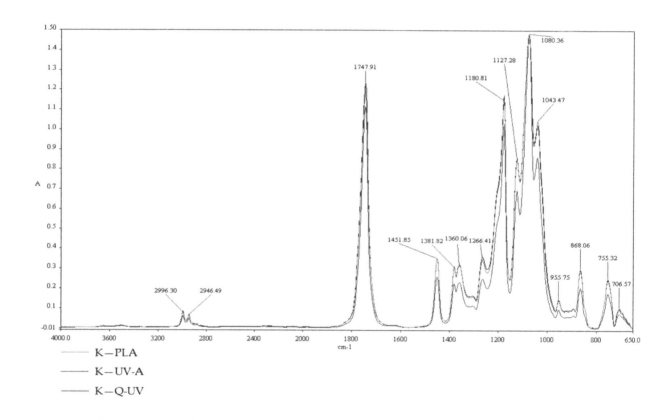

Figure 4. The FT-IR spectra of K—PLA, K—UV-A, K—Q-UV.

A representative spectrum of MHPC is shown in Figure 5. The peak at 3382.15 cm^{-1} was due to –OH vibration stretching. The symmetric stretching mode of methyl and hydroxyl propyl frequency was found in the range of 2918.71–2850.98 cm^{-1} in which all the –CH bonds extend and contract in phase. Symmetric vibrations were mainly displayed in the range of 1379.29 cm^{-1} and suggested cyclic anhydrides. The bands at 1040.36 and 1075.62 cm^{-1} were for the stretching vibration of ethereal C–O–C groups. These results were confirmed by Punitha S. et al. [40] and by Dong Ch. et al. [41].

The results demonstrated that accelerated irradiation had no effect on PLA film samples. The influence of UV-A and Q-SUN irradiation on hydrophilic MHPC was also not observed (Figure 5). However, with the UV-aging of the samples, the bend, at 1381.82, 1360.06, and 868.06 cm^{-1}, increased in intensity.

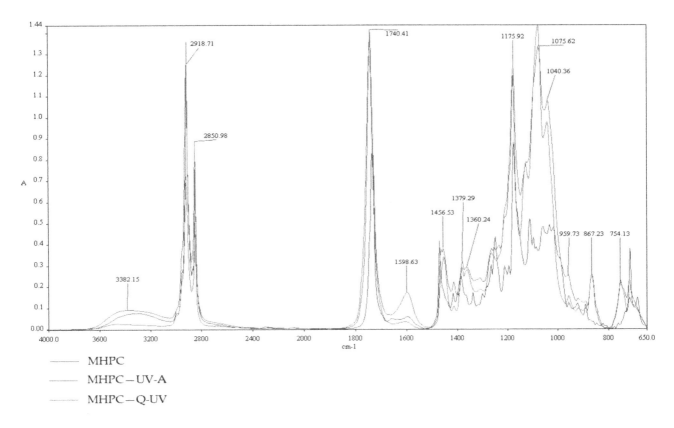

——— MHPC

——— MHPC—UV-A

——— MHPC—Q-UV

Figure 5. The FT-IR spectra of MHPC, MHPC—UV-A, MHPC—Q-UV.

A MHPC coating containing cocoa butter (CB—MHPC), a UV-A irradiated MHPC coating with cocoa butter (CB—MHPC—UVA) and a Q-SUN irradiated MHPC coating with hydrophobic additive (CB—MHPC—Q-UV) are presented in Figure 6. There are three regions viewed in the FT-IR spectroscopy that range from (1) 3600 to 2400 cm^{-1}; (2) 1800 to 1400 cm^{-1}; and (3) 1000 to 650 cm^{-1}. In the case of a 2851.05 cm^{-1} peak, this was noted and is consistent with absorption, stimulated by O=C–H double bonds. The 1747.53 cm^{-1} peak was stimulated by stretching vibration of carbonyl groups (C=O) from esters of triglycerides. The 1254.55 cm^{-1} peak was stimulated by C–O stretching vibration in ester. Alternatively, spectra peaks ranging from 1600 to 1400 cm^{-1}, were observed for a peak with C=C induced absorption. These peaks were only observed for a coating that was not irradiated. UV-A and Q-SUN irradiations caused the peaks to disappear. Different peak properties were shown at 937.89 and 720 cm^{-1}. The peaks were also observed in the case of a MHPC coating containing cocoa butter. The results were confirmed by Vesela A. et al. [42] and by Suparman et al. [43]. The presence of these peaks for a Q-SUN and an UV-A irradiated MHPC coatings was not noted. It was clearly confirmed that accelerated UV-A and Q-SUN irradiations altered the chemical composition of the CB—MHPC layer. It is tempting to suggest that accelerated irradiations cased the oxidation of the hydrophobic additive. Similar results were obtained for the MHPC coatings containing cocoa butter as a hydrophobic additive and *E. comosa* extract as an active substance (Figure 7). There are also three regions viewed in the FT-IR spectroscopy that range from (1) 3600 to 2400 cm^{-1}; (2) 1600 to 1400 cm^{-1}; and (3) 1000 to 650 cm^{-1}. In the case of coatings containing an active extract, the disappearance of the peaks after accelerated irradiation was also noted. This means that accelerated irradiations caused an unsaturated bond oxidation of the cocoa butter fatty acids. This led us to believe that ZnO nanoparticles have shielding properties, in contrast to the *E. comosa* extract, which did not shield the coatings against accelerated irradiation. Mizielińsska et al. [28] showed that nano ZnO shielded the MHPC layer against Q-SUN irradiation. This conclusion was confirmed by El-Feky O.M. et al. [31] who used ZnO nanoparticles as an additive for coatings as a protection against UV irradiation.

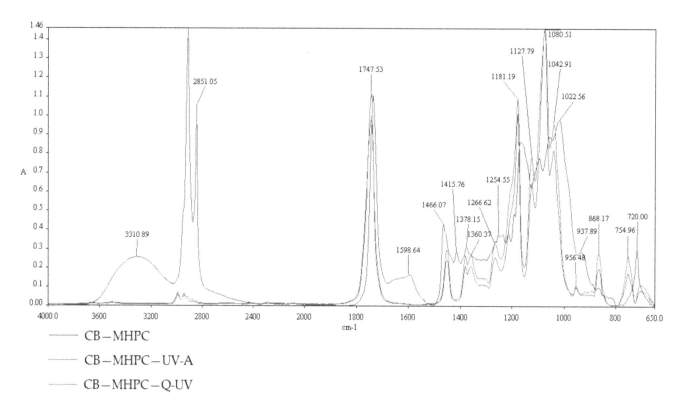

Figure 6. The FT-IR spectra of CB—MHPC, CB—MHPC—UV-A, CB—MHPC—Q-UV.

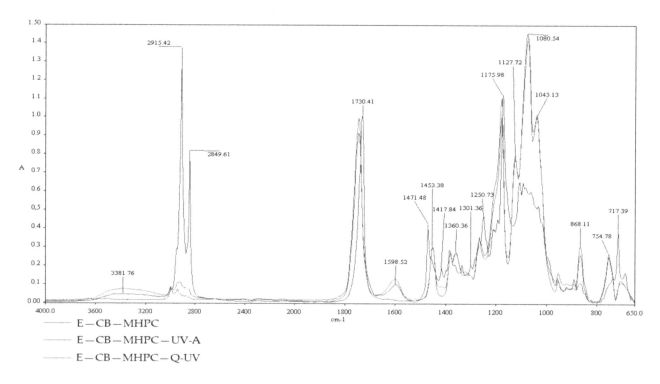

Figure 7. The FT-IR spectra of E—CB—MHPC, E—CB—MHPC—UV-A, E—CB—MHPC—Q-UV.

3.3. Barrier Characteristic

The results of this study demonstrated that the MHPC coating had no significant influence on the barrier properties of PLA films. As clearly seen below (Table 1). The accelerated UV-A and Q-SUN did not improve and/or lower the water permeability of PLA covered with MHPC. Cocoa butter as a hydrophobic additive decreased the water permeability of coatings from 76.88 ± 1.24 ($g/(m^2 \times h)$) to 41.38 ± 3.45 ($g/(m^2 \times h)$). It was observed that contrary to UV-A irradiation, Q-UV irradiation improved the barrier properties of the coated PLA. The addition of *E. comosa* extract into the coating decreased the water permeability of the MHPC coating with cocoa butter. It should be mentioned that UV-A and Q-UV irradiations increased the water permeability of the PLA covered with MHPC containing cocoa butter as a hydrophobic substance and *E. comosa* extract as an active substance.

Table 1. The water permeability of PLA films before and after irradiation.

Sample		WVTR ($g/(m^2 \times h)$)
	PLA	80.22 ± 1.78
Devoid of irradiation	MHPC	76.88 ± 1.24
	CB—MHPC	41.38 ± 3.45
	E—CB—MHPC	29.91 ± 2.28
	MHPC	76.45 ± 0.25
UV-A irradiated	CB—MHPC	48.34 ± 4.68
	E—CB—MHPC	60.31 ± 1.02
	MHPC	74.89 ± 2.76
Q-UV irradiated	CB—MHPC	40.23 ± 3.49
	E—CB—MHPC	63.26 ± 1.44

4. Discussion

Pollution is increasing day by day, and environmental laws are becoming more stringent. Therefore, companies need to transform to meet their traditional objectives of cost reduction and good product quality, but also make efforts to implement a green and innovative set of technologies [44,45]. There is a need to replace synthetic polymers with natural and biodegradable materials such as polylactic acid. PLA is the most commercial biodegradable polymer, showing many advantages. Hence, different methods have been suggested in order to improve its chemical–physical performance and to obtain antimicrobial properties of the films. With respect to synthetic polymers, the barrier behaviour showed a high gas and water vapour transmission that rendered this polymer unsuitable for several food packaging applications. To improve barrier characteristics and activate some kind of antimicrobial effect, a surface polymer modification, such as a coating, would need to be created [2,7–13,16,17]. Coatings containing active substances can be divided into those that migrate into the packed product and those that do not. The active compounds should inhibit the growth of microorganisms responsible for packed product spoilage and pathogenic microorganisms. An active packaging containing plant and spice extracts or essential oils has antimicrobial properties against many bacteria and fungi. Their natural components include antimicrobial phenolic compounds, aldehydes, ketones, alcohols, ethers, and hydrocarbons. [10,11,16,17]. Because *E. comosa extract* [25,26] was found to be active against *S. aureus*, *B. cereus*, and *B. atrophaeus*, it was used as an active substance in the coatings. Due to the high water vapour permeability [13,16,17] of the PLA films or MHPC carrier, a hydrophobic additive was also introduced into the coatings.

The results of the study showed that MHPC/cocoa butter coatings containing *E. comosa* extract reduced the number of *S. aureus*, *B. cereus*, and *B. atrophaeus* cells. The addition of cocoa butter into the coatings decreased the water vapour permeability of PLA. It was shown that accelerated UV-A and Q-SUN irradiations altered the chemical composition of the coatings containing cocoa butter. It was also observed that contrary to UV-A irradiation, Q-UV irradiation that caused the oxidation of the double bonds improved the barrier properties of the coated PLA. The addition of *E. comosa* extract into the coating decreased the water vapour permeability of the MHPC coating with cocoa butter. It should be mentioned that UV-A and Q-UV irradiations increased the water vapour permeability of covered PLA containing *E. comosa* extract.

Despite alterations made to the chemical composition of the layers, accelerated Q-SUN and UV-A irradiation had no influence on the antimicrobial properties of *E. comosa* extract coatings against *S. aureus*, and *B. cereus* (in contrast to barrier characteristic). Only Q-SUN irradiation decreased coating activity with an extract against *B. atrophaeus*, but this was negligible. Similar results were obtained in a previous study [28] that showed that accelerated Q-SUN and UV-A irradiation had no influence on the antimicrobial properties of coatings against *S. aureus* and *B. cereus*. The coatings used in this previous study contained ZnO nanoparticles that shielded the active MHPC coating during irradiation, which was confirmed by the authors [31,46]. This work showed that *E. comosa* extract did not shield the MHPC/cocoa butter coating during irradiation. Although the extract had no shielding properties and the chemical composition changed, the coatings were still active against Gram-positive bacteria.

The growing demand for increased fresh food shelf life as well as the need of protection against foodborne diseases urged the development of antimicrobial food packaging. Active food packaging may include oxygen scavengers, moisture absorbers, ultraviolet barriers, or compounds that deliver flavouring, antioxidant, or antimicrobial agents. In the context of increasing demand of multiple hurdle technology to achieve high food safety standards, the development of antimicrobial packaging systems is of great interest. The use of natural products, such as essential oils (EOs) or plant extracts, as food preserving agents, is being promoted given the current trend towards green consumerism. These extracts or oils often contain compounds such as polyphenols or terpenes with antimicrobial properties [47–54]. Ginja cherry stem extracts, extracts from green tea were used in food packaging [50,51].

It is known that Gram-positive bacteria such as *S. aureus*, *Bacillus* sp., and *Listeria monocytogenes* were often isolated from raw fish, meat, or ready-sliced ham and cheese. Responsible for foodborne diseases, *Clostridium* sp. was isolated from meat packed in vacuum. The packaging materials covered with coatings, that are active against Gram-positive microorganisms could be the solution for vacuum system packaging. Shakila R.J. et al. [49] evaluated antimicrobial properties against *S. aureus* and *Listeria monocytogenes* of coatings with different active additives (chitosan, clove, and pepper) in vacuum packed fish steaks. The authors extended the shelf-life of product from 4 to 8 days at 4 °C. Sandoval L.N. et al. [54] proved that the active packaging containing chitosan as an antimicrobial agent offered an alternative for the preservation of the quality of fresh cheese during storage, and increased shelf life, and more importantly, it inhibited the growth of *L. monocytogenes*. This study showed that *E. comosa* extract coatings were active against Gram-positive bacteria even after UV-aging. These coatings could be used as packaging material for vacuum-packed fish, meat, or cheese.

5. Conclusions

The food packaging industry is going to replace non-compostable polymers with biodegradable materials. It should be underlined that polylactic acid is a biodegradable biopolymer that has been commercialized as a good alternative substitute for synthetic polymers.

It should be underlined that the covering of PLA films with a MHPC carrier containing cocoa butter decreased water vapour permeability of the biopolymer. The additional advantage of the material was its antibacterial activity, obtained by introducing *E. comosa* extract as an active substance into the MHPC coating. Accelerated UV-A and Q-SUN irradiations altered the chemical composition of the coatings containing cocoa butter or cocoa butter and *E. comosa* extract. It caused an increase in the water vapour permeability of the coated PLA. It should be highlighted that the accelerated Q-SUN and UV-A irradiation had no influence on the antimicrobial properties of *E. comosa* extract coatings against *S. aureus* and *B. cereus*.

Due to the resistance of the coatings to UV-aging, and due to improved barrier characteristic of PLA, the coatings could be used to cover biopolymer films or boxes to obtain the active packaging material. The active coatings could extend the shelf life, the quality and freshness of food products.

Acknowledgments: Research funded by the West Pomeranian University of Technology Szczecin.

Author Contributions: Małgorzata Mizielińska conceived and designed the experiments, and wrote the paper; Małgorzata Mizielińska and Urszula Kowalska performed the microbiological tests; Małgorzata Mizielińska analyzed the data; Piotr Salachna prepared *E. comosa* bulbs extract; Łukasz Łopusiewicz performed FT-IR tests; Michał Jarosz performed barrier characteristic tests; Małgorzata Mizielińska analyzed the data; Małgorzata Mizielińska prepared reagents/materials; Małgorzata Mizielińska contributed analysis tools; Urszula Kowalska performed statistical analysis; Małgorzata Mizielińska analyzed the data.

References

1. Muller, J.; González-Martínez, C.; Chiralt, A. Combination of Poly(lactic) Acid and Starch for Biodegradable Food Packaging. *Materials* **2017**, *10*, 952. [CrossRef] [PubMed]
2. Siracusa, V.; Rosa, M.D.; Iordanskii, A.L. Performance of Poly(lactic acid) Surface Modified Films for Food Packaging Application. *Materials* **2017**, *10*, 850. [CrossRef] [PubMed]
3. Sánchez Aldana, D.; Duarte Villa, E.; De Dios Hernández, M.; Guillermo González Sánchez, G.; Rascón Cruz, Q.; Flores Gallardo, S.; Hilda Piñon Castillo, H.; Ballinas Casarrubias, L. Barrier Properties of Polylactic Acid in Cellulose Based Packages Using Montmorillonite as Filler. *Polymers* **2014**, *6*, 2386–2403. [CrossRef]
4. Musil, J. Flexible antibacterial coatings. *Molecules* **2017**, *22*, 813. [CrossRef] [PubMed]
5. Mizielińska, M.; Lisiecki, S.; Jędra, F.; Kowalska, U.; Tomczak, A. The barrier and the antimicrobial properties of polylactide films covered with exopolysaccharide layers synthesized by *Arthrobacter viscosus*. *Przem. Chem.* **2015**, *94*, 748–751. [CrossRef]
6. Chu, Z.; Zhao, T.; Li, L.; Fan, J.; Qin, Y. Characterization of Antimicrobial Poly (Lactic Acid)/Nano-Composite Films with Silver and Zinc Oxide Nanoparticles. *Materials* **2017**, *10*, 659. [CrossRef] [PubMed]
7. Mizielińska, M.; Lisiecki, S. Coating of polylactide films to generate their antimicrobial properties. *Przem. Chem.* **2015**, *94*, 752–755. [CrossRef]
8. Han, J.; Salmeri, S.; Le Tien, C.; Lacroix, M. Improvement of Water Barrier Property of Paperboard by Coating Application with Biodegradable Polymers. *J. Agric. Food Chem.* **2010**, *58*, 3125–3131. [CrossRef] [PubMed]
9. Torres-Moreno, M.; Torrescasana, E.; Salas-Salvadó, J.; Blanch, C. Nutritional composition and fatty acids profile in cocoa beans and chocolates with different geographical origin and processing conditions. *Food Chem.* **2015**, *166*, 125–132. [CrossRef] [PubMed]
10. Mizielińska, M.; Kowalska, U.; Pankowski, J.; Bieńkiewicz, G.; Malka, M.; Lisiecki, S. Coating the polyethylene films to generate the antibacterial properties. *Przem. Chem.* **2017**, *96*, 1317–1321. [CrossRef]
11. Mizielińska, M.; Ordon, M.; Pankowski, J.; Bieńkiewicz, G.; Malka, M.; Lisiecki, S.; Bartkowiak, A. Estimation of the antimicrobial properties of the coatings from industrial trials. *Przem. Chem.* **2017**, *96*, 1322–1324. [CrossRef]
12. Mizielińska, M.; Lisiecki, S.; Jotko, M.; Chodzyńska, I.; Bartkowiak, A. The antimicrobial properties of polylactide films covered with ZnO nanoparticles-containing layers. *Przem. Chem.* **2015**, *94*, 1000–1003. [CrossRef]
13. Bartkowiak, A.; Mizielińska, M.; Sumińska, P.; Romanowska-Osuch, A.; Lisiecki, S. Innovations in food packaging materials. In *Emerging and Traditional Technologies for Safe, Healthy and Quality Food*; Nedović, V., Raspor, P., Lević, J., Šaponjac, V.T., Barbosa-Cánovas, G.V., Eds.; Springer: Berlin/Heidelberg, Germany, 2016; pp. 383–412. [CrossRef]
14. Jafarzadeh, S.; Alias, A.K.; Ariffin, F.; Mahmud, S.; Najafi, A.; Ahmad, M. Fabrication and characterization of novel semolina-based antimicrobial films derived from the combination of ZnO nanorods and nanokaolin. *J. Food. Sci. Technol.* **2017**, *54*, 105–113. [CrossRef] [PubMed]
15. Wang, Y.; Ma, J.; Xu, Q.; Zhang, J. Fabrication of antibacterial casein-based ZnO nanocomposite for flexible coatings. *Mater. Des.* **2017**, *113*, 240–245. [CrossRef]
16. Kwiatkowski, P.; Giedrys–Kalemba, S.; Mizielińska, M.; Artur Bartkowiak, A. Antibacterial activity of rosemary caraway and fennel essential oils. *Herba Pol.* **2015**, *61*, 31–39. [CrossRef]
17. Kwiatkowski, P.; Giedrys–Kalemba, S.; Mizielińska, M.; Artur Bartkowiak, A. Modyfication of PLA foil surface by ethylocellulose and essential oils. *J. Microbiol. Biotechnol. Food Sci.* **2016**, *5*, 440–444. [CrossRef]

18. Taylor, J.L.S.; Van Staden, J. The effect of age, season and growth conditions on anti-inflammatory activity in *Eucomis autumnalis* (Mill) Chitt. plant extracts. *Plant Growth Regul.* **2011**, *34*, 39–47. [CrossRef]

19. Louw, C.A.M.; Regnier, T.J.C.; Korsten, L. Medicinal bulbous plants of South Africa and their traditional relevance in the control of infectious diseases. *J. Ethnopharmacol.* **2002**, *82*, 147–154. [CrossRef]

20. Salachna, P.; Zawadzińska, A. Effect of daminozide and flurprimidol on growth, flowering and bulb yield of *Eucomis autumnalis* (Mill.) Chitt. *Folia Hortic.* **2017**, *29*, 33–38. [CrossRef]

21. Masondo, N.A.; Finnie, J.F.; van Staden, J. Pharmacological potential and conservation prospect of the genus *Eucomis* (Hyacinthaceae) endemic to Southern Africa. *J Ethnopharmacol.* **2014**, *151*, 44–53. [CrossRef] [PubMed]

22. Salachna, P.; Grzeszczuk, M.; Wilas, J. Total phenolic content, phenolic content, photosynthetic pigment concentration and antioxidant activity of leaves and bulbs of selected *Eucomis L'Hér.* taxa. *Fresenius Environ. Bull.* **2015**, *24*, 4220–4225.

23. Bisi-Johnson, M.A.; Obi, C.L.; Hattori, T.; Oshima, Y.; Li, S.; Kambizi, L.; Eloff, J.N.; Vasaikar, S.D. Evaluation of the antibacterial and anticancer activities of some South African medicinal plants. *Complement. Altern. Med.* **2011**, 11–14. [CrossRef] [PubMed]

24. Motsei, M.L.; Lindsey, K.L.; van Staden, J.; Jäger, A.K. Screening of traditionally used South African plants for antifungal activity against *Candida albicans*. *J. Ethnopharmacol.* **2003**, *86*, 235–241. [CrossRef]

25. Salachna, P.; Mizielińska, M.; Soból, M. Exopolysaccharide Gellan Gum and Derived Oligo-Gellan Enhance Growth and Antimicrobial Activity in Eucomis Plants. *Polymers* **2018**, *10*, 242. [CrossRef]

26. Mizielińska, M.; Salachna, P.; Ordon, M.; Łopusiewicz, Ł. Antimicrobial activity of water and acetone extracts of some *Eucomis* taxa. *Asian Pac. J. Trop. Med.* **2017**, *10*, 892–895. [CrossRef] [PubMed]

27. Mizielińska, M.; Kowalska, U.; Jarosz, M.; Sumińska, P. A comparison of the effects of packaging containing nano ZnO or polylysine on the microbial purity and texture of cod (*Gadus morhua*) fillets. *Nanomaterials* **2018**, *8*, 158. [CrossRef] [PubMed]

28. Mizielińska, M.; Łopusiewicz, Ł.; Mężyńska, M.; Bartkowiak, A. The influence of accelerated UV-A and Q-SUN irradiation on the antimicrobial properties of coatings containing ZnO nanoparticles. *Molecules* **2017**, *22*, 1556. [CrossRef] [PubMed]

29. Nguyena, T.V.; Dao, P.H.; Khanh Linh Duong, K.L.; Duong, Q.H.; Vu, Q.T.; Nguyena, A.H.; Mac, V.P.; Lea, T.L. Effect of R-TiO$_2$ and ZnO nanoparticles on the UV-shielding efficiency of water-borne acrylic coating. *Prog. Org. Coat.* **2017**, *110*, 114–121. [CrossRef]

30. Marvizadeh, M.M.; Oladzadabbasabadi, N.; Nafchi, A.M.; Jokar, M. Preparation and characterization of bionanocomposite film based on tapioca starch/bovine gelatin/nanorod zinc oxide. *Int. J. Biol. Macromol.* **2017**, *99*, 1–7. [CrossRef] [PubMed]

31. El-Feky, O.M.; Hassan, E.A.; Fadel, S.M.; Hassan, M.L. Use of ZnO nanoparticles for protecting oil paintings on paper support against dirt, fungal attack, and UV aging. *J. Cult. Herit.* **2014**, *15*, 165–172. [CrossRef]

32. Kairyte, K.; Kadys, A.; Luksiene, Z. Antibacterial and antifungal activity of photoactivated ZnO nanoparticles in suspension. *J. Photochem. Photobiol. Biol.* **2013**, *128*, 78–84. [CrossRef] [PubMed]

33. *ASTM Standard Test Method for Determining the Activity of Incorporated Antimicrobial Agent(s) in Polymeric or Hydrophobic Materials*; E 2180-01; ASTM: West Conshohocken, PA, USA, 2002.

34. Nichols, M.; Boisseau, J.; Pattison, L.; Campbell, D.; Quill, J.; Zhang, J.; Smith, D.; Henderson, K.; Seebergh, J.; Berry, D.; et al. An improved accelerated weathering protocol to anticipate Florida exposure behavior of coatings. *J. Coat. Technol. Res.* **2013**, *10*, 153–173. [CrossRef]

35. DIN 53122-1 ISO 2528:1995. Available online: http://sklep.pkn.pl/pn-iso-2528-2000p.html (accessed on 7 June 2016).

36. ISO 2528:1995. Available online: http://sklep.pkn.pl/pn-iso-2528-2000p.html (accessed on 7 June 2016).

37. Seda Tıglı Aydın, R.; Akyol, E.; Hazer, B. Influence of Soybean Oil Blending with Polylactic Acid (PLA) Films: In Vitro and In Vivo Evaluation. *J. Am. Oil Chem. Soc.* **2017**, *94*, 413–424. [CrossRef]

38. Yingfeng, Z.; Yiqiang, W.; Jiyou, G.; Yanhua, Z. The UV Aging Properties of Maleic Anhydride Esterified Starch/Polylactic Acid Composites. *J. Wuhan Univ. Technol.* **2017**, *32*, 917–977. [CrossRef]

39. Van Cong, D.; Trang, N.T.T.; Giang, N.V.; Lam, T.D.; Hoang, T. Effect of TiO$_2$-Crystal Forms on the Photo-Degradation of EVA/PLA Blend Under Accelerated Weather Testing. *J. Electron. Mater.* **2016**, *45*, 2536–2546. [CrossRef]

40. Punitha, S.; Uvarani, R.; Panneerselvam, A.; Nithiyanantham, S. Physico-chemical studies on some saccharides in aqueous cellulose solutions at different temperatures—Acoustical and FTIR analysis. *J. Saudi Chem. Soc.* **2014**, *18*, 657–665. [CrossRef]

41. Dong, H.; Zhang, X.; Cai, H. Green synthesis of monodisperse silver nanoparticles using hydroxy propyl methyl cellulose. *J. Alloys Compd.* **2014**, *583*, 267–271. [CrossRef]

42. Vesela, A.; Barros, A.S.; Synytsya, A.; Delgadillo, I.; Copikova, J.; Manuel Coimbra, A. Infrared spectroscopy and outer product analysis for quantification of fat, nitrogen, and moisture of cocoa powder. *Anal. Chim. Acta* **2007**, *601*, 77–86. [CrossRef] [PubMed]

43. Suparman; Rahayu, W.S.; Sundhani, E.; Dwi Saputri, S. The use of Fourier Transform Infrared Spectroscopy (FTIR) and Gas Chromatography Mass Spectroscopy (GCMS) for Halal Authentication in Imported Chocolate with Various Variants. *J. Food. Pharm. Sci.* **2015**, *2*, 6–11.

44. Díaz-Tena, E.; Rodríguez-Ezquerro, A.; López de Lacalle Marcaide, L.N.; Gurtubay Bustinduy, L.; Elías Sáenz, A. A sustainable process for material removal on pure copper by use of extremophile bacteria. *J. Clean. Prod.* **2014**, *84*, 752–760. [CrossRef]

45. Díaz-Tena, E.; Barona, A.; Gallastegui, G.; Rodrigez, A.; López de Lacalle, L.N.; Elías, A. Biomachining: Metal etching viamicroorganisms. *Crit. Rev. Biotechnol.* **2017**, *37*, 323–332. [CrossRef] [PubMed]

46. Venkatesan, R.; Natesan Rajeswari, N. ZnO/PBAT nanocomposite films: Investigation on the mechanical and biological activity for food packaging. *Polym. Adv. Technol.* **2017**, *28*, 20–27. [CrossRef]

47. Valdés, A.; Ramos, M.; Beltrán, A.; Jiménez, A.; Garrigós, M.C. State of the Art of Antimicrobial Edible Coatings for Food Packaging Applications. *Coatings* **2017**, *7*, 56. [CrossRef]

48. Djenane, D.; Roncalés, P. Carbon Monoxide in Meat and Fish Packaging: Advantages and Limits. *Foods* **2018**, *7*, 12. [CrossRef] [PubMed]

49. Shakila, R.J.; Jeevithan, E.; Arumugam, V.; Jeyasekaran, G. Suitability of antimicrobial grouper bone gelatin films as edible coatings for vacuum-packaged fish steaks. *J. Aquat. Food Prod. Technol.* **2016**, *25*, 724–734. [CrossRef]

50. Jongberg, S.; Tørngren, M.A.; Skibsted, L.H. Protein Oxidation and Sensory Quality of Brine-Injected Pork Loins Added Ascorbate or Extracts of Green Tea or Maté during Chill-Storage in High-Oxygen Modified Atmosphere. *Medicines* **2018**, *5*, 7. [CrossRef] [PubMed]

51. Campos, D.; Piccirillo, C.; Pullar, R.C.; Castro, P.M.L.; Pintado, M.M.E. Characterization and antimicrobial properties of food packaging methylcellulose films containing stem extract of *Ginja cherry*. *J. Sci. Food. Agric.* **2014**, *94*, 2097–2103. [CrossRef] [PubMed]

52. Quesada, J.; Sendra, E.; Navarro, C.; Sayas-Barberá, E. Antimicrobial Active Packaging including Chitosan Films with *Thymus vulgaris* L. Essential Oil for Ready-to-Eat Meat. *Foods* **2016**, *5*, 57. [CrossRef] [PubMed]

53. Rahman, P.M.; Abdul Mujeeb, V.M.; Muraleedharan, K. Flexible chitosan-nano ZnO antimicrobial pouches as a new materialfor extending the shelf life of raw meat. *Int. J. Biol. Macromol.* **2017**, *97*, 382–391. [CrossRef] [PubMed]

54. Sandoval, L.N.; López, M.; Montes-Díaz, E.; Espadín, A.; Tecante, A.; Gimeno, M.; Shirai, K. Inhibition of Listeria monocytogenes in Fresh Cheese Using Chitosan-Grafted Lactic Acid Packaging. *Molecules* **2016**, *21*, 469 [CrossRef] [PubMed]

The Quality Evaluation of Postharvest Strawberries Stored in Nano-Ag Packages at Refrigeration Temperature

Cheng Zhang [1], Wenhui Li [1], Bifen Zhu [1], Haiyan Chen [1], Hai Chi [1], Lin Li [2], Yuyue Qin [1,*] and Jing Xue [3,*]

[1] Institute of Yunnan Food Safety, Kunming University of Science and Technology, Kunming 650500, China; 13136640259@163.com (C.Z.); 15559823733@163.com (W.L.); 18469189957@163.com (B.Z.); seacome@163.com (H.C.); 18468273140@163.com (H.C.)

[2] College of Food Sciences and Engineering, South China University of Technology, Guangzhou 510640, China; felinli@scut.edu.cn

[3] State Key Laboratory of Oral Diseases, Sichuan University, Chengdu 610041, China

* Correspondence: rabbqy@163.com (Y.Q.); aqua119@163.com (J.X.);

Abstract: Different percentages (0%, 1%, 5%, and 10%) of nano-Ag particles were added to polylactic acid (PLA) to make an active nanocomposite packaging film. Strawberries were packaged by the nanocomposite films and stored at 4 ± 1 °C for 10 days. The freshness of strawberries was assessed by regularly measuring the physicochemical properties of the strawberries in each packaging film. The difference in the freshness of strawberries was evaluated by determining the following parameter changes: weight loss, hardness, soluble solids, titratable acid, color, vitamin C, total phenol, free radical scavenging activity, peroxidase activity, and sensory evaluation. The results revealed that the active nanocomposite packaging film has better preservation effect when compared with pure PLA film. Its preservation effect is mainly reflected in the more effective reduction of vitamin C loss, delaying the decline of total phenols and 1-Diphenyl-2-picrylhydrazyl (DPPH) in strawberries. It also showed better physical properties. The results showed that the PLA nanocomposite packaging film could effectively preserve freshness of strawberries.

Keywords: strawberry; nano-Ag packaging; storage; quality change

1. Introduction

Strawberry (*Fragaria x ananassa* Duch.) is a perennial herb of the genus *Rosaceae*. Strawberry has a unique scent and is juicy. It is rich in vitamins, carotene, anthocyanins, and other nutrients. It is known as the Queen of fruits and widely welcomed by consumers. However, due to its delicate skin, it can easily cause surface crushing in picking and transportation projects. After picking, the strawberry is vigorously breathing and can easily cause spoilage of the fruit [1]. Strawberry is a non-climacteric fruit that can be harvested at maturity to obtain the best taste of the fruit. However, the best tasting period for strawberry is very short, so how to keep strawberries fresh is always the focus of research work. At present, the methods for keeping fresh strawberry mainly include low-temperature refrigeration and fresh-keeping technology [2], modified atmosphere packaging storage technology [3], UV shortwave ultraviolet radiation technology [4], chemical preservation [5] and so on. However, most of these treatments are expensive, time consuming, and may even damage the appearance of the strawberry.

Polylactic acid (PLA) is a biodegradable material and can be regenerated by bacterial fermentation through corn or sugar beets. Due to its low cost of acquisition and biodegradability, it has excellent biocompatibility. PLA is widely studied and used in packaging materials [6]. In addition, PLA has been approved by the US Food and Drug Administration (FDA) and European food regulations for use in food packaging. Thus, it is safe to contact with foodstuff as a packaging material. However, pure PLA is inferior to traditional petroleum-based packaging materials in many aspects such as toughness and thermal properties [7]. In general, pure PLA is polymerized or blended with other biologically active ingredients to improve the desired properties of PLA, such as nano-TiO_2 [8], nano-Ag, nano-SiO_2 [9], and nano-ZnO [10].

Nanotechnology is a high-tech science and technology that has been rapidly developed in recent years. It has great market prospects and application value. The addition of nanoparticles not only improves the physical properties of the material such as flexibility, light transmission, and plasticity. It has effective antifungal, microbiological and antiviral activity [11]. Studies have shown that Ag particles have antibacterial effects [12]. Therefore, nano-Ag particles have been widely used in food packaging, textiles, water filtration, and healthcare [13]. Li et al. modified active films by adding nano-Ag into LDPE films and applied them to the preservation of juices. The results show that nano-Ag composite packaging films have greater anti-microbial effects [14]. Therefore, this study embedded nano-Ag particles into PLA to investigate the effect of nano-Ag packaging films on the quality of strawberries stored at $4 \pm 1\,°C$. Then, the effect of nano-Ag PLA-based packaging films on fresh-keeping of strawberry was discussed.

2. Materials and Method

2.1. Materials

PLA (M_w = 280 kDa, M_w/M_n = 1.98) used in this experiment was obtained from Natureworks LLC (Blair, NE, USA). Acetyl tributyl citrate (ATBC) was purchased from Shanghai Macklin Biochemical Co., Ltd. (Shanghai, China). Nano-Ag was purchased from Wanjing New Material Co., Ltd. (Hangzhou, China). Dichloromethane, methanol, NaOH, hydrogen peroxide, Na_2CO_3 were obtained from Chengdu Kelong Chemical Co., Ltd. (Chengdu, China). DPPH, Guaiacol were purchased from Sigma (St. Louis, MO, USA). All the reagent were analytical reagent. Texture analyzer (Texture Exponent 32, Stable Micro Systems Ltd., London, UK), Colorimeter (WSC-S; Shanghai Precision Instrument Co., Ltd., Shanghai, China), Digital Refractometer (MZB 92, Shanghai Miqingke Industrial Testing Co., Ltd., Shanghai, China), PH meter (Merck, Barcelona, Spain), spectrophotometry (UV-1800, Mapada Instruments Co., Ltd., Shanghai, China), Centrifuged (TGL-16M, Xiangyi Centrifuge instruments Co., Ltd., Shanghai, China).

2.2. Preparation of Film and Sample

The PLA-based nanocomposite films were prepared using the solvent evaporation method according to Qin et al. [15] with some modifications. PLA containing 1 wt % of ATBC plasticizer and different percent of nano-Ag (0, 1, 5, and 10 wt % of 2 g PLA) were dissolved in 50 mL of dichloromethane. The solution was stirred by magnetic stirrer at room temperature for about 10 h. When the solution was completely blended, it was poured onto a 200 mm × 200 mm polytetrafluoroethylene plate and left to stand overnight in the well-ventilated place. Nano-Ag was incorporated into PLA as 0, 1, 5, and 10 wt % loading, and named as PLA, PLA/Ag1%, PLA/Ag5%, and PLA/Ag10%.

Fresh strawberry was harvested from Kunming strawberry greenhouse. About 70–80% ripeness (approximately 70–80% of strawberry retained the red color), same size, no pests, and no mechanical damage extrusion strawberries were selected as the test material. The strawberries were transported to the laboratory immediately after being harvested. Strawberry samples picked from the greenhouse were simply cleaned of soil on the surface and put directly into the bag. Note that water was not used to wash the surface of the strawberry; the soil and other dirt was simply shaken off. Alcohol was applied to both sides of the films and then they were placed on an aseptic table. The films were irradiated with UV light for 15 min to ensure film hygiene as well as to not affect the accuracy of subsequent microbial experiments. Ten fresh strawberries were randomly selected to per type of material package and stored at 4 ± 1 °C. A sample was taken every two days to determine the indicators during 10 days (on Days 0, 2, 4, 6, 8, and 10).

Refrigerated Storage

Strawberries were packaged with five kinds of active films (PLA, PLA/Ag1%, PLA/Ag5%, and PLA/Ag10%). Following treatment, strawberries were stored at 4 ± 1 °C for 10 days. The effectiveness of the treatments was evaluated by determining quality changes every two days. Then, combined with sensory evaluation, the acceptability of strawberry preservation was obtained.

2.3. Weight Loss (LS)

Three bags were selected randomly from the four active packages to determine the weight every two days, and compare the weight difference with the original fresh weight on the first day. Gravimetry was used to measure the weight loss, expressed as the percentage of the original weight. Equation (1) can be used to express weight loss:

$$\text{Weight Loss}(\%) = \frac{M_0 - M_1}{M_0} \times 100 \tag{1}$$

M_0 is the fresh weight of fruit on the first day, and M_1 is the measured weight on each sampling day [16].

2.4. Firmness Measurement

The firmness of the strawberry was determined using a texture analyzer (Texture Exponent 32, Stable Micro Systems Ltd., London, UK) equipped with 2 mm diameter cylindrical probe. Four different locations were used to measure firmness around the equatorial region on each fruit. The penetration depth of the probe into the sample was 5 mm and the crosshead speed of the texture analyzer as 2 mm/s. From the force vs. time curve, the maximum force firmness was expressed in N·cm^{-2}.

2.5. Surface Color

A colorimeter (WSC-S; Shanghai Precision Instrument Co., Ltd., Shanghai, China) was used to determine the surface color of the samples detected by measuring L-a-b values as L* (light/dark), a* (red/green), and b* (yellow/blue), and the results were expressed by hue angle. Hue angle was obtained using the equation hue angle (h = arctangent (b*/a*)), where 0° equals red/purple, 90° equals yellow, 180° equals bluish/green, and 270° equals blue [17]. Every sample was measured at three equidistant points, and three samples selected randomly from each package (three package were randomly chosen from the each density active package) were analyzed [18].

2.6. Soluble Solid Concentration (SSC) and Titratable Acidity (TA)

The sample was ground in a mortar and squeezed by hand to obtain juice. Measurement of SSC from fruit juices using a Digital Refractometer (MZB 92, Shanghai Miqingke Industrial Testing Co., Ltd., Shanghai, China). Titratable acidity of strawberry was determined according to the international standard ISO 750-1981. A certain weight of strawberry juice was diluted 100 times with distilled water and filtered to remove the pulp. Two grams of homogenate equal a one gram sample. The samples were heated in a 70–80 °C water bath for 30 min while shaking and cooled to room temperature. Phenolphthalein was used as an indicator. The acidity was measured using a 526 WWW pH meter (Merck, Barcelona, Spain) with a glass electrode. The acidity was measured by titration with 0.1 mol·L^{-1} NaOH to pH 8.1 and expressed as a percentage of citric acid.

2.7. Determination of Vitamin C

Vitamin C content in strawberries was determined by Spectrophotometric method. An appropriate amount of water was added to the sample to make the homogenate, which was centrifuged at $10,000 \times g$ for 10 min and then the supernatant was removed. The dilution times of the sample fluid depend on VC content and sample fluid color. In this study, 1:10 was used as the solid-to-liquid ratio to make the homogenate, as the color of the homogenate was too dark to be determined by spectrophotometry (UV-1800, Mapada Instruments Co., Ltd., Shanghai, China) [19]. Then, the homogenate was diluted 10 times with distilled water. Distilled water was used as a reference, and a series of VC standard solutions were prepared. The absorption spectrum curve of VC was plotted in the range of 220–320 nm to determine the maximum absorption wavelength. The absorbance was finally measured at 267 nm to draw a standard curve and then the vitamin C content of the sample was calculated from the standard curve [18].

2.8. Total Phenolics Content

Total phenolics content in strawberries were determined by the Folin–Ciocalteu method [20]. Briefly, the samples (10 g) were mixed with 10-fold 80% cold methanol (100 mL) at room temperature for 30 min. During this time, the strawberries were thoroughly ground in a methanol (100 mL) mortar and pestle. Then, the homogenate was centrifuged at $10,000 \times g$ for 20 min and filtered. The supernatant was taken and the residue was extracted twice. The supernatants were combined. The crude extract (1 mL) was mixed with 10-fold diluted 2 N Folin–Ciocalteu reagent (4.0 mL). The mixture was kept at room temperature for 5 min and then 4 mL of Na_2CO_3 (7.5% w/v) was added. Before the mixture was incubated for 1 h at room temperature, it was allowed to react in a vortex mixer. The Absorbance was measured by spectrophotometer at 765 nm. The result was expressed as mg·GAE/100 g FW. The amount of total phenolics in the strawberries was calculated using a gallic acid calibration curve.

2.9. 1-Diphenyl-2-Picrylhydrazyl (DPPH) Determination

DPPH free radical scavenging activity of sample extracts was determined following the method described by Cao et al. [21] to assess antioxidant capacity. Crude sample extraction method was the same as total phenol extraction method. Exactly 0.0394 g DPPH was weighed and diluted with 100 mL of 95% methanol to obtain 1 mmol/L DPPH liquor, which was diluted 12 times with methanol to form 0.12 mmol/L DPPH reaction solutions. Then, 0.1 mL sample solution and 2.9 mL DPPH reaction solution were put together in a 5 mL centrifuge tube. The mixture was shaken for 30 min in a dark place at room temperature (Absorbance$_{sample}$). The absorbance was measured at a wavelength of 517 nm using spectrophotometry. The absorbance of 2.9 mL DPPH reaction solution and 0.1 mL methanol mixture was measured (A_0). The absorbance of water in the DPPH reaction mixture was measured simultaneously under the same condition (A_1). $A_0 - A_1 =$ Absorbance$_{control}$. All assays were performed in triplicate. The result was calculated according to Equation (2):

$$\text{DPPH radical scavenging activity}(\%) = (1 - \frac{\text{Absorbance}_{sample}}{\text{Absorbance}_{control}}) \times 100\% \tag{2}$$

2.10. Pyrogallol Peroxidase Assay

The pyrogallol peroxidase (POD) activity was measured according to Chen et al. [22] with some modifications. The fresh fruits were crushed into coarse pieces; 5-g samples were homogenized for 3 min with 20 mL of pre-cooled sodium phosphate buffer (0.1 mol/L, pH 6.8) and 2 wt % of PVP in a cooled mortar and pestle. The homogenate was collected into the centrifuge tube and centrifuged

(TGL-16M, Xiangyi Centrifuge instruments Co., Ltd., Shanghai, China) at $10,000 \times g$ for 20 min. During centrifugation, the homogenate was kept at 4 ± 1 °C. Then, the supernatant was used as a crude enzyme solution. The reaction mixture consisted of 0.5 mL of crude enzyme solution, 3 mL of 25 mM guaiacol, and 200 μL of 0.5 M hydrogen peroxide (30%). Distilled water was used as a reference. The POD value was measured at 470 nm. The measurement was started after 15 s of reaction and then repeated three times every minute. One unit of enzyme activity was defined as an increase in absorbance of 0.001/min. The result was expressed as U per gram of fresh fruit (U·g^{-1} FW).

2.11. Sensory Evaluation of Strawberries

The quality assessment consisted of ten trained reviewers from Yunnan Institute of Food Safety, Kunming University of Science and Technology, Kunming, China. A nine-point hedonic scale based on color, texture, odor, and overall acceptability of samples was used to differentiate changes in sample quality, where 1 = inedible, 3 = poor, 5 = fair, 7 = very good, and 9 = excellent [23].

2.12. Microbiological Analysis

The total bacterial count of the sample was evaluated according to the plate count method. Briefly, 25 g of samples was aseptically transferred to 225 mL of a 0.85% (w/v) sterile physiological saline solution and homogenized. Serial decimal dilutions were prepared in sterile peptone water and poured onto plate count agar (PCA) plates. The total bacterial count was incubated on plate agar (Oxoid, London, UK) for 48 h at 30 °C. All counts were the average of the samples and expressed as log cfu/mL.

2.13. Statistical Analysis

The result was represented as means ± standard deviations and analyzed by analysis of variance (ANOVA) using SPSS (version 19.0, SPSS Inc., Chicago, IL, USA). Duncan's multiple-range test was used to determine significant differences at 95% confidence level.

3. Results and Discussion

3.1. Weight Loss

The weight loss value of strawberry samples during the preservation period is shown in Figure 1. During the preservation period, the weight loss of all packaged samples gradually increased, which might be related to the continuous loss of water from the strawberries to the surrounding environment. As shown in Figure 1, after two days of storage, the mass loss of the strawberry samples packed with PLA/Ag10%, PLA/Ag5%, and PLA/Ag1% film was significantly ($p < 0.05$) lower than that of the PLA films. This was because a certain amount of nano-Ag was incorporated in PLA matrix. Nano-Ag has an outstanding antimicrobial property. It can attach to the cell membranes and penetrate into bacteria, block the bacterial respiratory chain, and eventually kill the bacteria on the surface of the attached material. Thus, higher concentration nano-Ag package has higher antimicrobial property. The fruit packed in the high concentration nano-Ag package was less prone to spoil and deteriorate [24]. The weight loss of the fruit packaged by different concentration film was presented as: PLA film > PLA/Ag1% film > PLA/Ag5% film > PLA/Ag10% film.

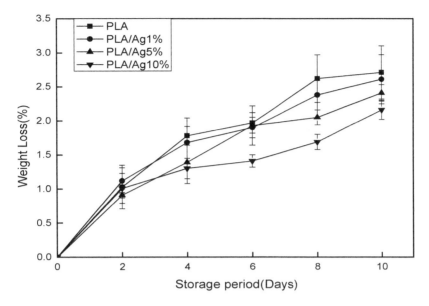

Figure 1. Effect of different concentration of nano-Ag active packages on weight loss of strawberry stored at $4 \pm 1\,^{\circ}C$ for 10 days. Data are presented as mean $\pm$ standard deviation.

3.2. Firmness

Firmness is one of the important indicators for strawberries. The change trend of the tissue firmness of strawberry samples packaged in different packaging films is shown in Figure 2. During the $4 \pm 1\,^{\circ}C$ preservation period, the firmness value of all samples showed a downward trend. After 10 days of storage, the PLA/Ag5% film package had the highest firmness value (40.96 g) and the PLA film package had the lowest firmness value (39.65 g). This might be because the water vapor permeability of nano-mixed membranes was higher than that of other membranes and nano-Ag has certain antibacterial properties. Low-permeability packaging films could increase the relative humidity inside the package to accelerate the softening of the strawberries. Studies have shown that, during the storage period, strawberry tissue becomes soft due to metabolic changes and moisture loss of the enzyme, which in turn reduces the firmness of the strawberry [25].

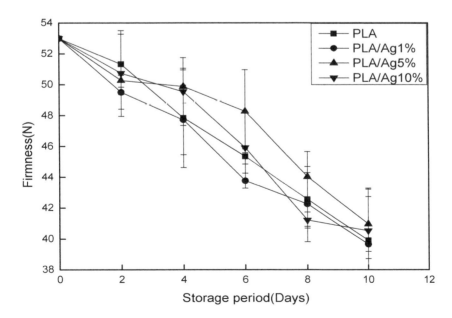

Figure 2. Effect of different concentration of nano-Ag active packages on firmness of strawberry stored at $4 \pm 1\,^{\circ}C$ for 10 days. Data are presented as mean $\pm$ standard deviation.

3.3. SSC and TA

The amount of SSC in strawberries could be used to assess the ripeness of strawberries and fully mature strawberries contain the most soluble solids [26]. In Figure 3, longer storage periods leads to greater soluble solids content in the four groups of strawberries. During the first two days of storage, there was no significant ($p > 0.05$) difference in SSC between PLA/Ag1% and PLA/Ag5% group. The increase in SSC value in the four groups of strawberries was because the starch in the strawberries continuously converted to soluble sugars as the storage period increased. PLA/Ag5% and PLA/Ag10% film had better inhibition of strawberry respiration and inhibition of metabolic enzyme activity than PLA film. Therefore, the soluble solids growth of strawberry coated with PLA/Ag5% and PLA/Ag10% film was slower.

The amount of TA in strawberries is closely related to their flavor, and the acid content decreases as the respiratory metabolism of strawberries increases. In Figure 4, the TA content in the four groups of strawberries decreased continuously with the extension of the storage period. In the PLA group, TA value rapidly decreased most in the strawberry, followed by the PLA/Ag1% group, and was slowest in the PLA/Ag5% and PLA/Ag10% groups. In the whole process of storage, the content of TA of nano-active packaging film is higher than that of pure polylactic acid packaging film, indicating that nano-modified film is conducive to delaying the decrease of TA content of fruits during storage. This is consistent with the findings of Li et al. [27]. The experimental results showed that the nano-composite membrane could inhibit the respiration of strawberry and slow down the consumption of acid in the physiological metabolic activities of strawberry, thus effectively slowing down the downward trend of titratable acid and extending the shelf life of the strawberries.

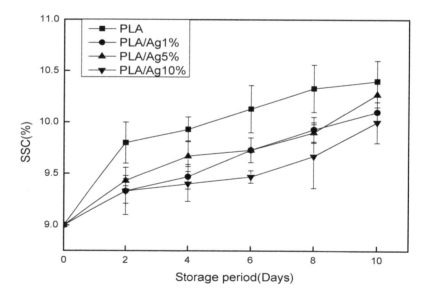

Figure 3. Effect of different concentration of nano-Ag active packages on soluble solid content of strawberry stored at $4 \pm 1\,°C$ for 10 days. Data are presented as mean $\pm$ standard deviation.

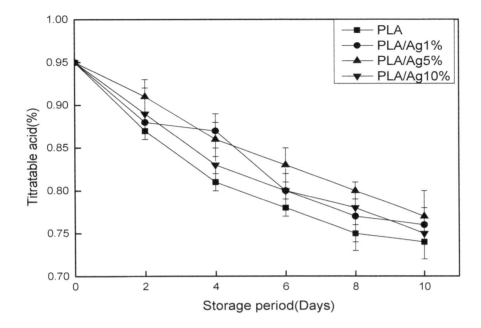

Figure 4. Effect of different concentration of nano-Ag active packages on titratable acid of strawberry stored at 4 ± 1 °C for 10 days. Data are presented as mean ± standard deviation.

3.4. Surface Color

During storage, discoloration mainly occurs on the surface of the strawberry and is mainly manifested as darkening or spoilage. This greatly reduces the nutritional value and sensory characteristics of strawberry itself, and seriously affects consumers' desire to purchase products. Among them, the enzymatic browning of the reaction of polyphenol compounds with oxygen catalyzed by the endogenous polyphenol oxidase (PPO) [28] in strawberry is a major problem. Hue angle was used to characterize the color change of strawberry surface. In Figure 5, the degree of coloration of the samples in each group of packaging films during storage gradually decreased. On Day 10 of storage, the hue angles of PLA, PLA/Ag1%, PLA/Ag5%, and PLA/Ag10% decreased by 18.31%, 16.31%, 13.47%, and 16.78%, respectively, when compared with the initial value. The color difference value of the fruit in the PLA active packaging film, in which the amount of nano-Ag was added at 5%, was lower than the initial value. The value of the hue angle was always higher than other groups, and the difference was significant ($p < 0.05$). The concentration of the active packaging film showed better protection of strawberry surface color than other concentrations. However, the result showed that, although the different processing and storage methods could improve the decline of the surface color of the strawberry, the change of the surface color of the strawberry could not be used as an indicator to effectively reflect the strawberry fruit quality [29].

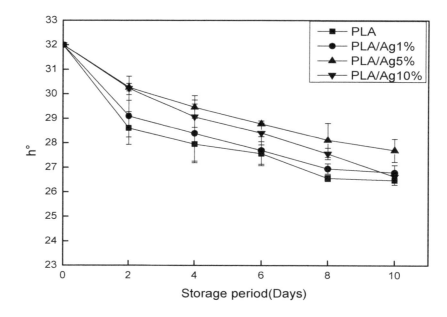

Figure 5. Effect of different concentration of nano-Ag active packages on the hue angle of strawberry stored at $4 \pm 1\,^{\circ}$C for 10 days. Data are presented as mean $\pm$ standard deviation.

3.5. Determination of Vitamin C

The antioxidant properties of vitamin C play an important role in the metabolism of plants and animals [30]. Since the human body cannot synthesize vitamin C, it can only meet the needs of the human body through diet or supplementary nutrients, and a very healthy way to supplement vitamin C is eating fruit. Thus, vitamin C plays a vital role in the preservation of food, especially in vegetable and fruits during the storage. As Mcerlain and Marson mentioned, improper preservation methods can accelerate the loss of vitamins in food [31]. Therefore, the change of VC content in strawberry in different content of nano-Ag packaging film can verify whether nano-Ag film package has a good preservation effect. Changes in vitamin C content of strawberries during storage are shown in Figure 6, in which the content of vitamin C decreases with time for all the samples, which might be attributed to its oxidation through OH radicals in the strawberries [32]. It could be seen in the figure that vitamin C content decreased linearly after storage. With the prolongation of storage time, the strawberry in all active packaging films gradually showed significant ($p < 0.05$) difference in vitamin C content compared with pure polylactic acid from Day 4 until the end of storage days. Among them, PLA/Ag1% is significantly ($p < 0.05$) different from PLA during the whole storage process. Over time, the vitamin C content of strawberries in 5% nano-Ag active packages was higher than other active packaging film concentrations during storage. On Day 10 of storage, the content of ascorbic acid in the pure PLA film and nano-Ag-added active packaging had a significant ($p < 0.05$) difference from the initial value, indicating that the nano-Ag active film can effectively reduce the loss of ascorbic acid in strawberry. This might be due to the addition of nano-Ag, changes in membrane permeability [18], inhibition of respiration, and delay in occurrence of ascorbic acid oxidation reaction. The ascorbic acid content of strawberry in the packaging of nano-Ag films with a low to high concentration decreased by 38.18%, 37.13%, 33.38%, and 42.93%, respectively, compared with the initial values.

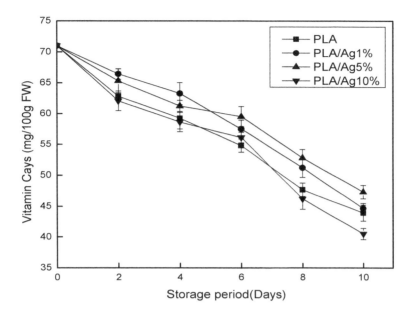

Figure 6. Effect of different concentration of nano-Ag active packages on the vitamin C level strawberry stored at 4 ± 1 °C for 10 days. Data are presented as mean $\pm$ standard deviation.

3.6. Pyrogallol Peroxidase (POD)

Enzymatic browning caused by oxidation of phenolic compounds is one of the most important causes of fruit color deterioration. POD is one of the oxidoreductases involved in enzymatic browning [33]. It might cause not only spoilage of the appearance of the fruit but also the weight loss. In Figure 7, there was no significant ($p < 0.05$) difference in pod content in each packaging film during the first two days of storage. The pod activity in all active packaging film packages gradually increased until the activity peaks on Day 6. The maximum pod activity values of PLA, PLA/Ag1%, PLA/Ag5% and PLA/Ag10% were 15.12, 15.55, 17.02 and 16.69 U/g FW, respectively. PLA and all active packaging films are significantly ($p < 0.05$) different. PLA/Ag5% film was higher than that of other concentrations of nano-active film and pure PLA film samples. Liu et al. [34] found that POD activity of mushrooms increased first and then decreased during storage. Moreover, the POD activity of mushrooms was always higher than that of the control group, so it was beneficial to prolong the storage period of mushrooms, which was consistent with the results of this study. This could be due to the antibacterial properties of the nanoparticles in the film, which could delay fruit senescence. The high activity of POD indicated that the sample had a low degree of tissue aging and that the fruits and vegetables were fresher. After Day 6, the activity gradually decreased. On Day 8, PLA10 was significantly lower than all packaging films. Then, there was no significant ($p < 0.05$) difference between the packaging films over time. The POD enzyme activity increased rapidly from the beginning of storage to Day 6, possibly due to the POD-induced enzyme action. When the fruit was adversely affected by the outside, the enzyme activity increased; and the higher the cellular activity, the greater the increased [35]. As the POD activity declined, the quality of the fruit began to deteriorate. Studies have shown that POD may be involved in the production of ethylene in fruits and can be used as an indicator of fruit ripening and senescence [36]. This might regulate the maturation of the strawberry and ultimately improve the sensory quality of the fruit in the nano-active packaging film. Considering fruit preservation effect and material cost savings, from Day 0 to Day 6, PLA/Ag5% film is better.

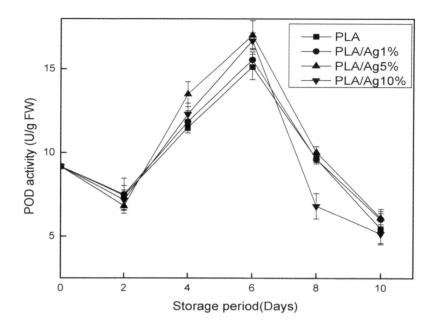

Figure 7. Effect of different concentration of nano-Ag active packages on the activity of POD of strawberry stored at 4 ± 1 °C for 10 days. Data are presented as mean ± standard deviation.

3.7. Total Phenolics Content

Total phenols are very important for the quality of strawberries, not only because of their contribution to the taste, but also polyphenols and anthocyanins affect the appearance of pigments in fruits and vegetables [37]. Phenols have anti-aging, anti-cancer, anti-inflammatory and anti-oxidation biological functions, making total phenols an important indicator to assess the fruit and vegetable preservation effect [38]. Changes in the total phenolic content of the sample during storage are shown in Figure 8. In Figure 8, the content of total phenols encapsulated by various contents of nano-Ag increased at first and then decreased. The total phenolic content of the pure PLA packaging film increased rapidly after storage began and peaked on Day 4. The content of total phenols in the nano-Ag packaging films with different contents rose slowly and peaked on Day 6. The peak value was significantly ($p < 0.05$) lower than the peak value of total phenolics in the pure PLA packaging film. From Day 6 to Day 10, the total phenolic content of strawberry in pure PLA packaging film and nano-Ag active packaging film showed a decreasing trend, but the total phenol content of nano-Ag active packaging film was higher than that of pure nano-Ag packaging film. The content is maintained at a high level and the downward trend tends to be gentle. Yang et al. showed that this might be due to the fact that nano-packaging inhibits the accumulation of anthocyanin in fruits [39]. The rapid accumulation of total phenols in the first four days might be due to the accumulation of other phenolic substances. Shin et al. found that changes in the total phenolic content of strawberries during storage are affected by the maturity at harvest [40]. The total phenolic content of fully ripened strawberries will continue to decrease during storage. The experimental samples maturation were all 70–80% but not fully matured. Therefore, the phenols gradually accumulated during the early storage and led to increased levels. Studies have also shown that higher concentrations of oxygen in the environment will accelerate the accumulation of total phenols in fruits [41]. The nano-Ag particles added to the PLA film change the oxygen and carbon dioxide transmission rate, which in turn reduces the oxygen content in the package [39]. Therefore, the nano-Ag active packaging film has lower total phenol content than the pure PLA packaging film, which can effectively delay the decay of the total phenol content. In addition, the total phenol content in the active packaging film does not increase with the increase of nano-Ag content in the active packaging film. In the low concentration range of the active film, the Ag concentration of 5% has higher total phenol content than other concentrations of the nano-Ag active packaging film. The overall trend of nano-Ag films with a concentration of 10% is higher than that of pure PLA films lower than that of low-concentration nano-Ag films. This might be because, as the nanoparticles are added, the nanoparticles are aggregated, the voids of the membrane are enlarged, and the respiration continuously causes the loss of phenolic content.

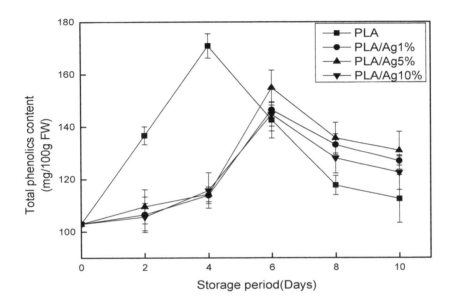

Figure 8. Effect of different concentration of nano-Ag active packages on total phenolics of strawberry stored at $4 \pm 1\,^{\circ}$C for 10 days. Data are presented as mean $\pm$ standard deviation.

3.8. 1-Diphenyl-2-Picrylhydrazyl (DPPH)

Antioxidant ability is one of the important indicators for evaluating fruits and vegetables. There are many methods to determine the antioxidant capacity of samples, such as ferric reduction antioxidant (FRAP) method, oxygen radical absorbance capacity (ORAC) method for measuring antioxidant capacity, and so on. However, the DPPH method has been widely used in the determination of the antioxidant capacity of fruits and vegetables and their extracts due to their sensitivity, rapidity, reliability, and simple method [42]. In Figure 9, the DPPH radical scavenging rates of the packaging films of strawberry during the whole storage period showed a trend of increasing first and then decreasing. On Day 4, the pure PLA film reached its peak first and then decreased rapidly. On Day 6, the DPPH radical scavenging rate of each group of nano-Ag active packaging films also peaked and then decreased and maintained at a relatively high level. There was a significant ($p < 0.05$) difference between the numerical value and the pure PLA film. Li and coworkers' [43] research results are consistent with the results of this study.

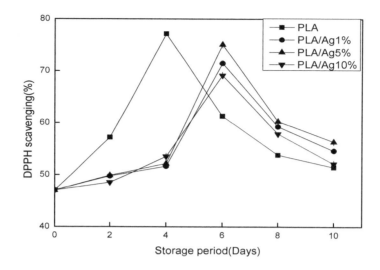

Figure 9. Effect of different concentration of nano-Ag active packages on antioxidant capacity measured by DPPH. Data are presented as mean ± standard deviation.

3.9. Microbiological Analysis

Total bacterial is one of the indicators for judging food hygiene. The change in the total number of bacteria during storage is shown in Figure 10. The total number of bacteria gradually increased throughout the storage period. There was no significant ($p > 0.05$) difference in the total number of bacteria between the PLA/Ag1% and PLA/Ag5% groups during the entire storage period. The total number of bacteria in the PLA/Ag10% wrapped strawberry samples was significantly ($p > 0.05$) lower than that in PLA. This may be because the incorporation of nano-Ag into PLA can effectively inhibit the growth of fruit microorganisms. This may be related to the antibacterial mechanism of nano-Ag. The antibacterial mechanism of nano-Ag is related to membrane damage caused by free radicals derived from the surface of nano-Ag [44]. Fan et al. [45] found that nano active film is better than PLA film for food preservation. Li and coworkers' [46] study shows that the higher content of ZnO composite film with 3 wt % content has the best preservation performance compared with 0 wt % and 1 wt %.The experimental results are consistent with the above studies.

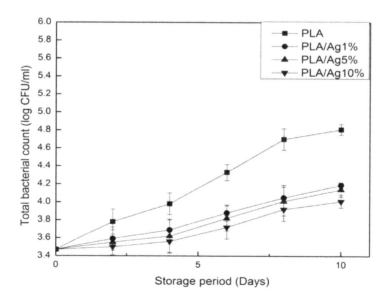

Figure 10. Effect of different concentration of nano-Ag active packages on the account of total bacterial. Data are presented as mean ± standard deviation.

3.10. Sensory Evaluation of Strawberries

The sensory evaluation of the strawberry was performed on a nine-point scale, and the odor, color, texture, and overall acceptability were evaluated. Table 1 lists the sensory scores for strawberries, with the values for all groups decreasing with increasing storage time. There was no significant ($p > 0.05$) difference in sensory scores of all packaged samples on Day 2. However, after eight days of storage, the odor and texture of the strawberry samples packed with PLA/Ag1%, PLA/Ag5%, and PLA/Ag10% were not affected. The overall acceptability score was significantly higher than that of the PLA film. At the end of the storage time, the overall acceptability scores of samples packaged with PLA/Ag1%, PLA/Ag5%, and PLA/Ag10% were still higher than 5, and the sample maintained proper quality characteristics considering the sensory parameters such as odor, color, and texture. After 10 days of storage, the overall acceptability of the sample packed with PLA/Ag5% film was highest, and it was not significantly ($p > 0.05$) different from that of PLA/Ag%1 and PLA/Ag10% films. Six points are the criteria for good product quality and marketability. The shelf life of strawberry samples packed with PLA, PLA/Ag1%, PLA/Ag%5 and PLA/Ag10% films was four, six, eight and six days, respectively. The results showed that PLA/Ag film can improve the quality of strawberry during refrigerated storage. The main reason is that the nanofiber membrane has a small pore size and a high porosity. It can prevent gas and water from passing through and can block foreign pollutants from infecting the sample. It also improves the hydrophilicity of the membrane and prevents condensation of water vapor on the inner surface of the membrane [47].

Table 1. Sensory evaluation of strawberries stored for 10 days at $4 \pm 1\ °C$ in different concentration of nano-Ag active packages.

Treatment	Odor	Color	Texture	Overall Acceptability
Day 0	9	9	9	9
Day 2				
PLA	8.18 ± 0.03 [ab]	7.87 ± 0.32 [a]	8.07 ± 0.17 [a]	7.94 ± 0.02 [a]
PLA/Ag1%	8.05 ± 0.18 [a]	8.06 ± 0.32 [a]	8.22 ± 0.04 [ab]	8.05 ± 0.59 [a]
PLA/Ag5%	8.29 ± 0.07 [b]	8.08 ± 0.05 [a]	8.28 ± 0.07 [b]	8.13 ± 0.11 [a]
PLA/Ag10%	8.22 ± 0.04 [ab]	8.07 ± 0.15 [a]	8.19 ± 0.04 [ab]	8.04 ± 0.02 [a]
Day 4				
PLA	7.19 ± 0.04 [a]	7.05 ± 0.39 [a]	6.98 ± 0.14 [a]	6.89 ± 0.27 [a]
PLA/Ag1%	7.37 ± 0.18 [a]	6.85 ± 0.22 [a]	7.05 ± 0.18 [a]	7.07 ± 0.01 [ab]
PLA/Ag5%	7.37 ± 0.13 [a]	7.51 ± 0.44 [a]	7.18 ± 0.01 [a]	7.2 ± 0.08 [b]
PLA/Ag10%	7.24 ± 0.02 [a]	7.19 ± 0.21 [a]	7.12 ± 0.08 [a]	7.12 ± 0.06 [ab]
Day 6				
PLA	$6.68 + 0.06$ [ab]	6.27 ± 0.21 [a]	6.45 ± 0.05 [a]	6.03 ± 0.4 [a]
PLA/Ag1%	6.67 ± 0.02 [ab]	6.31 ± 0.17 [a]	6.49 ± 0.01 [ab]	6.52 ± 0.02 [b]
PLA/Ag5%	6.76 ± 0.09 [b]	6.63 ± 0.02 [b]	6.55 ± 0.04 [b]	6.71 ± 0.05 [b]
PLA/Ag10%	6.59 ± 0.1 [a]	6.52 ± 0.15 [ab]	6.46 ± 0.04 [a]	6.4 ± 0.08 [ab]
Day 8				
PLA	6.22 ± 0.02 [ab]	5.17 ± 0.14 [a]	5.99 ± 0.27 [a]	4.78 ± 0.06 [a]
PLA/Ag1%	6.13 ± 0.16 [a]	5.38 ± 0.21 [a]	6.1 ± 0.07 [a]	5.9 ± 0.04 [ab]
PLA/Ag5%	6.29 ± 0.05 [b]	5.43 ± 0.21 [a]	6.12 ± 0.35 [a]	6.3 ± 0.56 [b]
PLA/Ag10%	6.16 ± 0.03 [ab]	5.34 ± 0.06 [a]	6.08 ± 0.05 [a]	4.82 ± 1.02 [a]
Day 10				
PLA	5.71 ± 0.23 [b]	5.01 ± 0.55 [a]	5.45 ± 0.12 [ab]	4.23 ± 0.57 [a]
PLA/Ag1%	5.05 ± 0.08 [a]	5.03 ± 0.09 [a]	5.36 ± 0.12 [a]	5.36 ± 0.24 [b]
PLA/Ag5%	5.66 ± 0.32 [ab]	5.14 ± 0.25 [a]	5.6 ± 0.03 [b]	5.73 ± 0.04 [b]
PLA/Ag10%	5.13 ± 0.49 [ab]	4.99 ± 0.02 [a]	5.46 ± 0.06 [ab]	5.35 ± 0.15 [b]

a–b Values followed by different letters in the same column were significantly different ($p < 0.05$), where a is the lowest value.

4. Conclusions

Compared with pure PLA film, nano-Ag active film could effectively reduce the weight loss rate of strawberries during storage, and delay the drop in hardness, soluble solids, and titratable acid content. High levels of ascorbic acid and total phenols were maintained during the late storage period. The strawberries had a higher antioxidant capacity in the later period of storage, thus delaying the aging of the fruit and providing a better preservation effect. Therefore, the nano-Ag active packaging film has a better preservation effect, and PLA/Ag5% film is slightly better than other films.

Author Contributions: Conceptualization, Y.Q. and L.L.; Methodology, H.C. (Haiyan Chen); Software, B.Z.; Validation, W.L and B.Z.; Formal Analysis, C.Z.; Investigation, H.C. (Hai Chi); Resources, Y.Q.; Data Curation, J.X.; Writing-Original Draft Preparation, C.Z.; Writing-Review & Editing, C.Z.; Visualization, Y.Q.; Supervision, J.X.; Project Administration, Y.Q.; Funding Acquisition, Y.Q.

References

1.   Pombo, M.A.; Dotto, M.C.; Martinez, G.A.; Civello, P.M. UV-C irradiation delays strawberry fruit softening and modifies the expression of genes involved in cell wall degradation. *Postharvest Biol. Technol.* **2009**, *51*, 141–148. [CrossRef]

2.   Han, C.; Zhao, Y.; Leonard, S.W.; Traber, M.G. Edible coatings to improve storability and enhance nutritional value of fresh and frozen strawberries (*Fragaria* × *ananassa*) and raspberries (*Rubus ideaus*). *Postharvest Biol. Technol.* **2004**, *33*, 67–78. [CrossRef]

3.   Almenar, E.; Hernandez-Munoz, P.; Lagaron, J.M.; Catala, R.; Gavara, R. Controlled atmosphere storage of wild strawberry fruit (*Fragaria vesca* L.). *J. Agric. Food Chem.* **2006**, *54*, 86–91. [CrossRef] [PubMed]

4.   Ju, Y.K.; Kim, H.J.; Lim, G.O.; Jang, S.A.; Song, K.B. The effects of aqueous chlorine dioxide or fumaric acid treatment combined with UV-C on postharvest quality of 'Maehyang' strawberries. *Postharvest Biol. Technol.* **2010**, *56*, 254–256.

5.   Chen, F.; Liu, H.; Yang, H.; Lai, S.; Cheng, X.; Xin, Y.; Yang, B.; Hou, H.; Yao, Y.; Zhang, S. Quality attributes and cell wall properties of strawberries (*Fragaria annanassa* Duch.) under calcium chloride treatment. *Food Chem.* **2011**, *126*, 450–459. [CrossRef]

6.   Bajpai, P.K.; Singh, I.; Madaan, J. Tribological behavior of natural fiber reinforced PLA composites. *Wear* **2013**, *297*, 829–840. [CrossRef]

7.   Eili, M.; Shameli, K.; Ibrahim, N.A.; Yunus, W.M. Degradability Enhancement of Poly(Lactic Acid) by Stearate-Zn_3Al LDH Nanolayers. *Int. J. Mol. Sci.* **2012**, *13*, 7938–7951. [CrossRef] [PubMed]

8.   Zapata, P.A.; Palza, H.; Delgado, K.; Rabagliati, F.M. Novel antimicrobial polyethylene composites prepared by metallocenic in situ polymerization with TiO_2-based nanoparticles. *J. Polym. Sci. Part A* **2012**, *50*, 4055–4062. [CrossRef]

9.   Hua, Y.Q.; Zhang, Y.Q.; Wu, L.B.; Huang, Y.Q.; Wang, G.Q. Mechanical and Optical Properties of Polyethylene Filled with Nano-SiO_2. *J. Macromol. Sci. Part B* **2005**, *44*, 149–159. [CrossRef]

10. Li, X.H.; Li, W.L.; Xing, Y.G.; Jiang, Y.H.; Ding, Y.L.; Zhang, P.P. Effects of Nano-ZnO Power-Coated PVC Film on the Physiological Properties and Microbiological Changes of Fresh-Cut "Fuji" Apple. *Adv. Mater. Res.* **2011**, *152–153*, 450–453. [CrossRef]

11. Brody, A.L. Noncomposite technology in food packaging. *Food Technol.* **2007**, *61*, 80–83.

12. Silver, S.; Le, T.P. Bacterial Heavy Metal Resistance: New Surprises. *Annu. Rev. Microbiol.* **1996**, *50*, 753–789. [CrossRef] [PubMed]

13. Azlin-Hasim, S.; Cruz-Romero, M.C.; Ghoshal, T.; Morris, M.A.; Cummins, E.; Kerry, J.P. Application of silver nanodots for potential use in antimicrobial packaging applications. *Innov. Food Sci. Emerg. Technol.* **2015**, *27*, 136–143. [CrossRef]

14. Emamifar, A.; Kadivar, M.; Shahedi, M.; Soleimanian-Zad, S. Effect of nanocomposite packaging containing Ag and ZnO on inactivation of *Lactobacillus plantarum* in orange juice. *Food Control* **2011**, *22*, 408–413. [CrossRef]

15. Qin, Y.Y.; Li, W.H.; Liu, D.; Yuan, M.L.; Li, L. Development of active packaging film made from poly(lactic acid) incorporated essential oil. *Prog. Org. Coat.* **2017**, *103*, 76–82. [CrossRef]

16. Qin, Y.Y.; Liu, D.; Wu, Y.; Yuan, M.L.; Li, L.; Yang, J.Y. Effect of PLA/PCL/cinnamaldehyde antimicrobial packaging on physicochemical and microbial quality of button mushroom (*Agaricus bisporus*). *Postharvest Biol. Technol.* **2015**, *99*, 73–79. [CrossRef]

17. Gong, Y.; Mattheis, J.P. Effect of ethylene and 1-methylcyclopropene on chlorophyll catabolism of broccoli florets. *Plant Growth Regul.* **2003**, *40*, 33–38. [CrossRef]

18. Qin, Y.Y.; Zhuang, Y.; Wu, Y.; Li, L. Quality evaluation of hot peppers stored in biodegradable poly(lactic acid)-based active packaging. *Sci. Hortic.* **2016**, *202*, 1–8. [CrossRef]

19. Khan, M.M.R.; Rahman, M.M.; Islam, M.S.; Begum, S.A. A simple UV-spectrophotometric method for the determination of vitamin C content in various fruits and vegetables at Sylhet area in Bangladesh. *J. Biol. Sci.* **2006**, *6*, 2238. [CrossRef]

20. Xu, G.H.; Liu, D.H.; Chen, J.C.; Ye, X.Q.; Ma, Y.Q.; Shi, J. Juice components and antioxidant capacity of citrus varieties cultivated in China. *Food Chem.* **2008**, *106*, 545–551. [CrossRef]

21. Larrauri, J.A.; Sanchez-Moreno, C.; Saura-Calixto, F. Effect of temperature on the free radical scavenging capacity of extracts from red and white grape pomace peels. *J. Agric. Food Chem.* **1998**, *46*, 2694–2697. [CrossRef]

22. Chen, X.N.; Ren, L.P.; Li, M.L.; Qian, J.; Fan, J.F.; Du, B. Effects of clove essential oil and eugenol on quality and browning control of fresh-cut lettuce. *Food Chem.* **2017**, *214*, 432–439. [CrossRef] [PubMed]

23. Velickova, E.; Winkelhausen, E.; Kuzmanova, S.; Alves, V.D.; Moldao-Martins, M. Impact of chitosan-beeswax edible coatings on the quality of fresh strawberries (*Fragaria ananassa* cv Camarosa) under commercial storage conditions. *LWT-Food Sci. Technol.* **2013**, *52*, 80–92. [CrossRef]

24. Shrivastava, S.; Bera, T.; Roy, A.; Singh, G.; Ramachandrarao, P.; Dash, D. Characterization of enhanced antibacterial effects of novel silver nanoparticles. *Nanotechnology* **2007**, *18*, 225103. [CrossRef]

25. Ahmadi-Afzadi, M.; Tahir, I.; Nybom, H. Impact of harvesting time and fruit firmness on the tolerance to fungal storage diseases in an apple germplasm collection. *Postharvest Biol. Technol.* **2013**, *82*, 51–58. [CrossRef]

26. Jiang, M.; Xiao-Jun, H.U.; Jiao, M.S.; Huang, S.J.; Qiu-Ping, L.U.; Qiao-Lin, F.U. Preservative effect of citric acid and nano-ZnO treatment on storage of mango at room temperature. *North. Hortic.* **2012**, *11*, 172–176.

27. Li, H.; Li, F.; Wang, L.; Sheng, J.; Xin, Z.; Zhao, L.; Xiao, H.; Zheng, Y.; Hu, Q. Effect of nano-packing on preservation quality of Chinese jujube (*Ziziphus jujuba* Mill. var. inermis (Bunge) Rehd). *Food Chem.* **2009**, *114*, 547–552. [CrossRef]

28. He, Q.; Luo, Y.G.; Chen, P. Elucidation of the mechanism of enzymatic browning inhibition by sodium chlorite. *Food Chem.* **2008**, *110*, 847–851. [CrossRef] [PubMed]

29. Ordidge, M.; García-Macías, P.; Battey, N.H.; Gordon, M.H.; John, P.; Lovegrove, J.A.; Vysini, E.; Wagstaffe, A.; Hadley, P. Development of colour and firmness in strawberry crops is UV light sensitive, but colour is not a good predictor of several quality parameters. *J. Sci. Food Agric.* **2012**, *92*, 1597–1604. [CrossRef] [PubMed]

30. Davey, M.W.; Van Montagu, M.; Inze, D.; Sanmartin, M.; Kanellis, A.; Smirnoff, N.; Benzie, I.J.J.; Strain, J.J.; Favell, D.; Fletcher, J. Plant L-ascorbic acid: Chemistry, function, metabolism, bioavailability and effects of processing. *J. Sci. Food Agric.* **2000**, *80*, 825–860. [CrossRef]

31. Mcerlain, L.; Marson, H.; Ainsworth, P.; Burnett, S.A. Ascorbic acid loss in vegetables: Adequacy of a hospital cook-chill system. *Int. J. Food Sci. Nutr.* **2001**, *52*, 205–211. [PubMed]

32. Hussain, P.R.; Dar, M.A.; Wani, A.M. Effect of edible coating and gamma irradiation on inhibition of mould growth and quality retention of strawberry during refrigerated storage. *Int. J. Food Sci. Technol.* **2012**, *47*, 2318–2324. [CrossRef]

33. Liu, Z.; Wang, X. Changes in color, antioxidant, and free radical scavenging enzyme activity of mushrooms under high oxygen modified atmospheres. *Postharvest Biol. Technol.* **2012**, *69*, 1–6. [CrossRef]

34. Chisari, M.; Barbagallo, R.N.; Spagna, G. Characterization of polyphenol oxidase and peroxidase and influence on browning of cold stored strawberry fruit. *J. Agric. Food Chem.* **2007**, *55*, 3469–3476. [CrossRef] [PubMed]

35. Junliang, L.; Chen, K.; Zhang, S. Study on the change in ethylene production, the activity of superoxide dismutase and peroxidase in Chinese gooseberry fruits during ripening. *J. Zhejiang Agric. Univ.* **1993**, *19*, 135–138.

36. Ku, H.S.; Yang, S.F.; Pratt, H.K. Ethylene production and peroxidase activity during tomato fruit ripening. *Plant Cell Physiol.* **1970**, *11*, 241–246. [CrossRef]

37. Mazza, G.; Brouillard, R. The mechanism of co-pigmentation of anthocyanins in aqueous solutions. *Phytochemistry* **1990**, *29*, 1097–1102. [CrossRef]

38. Bhat, R.; Stamminger, R. Impact of ultraviolet radiation treatments on the physicochemical properties, antioxidants, enzyme activity and microbial load in freshly prepared hand pressed strawberry juice. *Food Sci. Technol. Int.* **2015**, *21*, 354–363. [CrossRef] [PubMed]

39. Yang, F.M.; Li, H.M.; Li, F.; Xin, Z.H.; Zhao, L.Y.; Zheng, Y.H.; Hu, Q.H. Effect of Nano-Packing on Preservation Quality of Fresh Strawberry (*Fragaria ananassa* Duch. cv Fengxiang) during Storage at 4 degrees C. *J. Food Sci.* **2010**, *75*, C236–C240. [CrossRef] [PubMed]

40. Shin, Y.; Ryu, J.A.; Liu, R.H.; Nock, J.F.; Watkins, C.B. Harvest maturity, storage temperature and relative humidity affect fruit quality, antioxidant contents and activity, and inhibition of cell proliferation of strawberry fruit. *Postharvest Biol. Technol.* **2008**, *49*, 201–209. [CrossRef]

41. Ayala-Zavala, J.F.; Wang, S.Y.; Wang, C.Y.; Gonzalez-Aguilar, G.A. High oxygen treatment increases antioxidant capacity and postharvest life of strawberry fruit. *Food Technol. Biotechnol.* **2007**, *45*, 166–173.

42. Leong, L.P.; Shui, G. An investigation of antioxidant capacity of fruits in Singapore markets. *Food Chem.* **2002**, *76*, 69–75. [CrossRef]

43. Li, D.; Ye, Q.; Lei, J.; Luo, Z. Effects of nano-TiO_2-LDPE packaging on postharvest quality and antioxidant capacity of strawberry (*Fragaria ananassa* Duch.) stored at refrigeration temperature. *J. Sci. Food Agric.* **2016**, *97*, 1116. [CrossRef] [PubMed]

44. Kim, J.S.; Kuk, E.; Yu, K.N.; Kim, J.H.; Park, S.J.; Hu, J.L.; Kim, S.H.; Park, Y.K.; Yong, H.P.; Hwang, C.Y. Antimicrobial effects of silver nanoparticles. *Nanomed. Nanotechnol. Biol. Med.* **2007**, *3*, 95–101. [CrossRef] [PubMed]

45. Hussain, M.; Bensaid, S.; Geobaldo, F.; Saracco, G.; Russo, N. Photocatalytic Degradation of Ethylene Emitted by Fruits with TiO_2 Nanoparticles. *Ind. Eng. Chem. Res.* **2010**, *50*, 2536–2543. [CrossRef]

46. Fan, J.; Chu, Z.; Li, L.; Zhao, T.; Yin, M.; Qin, Y. Physicochemical Properties and Microbial Quality of *Tremella aurantialba* Packed in Antimicrobial Composite Films. *Molecules* **2017**, *22*, 500. [CrossRef] [PubMed]

47. Li, W.; Lin, L.; Yun, C.; Lan, T.; Chen, H.; Qin, Y. Effects of PLA Film Incorporated with ZnO Nanoparticle on the Quality Attributes of Fresh-Cut Apple. *Nanomaterials* **2017**, *7*, 207. [CrossRef] [PubMed]

Permissions

List of Contributors

Ludmila Motelica, Denisa Ficai and Ovidiu Cristian Oprea
Faculty of Applied Chemistry and Materials Science, University Politehnica of Bucharest, 060042 Bucharest, Romania

Ecaterina Andronescu and Anton Ficai
Faculty of Applied Chemistry and Materials Science, University Politehnica of Bucharest, 060042 Bucharest, Romania
Section of Chemical Sciences, Academy of Romanian Scientists, 050045 Bucharest, Romania

Durmuş Alpaslan Kaya
Department of Field Crops, Faculty of Agriculture, Hatay Mustafa Kemal University, 31030 Antakya Hatay, Turkey

Rachael Reid and Declan Bolton
Food Safety Department, Teagasc Food Research Centre, Ashtown, Dublin 15, Ireland

Andrey A. Tyuftin
Food Packaging Group, School of Food & Nutritional Sciences, University College Cork, College Road, Cork, Ireland

Joe P. Kerry, Séamus Fanning and Paul Whyte
School of Public Health, University College Dublin, Belfield, Dublin 4, Ireland

Velázquez-Contreras Friné
Campus de los Jerónimos, Universidad Católica San Antonio de Murcia, 135, 30107 Guadalupe, Murcia, Spain
Universidad Panamericana ESDAI, Álvaro del Portillo 49, Zapopan 45010, Jalisco, Mexico

Núñez-Delicado Estrella and Gabaldón José Antonio
Campus de los Jerónimos, Universidad Católica San Antonio de Murcia, 135, 30107 Guadalupe, Murcia, Spain

Acevedo-Parra Hector
Universidad Panamericana ESDAI, Álvaro del Portillo 49, Zapopan 45010, Jalisco, Mexico

Nuño-Donlucas Sergio Manuel
Departamento de Ingeniería Química, Universidad de Guadalajara, Blvd, Marcelino García Barragán 1421, Guadalajara 44430, Jalisco, Mexico

Jun Liang
State Key Laboratory of Food Nutrition and Safety, Tianjin University of Science & Technology, Tianjin 300222, China
College of Packaging and Printing Engineering, Tianjin University of Science & Technology, Tianjin 300222, China

Ruipeng Chen
State Key Laboratory of Food Nutrition and Safety, Tianjin University of Science & Technology, Tianjin 300222, China

Rui Wang
College of Packaging and Printing Engineering, Tianjin University of Science & Technology, Tianjin 300222, China

Maizatulnisa Othman, Haziq Rashid, Nur Ayuni Jamal, Sharifah Imihezri Syed Shaharuddin and Sarina Sulaiman
Department of Manufacturing and Materials Engineering, Faculty of Engineering, International Islamic University Malaysia, Gombak 50728, Selangor, Malaysia

H. Saffiyah Hairil
PERMATApintar College, National University Malaysia, Bangi 43600, Selangor, Malaysia

Khalisanni Khalid
Agri-Nanotechnology Program, Biotechnology and Nanotechnology Research Center, Malaysian Agricultural Research and Development Institute (Mardi), MARDI Headquarters, Persiaran MARDI-UPM, Serdang 43400, Selangor, Malaysia

Mohd Nazarudin Zakaria
Department of Biocomposite Technology, Faculty of Applied Sciences, Universiti Teknologi MARA, Shah Alam 40450, Selangor, Malaysia

Ana Maria Diez-Pascual
Department of Analytical Chemistry, Physical Chemistry and Chemical Engineering, Faculty of Sciences, Alcalá University, 28871 Madrid, Spain

Bi Foua Claude Alain Gohi
Biotechnology Institute, College of Chemical Engineering, Xiangtan University, Xiangtan 411105, China
School of Biological and Chemical Engineering, Panzhihua University, Panzhihua 617000, China

Yi Diao
School of Biological and Chemical Engineering, Panzhihua University, Panzhihua 617000, China

Hong-Yan Zeng, Xiao-Ju Cao and Kai-Min Zou
Biotechnology Institute, College of Chemical Engineering, Xiangtan University, Xiangtan 411105, China

Wenlin Shuai
College of Chemistry and Chemical Engineering, Xinjiang University, Urumqi 830046, China

Abla Alzagameem
Department of Natural Sciences, Bonn-Rhein-Sieg University of Applied Sciences, von-Liebig-Str. 20, D-53359 Rheinbach, Germany
Faculty of Environment and Natural Sciences, Brandenburg University of Technology BTU Cottbus-Senftenberg, Platz der Deutschen Einheit 1, D-03046 Cottbus, Germany

Stephanie Elisabeth Klein, Michel Bergs, Xuan Tung Do and Margit Schulze
Department of Natural Sciences, Bonn-Rhein-Sieg University of Applied Sciences, von-Liebig-Str. 20, D-53359 Rheinbach, Germany

Imke Korte, Sophia Dohlen, Carina Hüwe and Judith Kreyenschmidt
Rheinische Friedrich Wilhelms-University Bonn, Katzenburgweg 7-9, D-53115 Bonn, Germany

Birgit Kamm
Faculty of Environment and Natural Sciences, Brandenburg University of Technology BTU Cottbus-Senftenberg, Platz der Deutschen Einheit 1, D-03046 Cottbus, Germany
Kompetenzzentrum Holz GmbH, Altenberger Strasse 69, A- 4040 Linz, Austria

Michael Larkins
Department of Natural Sciences, Bonn-Rhein-Sieg University of Applied Sciences, von-Liebig-Str. 20, D-53359 Rheinbach, Germany
Department of Forest Biomaterials, North Carolina State University, 2820 Faucette Drive Biltmore Hall, Raleigh, NC 27695, USA

Tanpong Chaiwarit, Warintorn Ruksiriwanich and Pensak Jantrawut
Department of Pharmaceutical Sciences, Faculty of Pharmacy, Chiang Mai University, Chiang Mai 50200, Thailand

Kittisak Jantanasakulwong
School of Agro-Industry, Faculty of Agro-Industry, Chiang Mai University, Chiang Mai 50100, Thailand

Edwin A. Segura González
Universidad Interamericana de Panamá, Research Direction (DI-UIP 6338000), Av. Ricardo J. Alfaro, Panama City, Panama
Department of Materials Science and Engineering and Chemical Engineering, Instituto de Química y Materiales Álvaro Alonso Barba (IQMAA), Universidad Carlos III de Madrid, Leganés 28911, Madrid, Spain

Dania Olmos, Miguel Ángel Lorente and Javier González-Benito
Department of Materials Science and Engineering and Chemical Engineering, Instituto de Química y Materiales Álvaro Alonso Barba (IQMAA), Universidad Carlos III de Madrid, Leganés 28911, Madrid, Spain

Itziar Vélaz
Departamento de Química, Facultad de Ciencias, Universidad de Navarra, 31080 Pamplona, Navarra, Spain

Denis Mihaela Panaitescu, Cristian-Andi Nicolae and Augusta Raluca Gabor
National Institute for Research & Development in Chemistry and Petrochemistry-ICECHIM, Polymer Department, 202 Spl. Independentei, 060021 Bucharest, Romania

Eusebiu Rosini Ionita, Maria Daniela Ionita and Gheorghe Dinescu
National Institute for Laser, Plasma and Radiation Physics, Atomistilor 409, Magurele-Bucharest, 077125 Ilfov, Romania

Roxana Trusca
Science and Engineering of Oxide Materials and Nanomaterials, University Politehnica of Bucharest, 1-7 Gh. Polizu Street, 011061 Bucharest, Romania

Brindusa-Elena Lixandru
"Cantacuzino" National Medical-Military Institute for Research and Development, 103 Spl. Independentei, 050096 Bucharest, Romania

Irina Codita
"Cantacuzino" National Medical-Military Institute for Research and Development, 103 Spl. Independentei, 050096 Bucharest, Romania
Carol Davila University of Medicine and Pharmacy, Bulevardul Eroii Sanitari 8, 050474 Bucharest, Romania

Max Krepker, Nadav Nitzan, Ofer Prinz-Setter and Naama Massad-Ivanir
Department of Biotechnology and Food Engineering, Technion-Israel Institute of Technology, Haifa 3200003, Israel

Cong Zhang, Andrew Olah and Eric Baer
Center for Layered Polymeric Systems, Department of Macromolecular Science and Engineering, Case Western Reserve University, Cleveland, OH 44106-7202, USA

Ester Segal
Department of Biotechnology and Food Engineering, Technion-Israel Institute of Technology, Haifa 3200003, Israel
The Russell Berrie Nanotechnology Institute, Technion-Israel Institute of Technology, Haifa 3200003, Israel

Małgorzata Mizielińska, Urszula Kowalska, Łukasz Łopusiewicz and Michał Jarosz
Center of Bioimmobilisation and Innovative Packaging Materials, Faculty of Food Sciences and Fisheries, West Pomeranian University of Technology Szczecin, Janickiego 35, 71-270 Szczecin, Poland

Piotr Salachna
Department of Horticulture, West Pomeranian University of Technology, 3 Papieża Pawła VI Str., 71-434 Szczecin, Poland

Cheng Zhang, Wenhui Li, Bifen Zhu, Haiyan Chen, Hai Chi and Yuyue Qin
Institute of Yunnan Food Safety, Kunming University of Science and Technology, Kunming 650500, China

Lin Li
College of Food Sciences and Engineering, South China University of Technology, Guangzhou 510640, China

Jing Xue
State Key Laboratory of Oral Diseases, Sichuan University, Chengdu 610041, China

Index

A

Aerobic Conditions, 161, 212

Algae, 28, 74-81, 83

Anthocyanins, 11, 27, 30, 226, 236, 242

Antibacterial Activity, 2-5, 7, 10-13, 15-16, 19, 26, 30-32, 41-42, 58, 71, 83, 93, 96-99, 101-103, 110, 125, 127-130, 137-140, 148, 150-151, 156, 170, 172-175, 177, 189-190, 195, 200, 222-223

Antibacterial Efficiency, 190

Antibacterial Properties, 15, 27-28, 31, 42-43, 69, 71, 83-86, 105-106, 133, 170, 189-190, 192-193, 209-210, 223, 231, 235

Antifungal Activity, 44, 48-49, 55-56, 196-197, 204, 207, 224

Antimicrobial Activity, 2, 4-6, 9, 11, 15-17, 19-20, 22, 24-25, 30-32, 35, 75, 83-84, 86, 88-89, 96-99, 105-111, 114, 121, 125-127, 129, 132-137, 139, 143-145, 148-149, 151, 156, 170, 172-174, 192, 196, 204, 207-208, 211, 214-215, 224

Antimicrobial Properties, 7, 17, 23, 25, 28-29, 41, 43, 55, 59, 68, 71, 88, 99, 105, 110, 125, 130, 147, 156, 175, 197, 207, 210-216, 221-225

Antioxidant Capacity, 134, 229, 237-238, 240-242

Antiradical Activity, 132, 141

B

Bacteria, 1, 5-6, 16, 20, 22, 37, 41-42, 56, 61, 68, 71, 76, 81, 86-89, 96-103, 105, 107, 110-111, 113-114, 125-126, 128-130, 132-134, 136-140, 143-145, 148, 159, 161, 164, 166-169, 212, 214-215, 221-222, 225, 230, 238

Bacterial Cellulose, 17, 31-32, 59-61, 63-71, 172-173, 175, 185, 188, 190-191, 193-195

Bacterial Growth, 37, 57, 81, 98, 158-159, 161, 166-167, 169, 178, 211

Biodegradation, 57, 74, 78-80, 82-83, 194

Biofilm Development, 159, 161, 164, 166, 169

Bionanocomposites, 19, 28-29, 31, 86, 106, 146, 172, 191-192

C

Carvacrol, 4-5, 9, 15-16, 25, 27, 29, 44-58, 173, 196-209

Chitosan, 1-3, 6-17, 19, 22-33, 42, 57, 59-61, 63-73, 78, 82, 90, 94, 105-106, 108-111, 114-117, 119-121, 125-130, 132, 135-136, 140-145, 147-149, 153, 157, 174, 192-193, 208, 222, 225, 241

Coatings, 9-10, 22, 24-33, 36-38, 59-60, 71, 84, 86, 88, 99, 105-108, 134-135, 146-147, 156, 159, 170, 193, 208, 210-217, 219, 221-225, 240-241

Coextrusion, 197-198, 201-202, 206, 208

Composite Films, 4-5, 17, 27-30, 59-60, 65, 68-69, 71, 129, 141, 145-146, 223, 242

D

Diffraction, 109, 112, 139, 158, 160-161, 169

E

Epoxy, 86, 106, 134, 139, 199

Essential Oils, 1-5, 14-17, 20, 22, 24-25, 27, 29-31, 33, 41-42, 45, 53, 56-58, 82-83, 142, 151, 156, 173, 196, 206-207, 209, 211, 221-223

F

Fabrication, 22-23, 33, 71, 75, 87, 99, 108, 128, 187, 192, 196, 198, 223

Feedstock, 84, 133

Fermentation, 17, 45, 194, 210, 226

Film Formulations, 152-153, 155

Food Packaging, 1-5, 9, 14, 19-34, 36, 42-45, 51, 57-58, 60, 63, 69-70, 74-75, 77, 82, 84, 86, 106, 111, 132-133, 150, 155-156, 159, 170, 172-175, 190-194, 197, 207-210, 221-223, 225, 227, 240

Fossil Resources, 45, 84

Free Radicals, 24, 88-89, 142, 238

G

Geranium Oil, 91, 95, 97, 107

Glycerol, 9, 11-12, 16-17, 20, 28, 30, 37-38, 61, 63, 65, 71-72, 75-78, 83, 85, 150-152, 154

H

Halloysite Nanotubes, 23, 196-197, 202, 208-209

Hydroxypropylmethylcellulose, 132, 135, 147

I

Inhibition Zone, 49, 102-103, 121, 137-138, 143, 200

L

Lignin, 7, 12-13, 27, 31, 132-148

Limonene, 9, 15, 29, 58, 150-155

Loading Content, 150, 152, 154-155

Low Methoxy Pectin, 150-151, 154

M

Metal Oxide Nanoparticles, 84, 87, 89, 93, 96

Microbial Contamination, 45, 150

Multilayered Films, 196-199, 201, 204-205, 207

N

Nanocomposites, 23, 25-27, 32-34, 72, 84, 89-103, 105-108, 128-129, 134, 137, 158-161, 163-164, 166, 169-172, 174-185, 190-195, 207-209, 218

Nanocrystals, 11, 16-17, 27, 30, 32, 71, 134, 147, 191

Nanofibers, 5, 11, 22, 29, 32-33, 71-72, 146, 171-175, 178, 181, 183-188, 190, 209

O

Orange Oil, 150-156

Organosolv, 132, 135-137, 140, 143, 145-148

P

Pathogenic Microorganisms, 45, 59, 132, 196, 221

Pectin Film, 150-151, 153, 155

Peptidoglycan, 88, 97, 101, 137-138

Peroxidation, 84, 87-88, 105, 130

Phenolic Content, 24, 224, 236-237

Physical Properties, 28, 32-33, 59, 63, 83, 85, 89, 106, 159, 226-227

Plasma Treatment, 5, 26, 37, 97, 172, 174-176, 178-188, 190, 193

Polyhydroxybutyrate, 172, 192-194

Polylactic Acid, 1, 7, 28-29, 33, 75, 158-159, 171, 192, 210, 215, 218, 221-224, 226, 232, 234

Polymer Chains, 44, 56, 60, 183

Polymer Matrix, 51, 53, 56, 95, 153-155, 169, 173, 196, 204

Polymer Nanocomposites, 158-159, 171

Polymeric Materials, 26, 51, 53, 84, 105, 108, 196, 207

Polymorphs, 88, 161, 169

Polyurethane, 9, 29, 86, 92, 99, 102-103, 106-108, 146-148, 211

Preservation Effect, 226, 234-236, 240

R

Reactive Oxygen Species, 84, 86-87, 107, 126, 138

S

Scanning Electron Microscopy, 48, 52, 62, 77, 93, 112, 152, 154, 172, 176, 199

Shelf-life, 26, 28-29, 74-75, 77, 209, 222

T

Tensile Strength, 11, 17, 62-65, 74, 76-80, 82, 152-154, 172, 174, 184-185

Thermal Properties, 12, 33, 53, 57, 72, 86, 159, 162, 170, 172, 191-192, 227

Thermosetting Polymers, 84-86, 89, 105

Thymol, 9, 12, 23, 27-29, 31, 33, 44-58, 197, 207

U

Unsaturation, 85, 100

Urea Method, 109-111, 129

V

Vegetable Oils, 84-86, 105

W

Water Solubility, 45, 56, 140

Water Vapor Permeability, 3, 60, 231

Printed in the USA
CPSIA information can be obtained
at www.ICGtesting.com
JSHW060010020124
54623JS00006B/112